To my late parents,

Anatole and Ruth Solow

and to my wife of many years,

Audrey

# SUMMARY OF PROOF TECHNIQUES

| What to Conclude | How to Do It |
|---|---|
| $B$ | Work forward from $A$ and apply the backward process to $B$. |
| $NOT\ A$ | Work forward from $NOT\ B$ and backward from $NOT\ A$. |
| Some contradiction | Work forward from $A$ and $NOT\ B$ to reach a contradiction. |
| That there is the desired object | Guess, construct, and so on, the object. Then show that it has the certain property and that the something happens. |
| That the something happens | Work forward from $A$ and the fact that the object has the certain property. Also work backward from the something that happens. |
| $B$ | Work forward by specializing $A$ to one particular object having the certain property. |

# How to Read and Do Proofs

## An Introduction to Mathematical Thought Processes

Fifth Edition

Daniel Solow
Department of Operations
Weatherhead School of Management
Case Western Reserve University
Cleveland, OH 44106
e-mail: daniel.solow@case.edu
web: http://weatherhead.cwru.edu/solow/

**WILEY**

John Wiley & Sons, Inc.

| | |
|---|---|
| Publisher: | Laurie Rosatone |
| Associate Editor: | Shannon Corliss |
| Marketing Manager: | Sarah Davis |
| Production Manager: | Micheline Frederick |

This book was set in LaTeX by the author and printed and bound by Courier Westford. The cover was printed by Courier Westford.

ISBN 9780470392164

Printed in the United States of America

10  9  8  7  6  5  4  3  2

# Contents

# *Foreword*

In a related article, "Teaching Mathematics with Proof Techniques," the author has written, "The inability to communicate proofs in an understandable manner has plagued students and teachers in all branches of mathematics." All of those who have had the experience of teaching mathematics and most of those who have had the experience of trying to learn it must surely agree that acquiring an understanding of what constitutes a sound mathematical proof is a major stumbling block for the student. Many students attempt to circumvent the obstacle by avoiding it—trusting to the indulgence of the examiner not to include any proofs on the test. This collusion between student and teacher may avoid some of the unpleasant consequences—for both student and teacher—of the student's lack of mastery, but it does not alter the fact that a key element in mathematics, arguably its most characteristic feature, has not entered the student's repertoire.

Dr. Solow believes that it is possible to teach the student to understand the nature of a proof by systematizing it. He argues his case cogently with a wealth of detail and examples in this book, and I do not doubt that his ideas deserve attention, discussion, and, above all, experimentation. One of his principal aims is to teach the student how to read the proofs offered in textbooks. These proofs are, to be sure, not presented in a systematic form. Thus, much attention is paid—particularly in the appendices—to showing the reader how to recognize the standard ingredients of a mathematical argument in an informal presentation of a proof.

There is a valid analogy here with the role of the traditional algorithms in elementary arithmetic. It is important to acquire familiarity with them and to understand why they work and to what problems they could, in principle, be applied. But once all of this has been learned, one would not slavishly execute the algorithms in real-life situations (even in the absence of a calculator!). So, the author contends, it is with proofs. Understand and analyze their structure—and then you will be able to read and understand the more informal versions you will find in textbooks—and finally you will be able to create your own proofs. Dr. Solow is not claiming that mathematicians create their proofs by consciously and deliberately applying the "forward-backward method"; he is suggesting that we have a far better chance of teaching an appreciation of proofs by systematizing them, than by our present, rather haphazard procedure based on the hope that the students can learn this difficult art by osmosis.

One must agree with Dr. Solow that, in this country, students begin to grapple with the ideas of mathematical proof far too late in their student careers—the appropriate stage to be initiated into these ideas is, in the judgment of many, no later than eighth grade. However, it would be wrong for university and college teachers merely to excuse their own failures by a comforting reference to defects in the student's precollege education.

Today, mathematics is generally recognized as a subject of fundamental importance because of its ubiquitous role in contemporary life. To be used effectively, its methods must be understood properly—otherwise we cast ourselves in the roles of (inefficient) robots when we try to use mathematics, and we place undue strain on our naturally imperfect memories. Dr. Solow has given much thought to the question of how an understanding of a mathematical proof can be acquired. Most students today do not acquire that understanding, and Dr. Solow's plan to remedy this very unsatisfactory situation deserves a fair trial.

PETER HILTON

*Distinguished Professor of Mathematics*
*State University of New York*
*Binghamton, NY*

# Preface to the Student

After finishing my undergraduate degree, I began to wonder why learning theoretical mathematics had been so difficult. As I progressed through my graduate work, I realized that mathematics possessed many of the aspects of a game—a game in which the rules had been partially concealed. Imagine trying to play chess before you know how all of the pieces move! It is no wonder that so many students have had trouble with abstract mathematics.

This book describes some of the rules by which the game of theoretical mathematics is played. It has been my experience that virtually anyone who is motivated and who has a knowledge of high school mathematics can learn these rules. Doing so greatly reduces the time (and frustration) involved in learning abstract mathematics. I hope this book serves that purpose for you.

To play chess, you must first learn how the individual pieces move. Only after these rules have entered your subconscious can your mind turn its full attention to the more creative issues of strategy, tactics, and the like. So it appears to be with mathematics. Hard work is required in the beginning to learn the fundamental rules presented in this book. In fact, your goal should be to absorb this material so that it becomes second nature to you. Then you will find that your mind can focus on the creative aspects of mathematics. These rules are no substitute for creativity, and this book is not meant to teach creativity. However, I do believe that the ideas presented here provide you with the tools needed to express your creativity. Equally important is the fact that these tools enable you to understand and appreciate the creativity of others. To that end, much emphasis is placed on teaching you how to read "condensed" proofs as they are typically presented in textbooks, journal articles, and other mathematical literature. Knowing how to read and understand such proofs enables you to assimilate the material in any advanced mathematics course for which you have the appropriate prerequisite background. In fact, knowing how to read and understand condensed proofs gives you the ability to learn virtually any mathematical subject on your own, with enough time and effort.

You are about to learn a key part of the mathematical thought process. As you study the material and solve problems, be conscious of your own thought processes. Ask questions and seek answers. Remember, the only unintelligent question is the one that goes unasked.

DANIEL SOLOW

*Department of Operations*
*Weatherhead School of Management*
*Case Western Reserve University*
*Cleveland, OH*

# Preface to the Instructor

## The Objective of This Book

The inability to communicate proofs in an understandable manner has plagued students and teachers in all branches of mathematics. The result has been frustrated students, frustrated teachers, and, oftentimes, a watered-down course to enable the students to follow at least some of the material or a test that protects students from the consequences of this deficiency in their mathematical understanding.

One might conjecture that most students simply cannot understand abstract mathematics, but my experience indicates otherwise. What seems to have been lacking is a proper method for explaining theoretical mathematics. In this book I have developed a method for communicating proofs—a common language that professors can teach and students can understand. In essence, this book categorizes, identifies, and explains (at the student's level) the various techniques that are used repeatedly in virtually all proofs.

Once the students understand the techniques, it is then possible to explain any proof as a sequence of applications of these techniques. In fact, it is advisable to do so because the process reinforces what the students have learned in the book.

Explaining a proof in terms of its component techniques is not difficult, as is illustrated in the examples of this book. Before each "condensed" proof is an analysis explaining the methodology, thought processes, and techniques that are used. Teaching proofs in this manner requires nothing more than preceding each step of the proof with an indication of which technique is about to be used and why.

When discussing a proof in class, I actively involve the students by soliciting their help in choosing the techniques and designing the proof. I have been pleasantly surprised by the quality of their comments and questions. It has been my experience that once students become comfortable with the proof techniques, their minds tend to address the more important issues of mathematics, such as why a proof is done in a particular way and why the piece of mathematics is important in the first place. This book is not meant to teach creativity, but I do believe that learning the techniques presented here frees the student's mind to focus on the creative aspects. I have also found that, by using this approach, it is possible to teach subsequent mathematical material at a more sophisticated level without losing the students.

In any event, the message is clear. I am suggesting that there are many benefits to be gained by teaching mathematical thought processes in addition to mathematical material. This book is designed to be a major step in the right direction by making abstract mathematics understandable and enjoyable to the students and by providing you with a method for communicating with them.

## What's New in the Fifth Edition

The main change in the fifth edition is a complete revision and expansion of the exercises in the main body of the text. This book now contains exercises that are appropriate for all levels of undergraduate students. As in the fourth edition, all exercises marked with a $B$ have completely worked-out solutions in the back of the book; those marked with a $W$ have complete solutions on the web at http://www.wiley.com/college/solow/; and the rest have solutions provided in the accompanying Solutions Manual that only instructors can obtain from the foregoing website. Exercises marked with a * symbol and whose solutions are not available to students are considered relatively more challenging or time consuming.

Other changes in this edition include the following:

1. A discussion in Chapter 1 of the need to identify the hypothesis and conclusion when the proposition is not stated in the standard form, "If $A$, then $B$." Several examples are given to illustrate how this is done and appropriate exercises are included.

2. An extended and more complete discussion in Chapter 3 of how to use a previously proved proposition in both the forward and backward processes.

3. A discussion in Chapter 5 of the equivalence of the statements, "For all objects $X$ with a certain property, something happens" and "If $X$ is

an object with a certain property, then $X$ satisfies the something that happens."

4. Replacing the previous Chapters 11 and 12 with new Chapters 11 through 14 in order to devote a separate self-contained chapter with exercises to each of the following techniques, respectively: uniqueness, induction, either/or, and max/min methods.

5. The inclusion of several final examples of how to read and do proofs in the summary Chapter 15 that serve to unify the student's knowledge of the various proof techniques.

Although these changes seem to make it even easier for students to understand proofs, I have still found no substitute for actively teaching the material in class instead of having the students read the material on their own. This active interaction has proved eminently beneficial to both student and teacher, in my case.

DANIEL SOLOW

*Department of Operations*
*Weatherhead School of Management*
*Case Western Reserve University*
*Cleveland, OH*

# Acknowledgments

## Acknowledgments for the First Edition

For helping to get this work known in the mathematics community, my deepest gratitude goes to Peter Hilton, an outstanding mathematician and educator. I also thank Paul Halmos, whose support greatly facilitated the dissemination of the knowledge of the existence of this book and teaching method. I am also grateful for discussions with Gail Young and George Polya.

Regarding the preparation of the manuscript for the first edition, no single person had more constructive comments than Tom Butts. He not only contributed to the mathematical content but also corrected many of the grammatical and stylistic mistakes in a preliminary version. I suppose that I should thank his mother for being an English teacher. I also acknowledge Charles Wells for reading and commenting on the first handwritten draft and for encouraging me to pursue the project further. Many others made substantive suggestions, including Alan Schoenfeld, Samuel Goldberg, and Ellen Stenson.

Of all the people involved in this project, none deserves more credit than my students. It is because of their voluntary efforts that the first edition of this book was prepared in such a short time. Thanks especially to John Democko for acting in the capacity of senior editor while concurrently trying to complete his Ph.D. program. Also, I appreciate the help that I received from Michael Dreiling and Robert Wenig in data basing the text and in preparing the exercises. Michael worked on this project almost as long as I did. A special word of thanks goes to Greg Madey for coordinating the second rewriting of the document, for adding useful comments, and, in general, for keeping me organized. His responsibilities were subsequently assumed by Robin Symes.

In addition, I am grateful to Ravi Kumar for the long hours he spent on the computer preparing the final version of the manuscript, to Betty Tracy and Martha Bognar for their professional and flawless typing assistance, and to Virginia Benade for her technical editing. I am also indebted to my class of 1981 for preparing the solutions to the exercises.

I thank the following professors for refereeing the original manuscript and for recommending its publication: Alan Tucker, David Singer, Howard Anton, and Ivan Niven.


## Acknowledgments for the Second Edition

Most of the credit for the improvements in the second edition goes to the students who have taught me so much about learning mathematics. I am also grateful for the many comments and suggestions I received from colleagues over the years and also from a questionnaire circulated by John Wiley.

On the technical end, I thank Dawnn Strasser for her most professional job of typing the preliminary version of the manuscript. I also am grateful to the Weatherhead School of Management at Case Western Reserve University for the use of their excellent word-processing facilities.

Finally, I am grateful to my wife, Audrey, for her help in proofreading and for her patience during yet another of my projects.


## Acknowledgments for the Third Edition

The third edition was typeset using LATEX with thanks to Donald Knuth and also Amy Hendrickson, whose macro package was a delight to use. Serge Karalli spent many hours helping me revise old exercises and develop new ones. Tom Engle created professional electronic drawings of the figures. I also thank John Wiley & Sons for providing me with the support I needed to accomplish this revision. It was also a pleasure to work with my editor, Kimberly Murphy, and her staff, especially Ken Santor.


## Acknowledgments for the Fourth Edition

I am especially grateful for the thorough and thoughtful comments of Richard Delaware at the University of Missouri, Kansas City, many of which have been incorporated in this edition. It has also been a pleasure to work with my editor, Laurie Rosatone, at John Wiley, who I also worked with many eons ago when she was at Addison-Wesley.

## Acknowledgments for the Fifth Edition

I thank the following reviewers for their useful comments in guiding the changes to this fifth edition: Jeff Dodd, Mary Bradley, Kimberly J. Presser, Ray Rosenstrater, and Richard Delaware. I am grateful to the Naval Postgraduate School where, as a visiting researcher, I was able to complete much of the work on this revision. It was there that I met Lieutenant Commander Jay Foraker, Eng Yau Pee, Andrew Swedberg, Eric Tollefson, Anthony Tvaryanas, and Roger Vaughn, who were dedicated enough to work through some of the new exercises that are included in this edition. I am especially grateful to Edward Agarwala at Case who prepared a first draft of the Solutions Manual for this edition and, in so doing, improved the quality of some of the exercises. Thanks also to Ralph Grimaldi from the Rose-Hulman Institute of Technology who, as an accuracy checker, found numerous minor mistakes that led to a cleaner version of the text and Solutions Manual. I also appreciate the help and support I received from my editors at John Wiley: Laurie Rosatone and Shannon Corliss.

D. S.

# 1

## *The Truth of It All*

The objective of mathematicians is to discover and to communicate certain truths. *Mathematics* is the language of mathematicians, and a *proof* is a method of communicating a mathematical truth to another person who also "speaks" the language. A remarkable property of the language of mathematics is its precision. Properly presented, a proof contains no ambiguity—there will be no doubt about its correctness. Unfortunately, many proofs that appear in textbooks and journal articles are presented for someone who already knows the language of mathematics. Thus, to understand and present a proof, you must learn a new language, a new method of thought. This book explains much of the basic grammar, but as in learning any new language, a lot of practice is needed to become fluent.

## 1.1  THE OBJECTIVES OF THIS BOOK

The approach taken here is to categorize and to explain the various **proof techniques** that are used in *all* proofs, regardless of the subject matter. One objective is to teach you how to read and understand a written proof by identifying the techniques that are used. Learning to do so enables you to study almost any mathematical subject on your own—a desirable goal in itself.

A second objective is to teach you to develop and to communicate your own proofs of known mathematical truths. Doing so requires you to use a certain amount of creativity, intuition, and experience. Just as there are many ways to express the same idea in any language, so there are different proofs for the same mathematical fact. The techniques presented here are designed to get you started and to guide you through a proof. Consequently, this book describes not only *how* the techniques work but also *when* each one is likely to be used and *why*. Often you will be able to choose a correct technique based on the form of the problem under consideration. Therefore, when attempting to create your own proof, *learn to select a technique consciously* before wasting hours trying to figure out what to do. The more aware you are of your thought processes, the better.

The ultimate objective, however, is to use your newly acquired skills and language to discover and communicate previously unknown mathematical truths. The first step in this direction is to reach the level of being able to read proofs and develop your own proofs of already-known facts. This alone will give you a much deeper and richer understanding of the mathematical universe around you.

Anyone with a good knowledge of high school mathematics can read this book. Advanced students who have seen proofs before can read the first two chapters, skip to the summary Chapter 15, and subsequently read any of the appendices to see how all the techniques fit together in a specific mathematical subject. Each chapter on a particular technique also contains a brief summary at the end that describes how and when to use the technique. The remainder of this chapter explains the types of relationships to which proofs are applied. Additional books on proofs and advanced mathematical reasoning are listed in the bibliography at the end of this book.

## 1.2   WHAT IS A PROOF?

In mathematics, a **statement** is a sentence that is either true or false. Some examples follow:

1. Two parallel lines in a plane have the same slope.

2. $1 = 0$.

3. $3x = 5$ and $y = 1$.

4. $x \not> 0$ ($x$ is not greater than 0).

5. There is an angle $t$ such that $\cos(t) = t$.

Observe that statement (1) is always true, (2) is always false, and statements (3) and (4) are either true or false, depending on the value of a variable. For this reason, (3) and (4) are called **conditional statements**.

It is perhaps not as obvious that statement (5) is always true. It therefore becomes necessary to have some method for *proving* that such statements are true. In other words, a **proof** is a convincing argument expressed in the

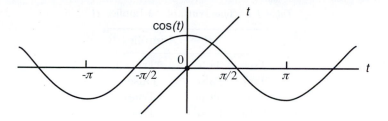

*Fig. 1.1* A proof that there is an angle $t$ such that $\cos(t) = t$.

language of mathematics. In this and other books, proofs are often given for what seem to be obviously true statements. One reason for doing so is to provide examples that are easy to follow so that you can eventually prove more difficult statements. Another reason is that some apparently "obvious" statements are, in fact, false. You will know that a statement is true only when you have *proved* it to be true.

A proof should contain enough mathematical details to be convincing to the person(s) to whom the proof is addressed. A proof of statement (5) that is meant to convince a mathematics professor might consist of nothing more than Figure 1.1; whereas a proof directed toward a high school student would require more details, perhaps even the definition of cosine. Your proofs should contain enough details to be convincing to someone else at your own mathematical level (for example, a classmate). It is the lack of sufficient detail that often makes a proof difficult to read and understand. One objective of this book is to teach you to decipher "condensed" proofs that are likely to appear in textbooks and other mathematical literature.

Given two statements $A$ and $B$, each of which may be either true or false, a fundamental problem of interest in mathematics is to show that the following statement—called an **implication**—is true:

If $A$ is true, then $B$ is true.

One reason for wanting to prove that an implication is true is when $B$ is a statement that you would like to be true but whose truth is not easy to verify. In contrast, suppose that $A$ is a statement whose truth is relatively easy to verify. If you have proved that "If $A$ is true, then $B$ is true," and if you can verify that $A$ is in fact true, then you will know that $B$ is true.

A proof is a formal method for convincing yourself (and others) that "If $A$ is true, then $B$ is true." To do a proof, you must know exactly what it means to show this. Statement $A$ is called the **hypothesis** and $B$ the **conclusion**. For brevity, the statement "If $A$ is true, then $B$ is true" is shortened to "If $A$, then $B$" or simply "$A$ **implies** $B$." Mathematicians have developed a symbolic shorthand notation. For instance, a mathematician would write "$A \Rightarrow B$" instead of "$A$ implies $B$." For the most part, textbooks do not use the symbolic notation, but teachers often do, and eventually you might find

Table 1.1  The Truth of "$A$ Implies $B$."

| $A$ | $B$ | $A$ implies $B$ |
|-----|-----|-----------------|
| True | True | True |
| True | False | False |
| False | True | True |
| False | False | True |

it useful, too. Therefore, the appropriate symbols are included in this book but are not used in the proofs. A complete list of symbols is presented in the glossary at the end of this book.

The conditions under which "$A$ implies $B$" are true depend on whether $A$ and $B$ themselves are true. Thus, there are four possible cases to consider:

1. $A$ is true and $B$ is true.

2. $A$ is true and $B$ is false.

3. $A$ is false and $B$ is true.

4. $A$ is false and $B$ is false.

Suppose, for example, that your friend made the statement, "If you study hard, then you will get a good grade." Here, statement $A$ is "you study hard" and $B$ is "you will get a good grade." To determine when the statement "$A$ implies $B$" is false, ask yourself in which of the four foregoing cases you would be willing to call your friend a liar. In the first case (that is, when you study hard and you get a good grade), your friend has told the truth. In the second case, you studied hard, and yet you did not get a good grade, as your friend said you would. Here your friend has not told the truth. In cases 3 and 4, you did not study hard. You would not want to call your friend a liar in these cases because your friend said that something would happen only when you did study hard. Thus, the statement "$A$ implies $B$" is true in each of the four cases except the second one, as summarized in Table 1.1.

Table 1.1 is an example of a **truth table**, which is a method for determining when a complex statement (in this case, "$A$ implies $B$") is true by examining all possible truth values of the individual statements (in this case, $A$ and $B$). Other examples of truth tables appear in Chapter 3.

According to Table 1.1, when trying to show that "$A$ implies $B$" is true, you might attempt to determine the truth of $A$ and $B$ individually and then use the appropriate row of the table to determine the truth of "$A$ implies $B$." For example, to determine the truth of the statement,

$$\text{if } 1 < 2, \text{ then } 4 < 3,$$

you can easily see that the hypothesis $A$ (that is, $1 < 2$) is true and the conclusion $B$ (that is, $4 < 3$) is false. Thus, using the second row of Table 1.1

(corresponding to $A$ being true and $B$ being false) you can conclude that in this case the statement "$A$ implies $B$" is false. Similarly, the statement,

$$\text{if } 2 < 1, \text{ then } 3 < 4,$$

is true according to the third row of the table because $A$ (that is, $2 < 1$) is false and $B$ (that is, $3 < 4$) is true.

Now suppose you want to prove that the following statement is true:

$$\text{if } x > 2, \text{ then } x^2 > 4.$$

The difficulty with using Table 1.1 for this example is that you cannot determine whether $A$ (that is, $x > 2$) and $B$ (that is, $x^2 > 4$) are true or false because the truth of the conditional statements $A$ and $B$ depend on the variable $x$, whose value is not known. Nonetheless, you can still use Table 1.1 by reasoning as follows:

> Although I do not know the truth of $A$, I do know that $A$ must be either true or false. Let me assume, for the moment, that $A$ is false (subsequently, I will consider what happens when $A$ is true). When $A$ is false, either the third or the fourth row of Table 1.1 is applicable and, in either case, the statement "$A$ implies $B$" is true—thus I would be done. Therefore, I need only consider the case in which $A$ is true.

> When $A$ is true, either the first or the second row of Table 1.1 is applicable. However, because I want to prove that "$A$ implies $B$" is true, I need to be sure that the first row of the truth table is applicable, and this I can do by establishing that $B$ is true.

From the foregoing reasoning, when trying to prove that "$A$ implies $B$" is true, *you can assume that $A$ is true; your job is to conclude that $B$ is true.*

Note that a proof of the statement "$A$ implies $B$" is not an attempt to verify whether $A$ and $B$ themselves are true but rather to show that $B$ is a logical result of having assumed that $A$ is true. Your ability to show that $B$ is true depends on the fact that you have assumed $A$ to be true; ultimately, you have to discover the relationship between $A$ and $B$. Doing so requires a certain amount of creativity. The techniques presented here are designed to get you started and guide you along the path.

The first step in doing a proof is to identify the hypothesis $A$ and the conclusion $B$. This is easy to do when the implication is written in the form "If $A$, then $B$" because everything after the word "if" and before the word "then" is $A$ and everything after the word "then" is $B$. Unfortunately, implications are not always written in this specific form. In such cases, everything that you are assuming to be true is the hypothesis $A$; everything that you are trying to prove is true is the conclusion $B$. You might have to interpret the meaning of symbols from the context in which they are used and even introduce your own notation sometimes. Consider the following examples.

**Example 1:** The sum of the first $n$ positive integers is $n(n+1)/2$.

> **Hypothesis:** $n$ is a positive integer. (Note that this is implied for the statement to make sense.)
> **Conclusion:** The sum of the first $n$ positive integers is $n(n+1)/2$.

**Example 2:** The quadratic equation $ax^2 + bx + c = 0$ has two real roots provided that $b^2 - 4ac > 0$, where $a \neq 0$, $b$, and $c$ are given real numbers.

> **Hypothesis:** $a$, $b$, and $c$ are real numbers with $a \neq 0$ and $b^2 - 4ac > 0$.
> **Conclusion:** The quadratic equation $ax^2 + bx + c = 0$ has two real roots.

**Example 3:** Two lines tangent to the endpoints of the diameter of a circle are parallel.

> **Hypothesis:** $L_1$ and $L_2$ are two lines that are tangent to the endpoints of the diameter of a circle.
> **Conclusion:** $L_1$ and $L_2$ are parallel.

**Example 4:** There is a real number $x$ such that $x = 2^{-x}$.

> **Hypothesis:** None, other than your previous knowledge of mathematics.
> **Conclusion:** There is a real number $x$ such that $x = 2^{-x}$.

Before starting a proof, always be clear what you are assuming—that is, the hypothesis $A$—and what you are trying to show—that is, the conclusion $B$.

## Summary

Hereafter, $A$ and $B$ are statements that are either true or false. The problem is to prove that "$A$ implies $B$" is true. According to Table 1.1, after identifying the hypothesis $A$ and conclusion $B$, you should:

1. Assume that $A$ is true.

2. Use this assumption to reach the conclusion that $B$ is true.

## Exercises

**Note:** Solutions to those exercises marked with a $B$ are in the back of this book. Solutions to those exercises marked with a $W$ are located on the web at http://www.wiley.com/college/solow/.

$^B$**1.1**   Which of the following are mathematical statements?

a. $ax^2 + bx + c = 0$.
b. $(-b + \sqrt{b^2 - 4ac})/(2a)$.

c. Triangle $XYZ$ is similar to triangle $RST$.
d. $3 + n + n^2$.

e. For every angle $t$, $\sin^2(t) + \cos^2(t) = 1$.

**1.2**    Which of the following are mathematical statements?

   a. There is an even integer $n$ that, when divided by 2, is odd.

   b. {integers $n$ such that $n$ is even}.

   c. If $x$ is a positive real number, then $\log_{10}(x) > 0$.

   d. $\sin(\pi/2) < \sin(\pi/4)$.

$^{B}$**1.3**    For each of the following problems, identify the hypothesis (what you can assume is true) and the conclusion (what you are trying to show is true).

   a. If the right triangle $XYZ$ with sides of lengths $x$ and $y$ and hypotenuse of length $z$ has an area of $z^2/4$, then the triangle $XYZ$ is isosceles.

   b. $n^2$ is an even integer provided that $n$ is an even integer.

   c. Let $a$, $b$, $c$, $d$, $e$, and $f$ be real numbers. You can solve the two linear equations $ax + by = e$ and $cx + dy = f$ for $x$ and $y$ when $ad - bc \neq 0$.

**1.4**    For each of the following problems, identify the hypothesis (what you can assume is true) and the conclusion (what you are trying to show is true).

   a. $r$ is irrational if $r$ is a real number that satisfies $r^2 = 2$.

   b. If $p$ and $q$ are positive real numbers with $\sqrt{pq} \neq (p+q)/2$, then $p \neq q$.

   c. Let $f(x) = 2^{-x}$ for any real number $x$. Then $f(x) = x$ for some real number $x$ with $0 \leq x \leq 1$.

$^{W}$**1.5**    For each of the following problems, identify the hypothesis (what you can assume is true) and the conclusion (what you are trying to show is true).

   a. Suppose that $A$ and $B$ are sets of real numbers with $A \subseteq B$. For any set $C$ of real numbers, it follows that $A \cap C \subseteq B \cap C$.

   b. For a positive integer $n$, define the following function:

$$f(n) = \begin{cases} n/2, & \text{if } n \text{ is even} \\ 3n + 1, & \text{if } n \text{ is odd} \end{cases}$$

Then for any positive integer $n$, there is an integer $k > 0$ such that $f^k(n) = 1$, where $f^k(n) = f^{k-1}(f(n))$, and $f^1(n) = f(n)$.

   c. When $x$ is a real number, the minimum value of $x(x - 1) \geq -1/4$.

$^{B}$**1.6**    "If I do not get my car fixed, I will miss my job interview," says Jack. Later, you come to know that Jack's car was repaired but that he missed his job interview. Was Jack's statement true or false? Explain.

**1.7**    "If I get my car fixed, I will not miss my job interview," says Jack. Later, you come to know that Jack's car was repaired but that he missed his job interview. Was Jack's statement true or false? Explain.

**1.8**    Suppose someone says to you that the following statement is true: "If Jack is younger than his father, then Jack will not lose the contest." Did Jack win the contest? Why or why not? Explain.

$^{B}$**1.9**    Determine the conditions on the hypothesis $A$ and conclusion $B$ under which the following statements are true and false and give your reason.

    a.   If $2 > 7$, then $1 > 3$.                b.   If $x = 3$, then $1 < 2$.

**1.10**    Determine the conditions on the hypothesis $A$ and conclusion $B$ under which the following statements are true and false and give your reason.

    a.   If $2 < 7$, then $1 < 3$.                b.   If $x = 3$, then $1 > 2$.

$^{B}$**1.11**    If you are trying to prove that "$A$ implies $B$" is true and you know that $B$ is false, do you want to show that $A$ is true or false? Explain.

**1.12**    By considering what happens when $A$ is true and when $A$ is false, it was decided that to prove the statement "$A$ implies $B$" is true, you can assume that $A$ is true and your goal is to show that $B$ is true. Use the same type of reasoning to derive another approach for proving that "$A$ implies $B$" is true by considering what happens when $B$ is true and when $B$ is false.

$^{B}$**1.13**    Using Table 1.1, prepare a truth table for "$A$ implies $(B$ implies $C)$."

$^{W}$**1.14**    Using Table 1.1, prepare a truth table for "$(A$ implies $B)$ implies $C$."

**1.15**    Using Table 1.1, prepare a truth table for "$B \Rightarrow A$." Is this statement true under the same conditions for which "$A \Rightarrow B$" is true?

$^{B}$**1.16**    Suppose you want to show that $A \Rightarrow B$ is false. According to Table 1.1, how should you do this? What should you try to show about the truth of $A$ and $B$? (Doing this is referred to as a **counterexample** to $A \Rightarrow B$.)

$^{B}$**1.17**    Apply your answer to Exercise 1.16 to show that each of the following statements is false by constructing a counterexample.

    a. If $x > 0$, then $\log_{10}(x) > 0$.

    b. If $n$ is a positive integer, then $n^3 \geq n!$ (where $n! = n(n-1)\cdots 1$).

*__1.18__    Apply your answer to Exercise 1.16 to show that each of the following statements is false by constructing a counterexample.

    a. If $n$ is a positive integer, then $3^n \geq n!$ (where $n! = n(n-1)\cdots 1$).

    b. If $x$ is a positive real number between 0 and 1, then the first three decimal digits of $x$ are not equal to the first three decimal digits of $2^{-x}$.

# 2

## The Forward-Backward Method

The purpose of this chapter is to describe one of the fundamental proof techniques: the **forward-backward method**. Special emphasis is given to the material of this chapter because all other proof techniques rely on this method.

Recall from Chapter 1 that, when proving "$A$ implies $B$," you can assume that $A$ is true and you must somehow use this information to reach the conclusion that $B$ is true. In attempting to reach the conclusion that $B$ is true, you will go through a **backward process**. When you make specific use of the information contained in $A$, you will go through a **forward process**. Both of these processes are described in detail now using the following example.

**Proposition 1** *If the right triangle $XYZ$ with sides of lengths $x$ and $y$ and hypotenuse of length $z$ has an area of $z^2/4$, then the triangle $XYZ$ is isosceles (see Figure 2.1).*

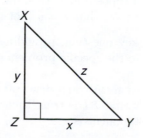

*Fig. 2.1*   The right triangle $XYZ$.

## 2.1    THE BACKWARD PROCESS

In the backward process you begin by asking, "How or when can I conclude that the statement $B$ is true?" The very manner in which you phrase this question is critical because you must eventually provide an answer. You should pose the question in an abstract way. To illustrate, consider the following conclusion for Proposition 1:

**B:** The triangle $XYZ$ is isosceles.

A correct abstract question for this statement $B$ is:

"How can I show that *a* triangle is isosceles?"

While it is true that you want to show that the specific triangle $XYZ$ is isosceles, by asking the *abstract* question, you call on your general knowledge of triangles, clearing away irrelevant details (such as the fact that the triangle is called $XYZ$ instead of $ABC$), thus allowing you to focus on those aspects of the problem that really seem to matter. The question obtained from statement $B$ in such problems is referred to here as the **key question**. A properly posed key question should contain no symbols or other notation (except for numbers) from the specific problem under consideration. The key to many proofs is formulating a correct key question.

Once you have posed the key question, the next step in the backward process is to provide an answer. Returning to Proposition 1, how can you show that a triangle is isosceles? Certainly one way is to show that two of its sides have equal length. Referring to Figure 2.1, you should show that $x = y$. Observe that answering the key question is a two-phase process:

**How to Answer a Key Question**

1. First, give an abstract answer that contains no symbols from the specific problem. (For example, to show that *a* triangle is isosceles, show that two of *its* sides have equal length.)

2. Next, apply this abstract answer to the specific problem using appropriate notation. [For Proposition 1, to show that two of *its* sides have equal length means to show that $x = y$ (not that $x = z$ or $y = z$)].

The process of asking the key question, providing an abstract answer, and then applying that answer to the specific problem constitutes one step of the backward process.

The backward process has given you a new statement, $B1$, with the property that, if you could show that $B1$ is true, then $B$ would be true. For the current example, the new statement is

**B1:** $x = y$.

If you can show that $x = y$, then the triangle $XYZ$ is isosceles.

Once you have the statement $B1$, all of your efforts must now be directed toward reaching the conclusion that $B1$ is true, for then it will follow that $B$ is true. How can you show that $B1$ is true? Eventually you will have to make use of the assumption that $A$ is true. When solving this problem, you would most likely do so now, but for the moment let us continue working backward from the new statement $B1$. This will illustrate some of the difficulties that can arise in the backward process. Can you pose a key question for $B1$?

Because $x$ and $y$ are the lengths of two sides of a triangle, a reasonable key question is, "How can I show that the lengths of two sides of a triangle are equal?" A second perfectly reasonable key question is, "How can I show that two real numbers are equal?" After all, $x$ and $y$ are also real numbers. One of the difficulties that can arise in the backward process is the possibility of more than one key question. Choosing the correct one is more of an art than a science. In fortunate circumstances, there will be only one obvious key question; in other cases, you may have to proceed by trial and error. This is where your intuition, insight, creativity, experience, diagrams, and graphs can play an important role. One general guideline is to let the information in $A$ (which you are assuming to be true) help you to choose the key question, as is done in this case.

Regardless of which question you finally settle on, the next step is to provide an answer, first in the abstract and then for the specific problem. Can you do this for the two foregoing key questions posed for $B1$? For the first one, you might show that two sides of a triangle have equal length by showing that the angles opposite them are equal. For the triangle $XYZ$ in Figure 2.1, this would mean you have to show that angle $X$ equals angle $Y$. A cursory examination of the contents of statement $A$ does not seem to provide much information concerning those angles of triangle $XYZ$. For this reason, the other key question is chosen, namely,

**Key Question:** "How can I show that two real numbers
(namely, $x$ and $y$) are equal?"

One answer to this question is to show that the difference of the two numbers is 0. Applying this answer to the specific statement $B1$ means you would have to show that $x - y = 0$. Unfortunately, there is another perfectly acceptable answer: show that the first number is less than or equal to the second number and also that the second number is less than or equal to the first number. Applying this answer to the specific statement $B1$, you would have to show that $x \leq y$ and $y \leq x$. Thus, a second difficulty can arise in the backward process: even if you choose the correct key question, there may be more than one answer. Moreover, you might choose an answer that will not permit you to complete the proof. For instance, associated with the key question, "How can I show that a triangle is isosceles?" is the answer, "Show that the triangle is equilateral." Of course it is impossible to show that triangle $XYZ$ in Proposition 1 is equilateral because one of its angles is 90 degrees.

Returning to the key question associated with $B1$, "How can I show that two real numbers (namely, $x$ and $y$) are equal?" suppose, for the sake of argument, that you choose the answer of showing that their difference is 0. Once again, the backward process has given you a new statement, $B2$, with the property that, if you could show that $B2$ is true, then in fact $B1$ would be true, and hence so would $B$. Specifically, the new statement is:

**B2:** $x - y = 0$.

Now all of your efforts must be directed toward reaching the conclusion that $B2$ is true. You must ultimately make use of the information in $A$ but, for the moment, let us continue the backward process applied to $B2$.

One key question associated with $B2$ is,

> **Key Question:** "How can I show that the difference of two
> real numbers is 0?"

After some reflection, it may seem that there is no reasonable answer to this question. Yet another problem can arise in the backward process: the key question might have no apparent answer. Do not despair—all is not lost. Remember that, when proving "$A$ implies $B$," you are allowed to assume that $A$ is true. It is now time to make use of this fact.

## 2.2   THE FORWARD PROCESS

The forward process involves deriving from the statement $A$, which you assume is true, some other statement, $A1$, that you know is true as a result of $A$ being true. It should be emphasized that the statements derived from $A$ are not haphazard. Rather, they are directed toward linking up with the last statement obtained in the backward process. Let us return to Proposition 1, keeping in mind that the last statement in the backward process is,

**B2:** $x - y = 0$.

For Proposition 1, you can assume that the following hypothesis is true:

**A:** "The right triangle $XYZ$ with sides of length $x$ and $y$ and
hypotenuse of length $z$ has an area of $z^2/4$."

One fact that you know (or should know) as a result of $A$ being true is that $xy/2 = z^2/4$, because the area of a triangle is one-half the base times the height—in this case, $xy/2$. So you have obtained the new statement,

**A1:** $xy/2 = z^2/4$.

Another useful statement follows from $A$ by the Pythagorean theorem because $XYZ$ is a right triangle, so you also have:

*Table 2.1* Proof of Proposition 1.

| Statement | Reason |
|---|---|
| $A$: Area of $XYZ$ is $z^2/4$ | Given |
| $A1$: $xy/2 = z^2/4$ | Area = (base)(height)/2 |
| $A2$: $x^2 + y^2 = z^2$ | Pythagorean theorem |
| $A3$: $xy/2 = (x^2 + y^2)/4$ | Substitute $A2$ into $A1$ |
| $A4$: $x^2 - 2xy + y^2 = 0$ | From $A3$ by algebra |
| $A5$: $(x - y)^2 = 0$ | Factor $A4$ |
| $B2$: $x - y = 0$ | From $A5$ by algebra |
| $B1$: $x = y$ | Add $y$ to both sides of $B2$ |
| $B$: $XYZ$ is isosceles | Because $B1$ is true |

**A2:** $x^2 + y^2 = z^2$.

With the forward process, you can also combine and use the new statements to produce more true statements. For instance, it is possible to combine $A1$ and $A2$ by replacing $z^2$ in $A1$ with $x^2 + y^2$ from $A2$, obtaining the statement,

**A3:** $xy/2 = (x^2 + y^2)/4$.

One of the problems with the forward process is that it is also possible to generate useless statements, for instance, "angle $X$ is less than 90 degrees." While there are no specific guidelines for producing new statements, keep in mind that the forward process is directed toward obtaining the statement $B2$: $x - y = 0$, which was the last one derived in the backward process. The fact that $B2$ does not contain the quantity $z^2$ is the reason that $z^2$ was eliminated from $A1$ and $A2$ to produce $A3$.

Continuing with the forward process, you should attempt to make $A3$ look more like $B2$ by rewriting. For instance, you can multiply both sides of $A3$ by 4 and subtract $2xy$ from both sides to obtain,

**A4:** $x^2 - 2xy + y^2 = 0$.

Factoring $A4$ yields

**A5:** $(x - y)^2 = 0$.

One of the most common steps in the forward process is to rewrite statements in different forms, as is done in obtaining $A4$ and $A5$. For Proposition 1, the final step in the forward process (and in the entire proof) is to realize from $A5$ that, if the square of a number (namely, $x - y$) is 0, then the number itself is 0, thus obtaining precisely the statement $B2 : x - y = 0$. The proof is now complete because you have successfully used the assumption that $A$ is true to reach the conclusion that $B2$, and hence $B$, is true. The steps and reasons are summarized in Table 2.1.

It is interesting to note that the forward process ultimately produced the elusive answer to the key question associated with $B2$, "How can I show that the difference of two real numbers is 0?" In this case, the answer is to show that the square of the difference is 0 (see $A5$ in Table 2.1).

## 2.3   READING PROOFS

In general it is not practical to write the entire thought process that goes into a proof, for this requires too much time, effort, and space. Rather, a highly condensed version is usually presented and often makes little or no reference to the backward process. For Proposition 1, the condensed proof might go something like this.

**Proof of Proposition 1.**   From the hypothesis and the formula for the area of a right triangle, the area of $XYZ = xy/2 = z^2/4$. By the Pythagorean theorem, $x^2 + y^2 = z^2$ and, on substituting $x^2 + y^2$ for $z^2$ and performing some algebraic manipulations, one obtains $(x - y)^2 = 0$. Hence $x = y$, and so the triangle $XYZ$ is isosceles.   $\square$

In the foregoing proof, the sentence, "From the hypothesis . . ." indicates that the author is working forward. Also, the $\square$ or some equivalent symbol is usually used to indicate the end of the proof. Sometimes the letters Q.E.D. are used as they stand for the Latin words *quod erat demonstrandum*, meaning "which was to be demonstrated."

Sometimes the condensed proof is partly backward and partly forward. For example:

**Proof of Proposition 1.**    The statement is proved by establishing that $x = y$, which in turn is done by showing that $(x-y)^2 = x^2 - 2xy + y^2 = 0$. But the area of the triangle is $xy/2 = z^2/4$ so that $2xy = z^2$. By the Pythagorean theorem, $z^2 = x^2 + y^2$ and hence $x^2 + y^2 = 2xy$, or, $x^2 - 2xy + y^2 = 0$.   $\square$

The proof can also be written entirely from the backward process. Although slightly unnatural, this version is worth seeing.

**Proof of Proposition 1.**    To reach the conclusion, it is shown that $x = y$ by verifying that $(x - y)^2 = x^2 - 2xy + y^2 = 0$, or equivalently, that $x^2 + y^2 = 2xy$. This is established by showing that $2xy = z^2$, for the Pythagorean theorem states that $x^2 + y^2 = z^2$. To see that $2xy = z^2$, or equivalently, that $xy/2 = z^2/4$, note that $xy/2$ is the area of the triangle and is equal to $z^2/4$ by hypothesis, thus completing the proof.   $\square$

Proofs found in research articles are very condensed, giving little more than a hint of how to do the proof. For example:

**Proof of Proposition 1.** The hypothesis together with the Pythagorean theorem yield $x^2 + y^2 = 2xy$ and hence $(x - y)^2 = 0$. Thus the triangle is isosceles, as required. □

Note that the word "hence" effectively conceals the reason that $(x - y)^2 = 0$. Is that reason algebraic manipulation (as we know it is) or something else?

### Reasons Why Reading a Condensed Proof Is Challenging

From these examples, you can see that there are several reasons why reading a condensed proof is challenging:

1. The steps of the proof are not always presented in the same order in which they were performed when the proof was done (see, for example, the first condensed proof of Proposition 1 given above).

2. The names of the techniques are often omitted (for instance, in the foregoing condensed proofs, no mention is made of the forward and backward processes or the key question).

3. Several steps of the proof are often combined into a single statement with little or no explanation (as is done in the last condensed proof of Proposition 1 given above).

### Steps for Reading a Condensed Proof

You should strive toward the ability to read and to dissect a condensed proof. To do so, you have to discover the thought processes that went into the proof, which you can do as follows.

1. Determine which techniques are used (because the forward-backward method is not the only one available).

2. Verify all of the steps involved by filling in missing details.

The more condensed the proof, the harder this process is. When an author writes, "It is easy to see that ...," or "Clearly, ..., " you can assume it will take you quite some time to fill in the missing details. In this book, reading proofs is made easier because preceding each condensed proof is an **analysis of proof** that describes the techniques, methodology, and reasoning involved in doing the proof. Additional suggestions and practice on reading condensed proofs are given throughout the book and in various exercises.

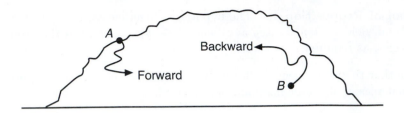

*Fig. 2.2*   Finding a needle in a haystack.

## Summary

To use the forward-backward method for proving that "*A* implies *B*," begin with the statement *B* that you are trying to conclude is true. Through the backward process of asking and answering the key question, derive a new statement, *B*1, with the property that, if *B*1 is true, then so is *B*. All efforts are now directed toward establishing that *B*1 is true. To that end, apply the backward process to *B*1, obtaining a new statement, *B*2, with the property that, if *B*2 is true, then so is *B*1 (and hence *B*). The backward process is motivated by the fact that *A* is assumed to be true. Continue working backward until either you obtain the statement *A* (in which case the proof is finished) or you can no longer pose and/or answer the key question fruitfully. In the latter case, it is time to start the forward process, in which you derive a sequence of statements from *A* that are necessarily true as a result of assuming that *A* is true. Remember that the goal of the forward process is to obtain precisely the last statement you have in the backward process, at which time the proof is complete.

These two processes are easily remembered by thinking of the statement *B* as a needle in a haystack. When you work forward from the assumption that *A* is true, you start somewhere on the outside of the haystack and try to find the needle. In the backward process, you start at the needle and try to work your way out of the haystack toward the statement *A* (see Figure 2.2).

Another way of remembering the forward-backward method is to think of a maze in which *A* is the starting point and *B* is the ending point (see Figure 2.3). You may have to alternate several times between the forward and backward processes because there are likely to be several false starts and blind alleys.

As a general rule, the forward-backward method is probably the first technique to try on a problem unless you have reason to use a different approach based on certain keywords that can arise in *A* and *B*, as you will learn. In any case, you will gain much insight into the relationship between *A* and *B*.

To read a condensed proof, you have to discover the thought processes that went into the proof. To do so, determine which techniques are used and verify all of the steps involved by filling in missing details.

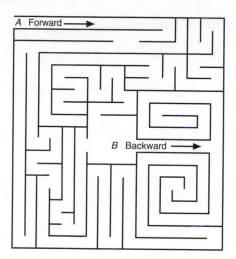

*Fig. 2.3*  The maze.

## Exercises

**Note:** Solutions to those exercises marked with a $B$ are in the back of this book. Solutions to those exercises marked with a $W$ are located on the web at http://www.wiley.com/college/solow/.

**Note:** All proofs should contain an analysis of proof and a condensed version. Definitions for all mathematical terms are provided in the glossary at the end of the book.

$^W$**2.1**    Explain the difference between the forward and backward processes. Describe how each one works and what can go wrong. How are the two processes related to each other?

**2.2**    When asking the key question, should you look at the last statement in the forward or backward process? When answering the key question, should you be guided by the last statement in the forward or backward process?

$^B$**2.3**    Suppose you are trying to prove that, "If $L_1$ and $L_2$ are two lines that are tangent to the endpoints of a diameter of circle, then $L_1$ and $L_2$ are parallel." What, if anything, is wrong with each of the following key questions?

    a. How can I show that $L_1$ and $L_2$ are two tangent lines to the endpoints of the diameter of a circle?

    b. How can I show that two lines are tangent to the endpoints of the diameter of a circle?

    c. How can I show that $L_1$ and $L_2$ are parallel?

    d. How can I show that two lines are parallel?

$^W$**2.4**    Consider the problem of proving that, "If $x$ is a real number, then the maximum value of $-x^2 + 2x + 1$ is greater than or equal to 2." What, if anything, is wrong with each of the following key questions?

a.  How can I show that the maximum value of a parabola is greater than or equal to a number?

b.  How can I show that a number is less than or equal to the maximum value of a polynomial?

c.  How can I show that the maximum value of the function $f$ defined by $f(x) = -x^2 + 2x + 1$ is greater than or equal to a number?

d.  How can I show that a number is less than or equal to the maximum of a quadratic function?

**2.5**    Suppose you are trying to prove that, "If $a$, $b$, and $c$ are integers for which $a$ divides $b$ and $b$ divides $c$, then $a$ divides $c$." What, if anything, is wrong with each of the following key questions?

a.  How can I show that $a$ divides $b$ and $b$ divides $c$?

b.  How can I show that $a$ divides $c$?

c.  How can I show that an integer divides another integer?

**2.6**    Suppose you are trying to prove that, "If $y = m_1 x + b_1$ and $y = m_2 x + b_2$ are two parallel lines in a plane, then $m_1 = m_2$." What, if anything, is wrong with each of the following key questions?

a.  How can I show that $m_1 = m_2$?

b.  How can I show that two lines are parallel?

c.  How can I show that two real numbers are equal?

**2.7**    Suppose you are trying to prove that, "If $a_0, \ldots, a_n$ are $n$ given real numbers, then the function $f(x) = (a_0 + a_1 x^1 + \cdots + a_n x^n)^2$ is a polynomial." What, if anything, is wrong with each of the following key questions?

a.  How can I show that a function is a polynomial?

b.  How can I show that the square of a function is a polynomial?

c.  How can I show the square of a polynomial is a polynomial?

**2.8**   Consider the problem of proving that, "If

$$
\begin{aligned}
R &= \{\text{real numbers } x : x^2 - x \le 0\}, \\
S &= \{\text{real numbers } x : -(x-1)(x-3) \le 0\}, \quad \text{and} \\
T &= \{\text{real numbers } x : x \ge 1\},
\end{aligned}
$$

then $R$ intersect $S$ is a subset of $T$." What, if anything, is wrong with each of the following key questions?

a. How can I show that a set is a subset of another set?

b. How can I show that $R$ intersect $S$ is a subset of $T$?

c. How can I show that every point in $R$ intersect $S$ is greater than or equal to 1?

d. How can I show that the intersection of two sets has a point in common with another set?

$^B$**2.9**   For the key question, "How can I show that a positive integer is prime (that is, can only be divided evenly by 1 and itself)?" what is wrong with the following answer: "Show that the integer is odd."

**2.10**   For the key question, "How can I show that two lines in a plane are parallel?" which of the following answers is incorrect? Explain.

a. Show that the slopes of the two lines are the same.

b. Show that each of the two lines is parallel to a third line.

c. Show that each of the two lines is perpendicular to a third line in the plane.

d. Show that the lines are on opposite sides of a quadrilateral.

**2.11**   For the key question, "How can I show that two nonzero integers are equal?" which of the following answers is incorrect? Explain.

a. Show that the ratio of the two integers is 1.

b. Show that the first integer is $\le$ the second integer and that the second integer is $\le$ the first integer.

c. Show that the squares of the two integers are equal.

d. Show that the absolute value of their difference is less than 1.

**2.12**   How does your answer to the previous exercise change if the key question is, "How can I show that two real numbers are equal?" (Assume that the word "integer" is changed to "real" in each answer in the previous exercise.)

$^B$**2.13**    List at least two key questions for each of the following problems. Be sure your questions contain no symbols or notation from the specific problem.

a. If $L_1$ and $L_2$ are tangent lines to a circle $C$ at the two endpoints $e_1$ and $e_2$ of a diameter $d$, respectively, then $L_1$ and $L_2$ are parallel.

b. If $f$ and $g$ are polynomials, then the function $f + g$ is a polynomial.

**2.14**    List at least two key questions for each of the following problems. Be sure your questions contain no symbols or notation from the specific problem.

a. If $n$ is an even integer, then $n^2$ is an even integer.

b. If $n$ is a given integer satisfying $-3n^2 + 2n + 8 = 0$, then $2n^2 - 3n = -2$.

**2.15**    List at least two key questions for each of the following problems. Be sure your questions contain no symbols or notation from the specific problem.

a. If $a$ and $b$ are real numbers, then $a^2 + b^2 \leq (a + b)^2$.

b. If $y = m_1 x + b_1$ and $y = m_2 x + b_2$ are the equations of two lines for which $m_1(m_2) = -1$, then the two lines are perpendicular.

**2.16**    List at least two key questions for each of the following problems. Be sure your questions contain no symbols or notation from the specific problem.

a. If $RST$ is an isosceles triangle with sides $RS$, $ST$, and $TR$, and $SU$ is a perpendicular bisector of $RT$, then $\overline{RS} = \overline{ST}$.

b. If $R$ and $T$ are the sets in Exercise 2.8, then $R$ intersect $T \neq \emptyset$.

$^B$**2.17**    List at least three answers to each of the following key questions.

a. How can I show that two real numbers are equal?

b. How can I show that two sets are equal?

$^W$**2.18**    List at least three answers to each of the following key questions.

a. How can I show that two lines in a plane are parallel?

b. How can I show that two triangles are congruent?

**2.19**    List at least two answers to each of the following key questions.

a. How can I show that two lines in a plane are perpendicular?

b. How can I show that a triangle is equilateral?

**2.20**   List at least two answers to each of the following key questions.

a. How can I show that a positive integer $> 1$ is not prime?

b. How can I show that a set is a subset of another set?

**2.21**   List at least three answers to each of the following key questions.

a. How can I show that a positive integer is less than another positive integer?

b. How can I show that an integer is even?

$^B$**2.22**   For the following implication, (1) pose a key question, (2) answer the question abstractly, and (3) apply your answer to the specific problem: "If $a$, $b$, and $c$ are real numbers for which $a > 0$, $b < 0$, and $b^2 - 4ac = 0$, then the solution to the equation $ax^2 + bx + c = 0$ is positive."

$^W$**2.23**   For the following implication, (1) pose a key question, (2) answer the question abstractly, and (3) apply your answer to the specific problem: "If $SU$ is a perpendicular bisector of $RT$, and $\overline{RS} = 2\overline{RU}$, then triangle $RST$ is equilateral (see the figure below)."

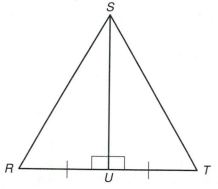

**2.24**   For the following implication, (1) pose a key question, (2) answer the question abstractly, and (3) apply your answer to the specific problem: "If $a, b, c, d, e$, and $f$ are real numbers with $b \neq 0$ and $d \neq 0$ and $a/b = c/d$, then the two lines $ax + by = e$ and $cx + dy = f$ are parallel."

**2.25**   For the following implication, (1) pose a key question, (2) answer the question abstractly, and (3) apply your answer to the specific problem: "If $d$ is an integer $> 2$, then the equation $x^2 + 3x + d = 0$ has no real solution."

$^B$**2.26**   Suppose you are trying to prove that, "If $R$ is a subset of $S$ and $S$ is a subset of $T$, then $R$ is a subset of $T$." What is wrong with the following statement in the forward process: "Because $R$ is a subset of $T$, it follows that every element of $R$ is also an element of $T$."

**2.27**    Suppose you are trying to prove that, "If $R$ is a subset of $S$ and $S$ is a subset of $T$, then $R$ is a subset of $T$." What is wrong with the following statement in the forward process: "Because $R$ is a subset of $S$, it follows that every element of $S$ is also an element of $R$."

*2.28    Identify the error in the forward steps of the following condensed proof of the statement, "If $n$ is a positive integer for which $n^2 < 2^n$, then $(n+1)^2 < 2^{n+1}$." Then modify the hypothesis so that the resulting statement is true.

> **Proof.** Now $(n + 1)^2 = n^2 + 2n + 1$ and, because $n^2 < 2^n$, it follows that $n^2 + 2n + 1 < 2^n + 2n + 1$. Finally, $2n + 1 < 2^n$ and so $(n + 1)^2 = n^2 + 2n + 1 < 2^n + 2^n = 2^{n+1}$.  □

$^B$**2.29**    For each of the following hypotheses, list at least two statements that are a result of applying the forward process one step from the hypothesis.

a. The real number $x$ satisfies $x^2 - 3x + 2 < 0$.

b. The sine of angle $X$ in triangle $XYZ$ of Figure 2.1 on page 9 is $1/\sqrt{2}$.

c. The circle $C$ consists of all values for $x$ and $y$ that satisfy the equation $(x - 3)^2 + (y - 2)^2 = 25$.

**2.30**    For each of the following hypotheses, list at least two statements that are a result of applying the forward process one step from the hypothesis.

a. The rectangle $ABCD$ is a square.

b. For the integer $n$, $n^2 - 1$ is odd.

c. The line $y = 3x - 1$ is tangent to the function $x^2 + x$.

$^B$**2.31**    Consider the problem of proving that, "If $x$ and $y$ are real numbers such that $x^2 + 6y^2 = 25$ and $y^2 + x = 3$, then $y = 2$." In working forward from the hypothesis, which of the following is not valid? Explain.

a.    $y^2 = 3 - x$.                         b.    $y^2 = 25/6 - (x/\sqrt{6})^2$.

c.    $(3 - y^2)^2 + 6y^2 - 25 = 0$.         d.    $x + 5 = -6y^2/(x - 5)$.

**2.32**    Consider the problem of proving that, "If $n$ is a positive integer, $a$ and $b$ are the lengths of the legs of a right triangle, and $c$ is the length of the hypotenuse, then $c^n > a^n + b^n$." In working forward from the hypothesis, which of the following is not valid? Explain.

a.    $c^2 = a^2 + b^2$.                      b.    $c > b$.

c.    $c^{n-2} > b^{n-2}$.                    d.    $c^n = c^2 c^{n-2}$.

*2.33   What, if anything, is wrong with the following proof that, "If $a$, $b$, and $c$ are real numbers for which $b - c = a$, then $\sqrt{b^2 - 4ac} = b - 2a$"?

> **Proof.** From the hypothesis that $b - c = a$, it follows that $4ab - 4ac = 4a^2$. But then $b^2 - 4ac = b^2 - 4ab + 4a^2$, that is, $b^2 - 4ac = (b - 2a)^2$. Taking the square root of both sides yields the desired conclusion that $\sqrt{b^2 - 4ac} = b - 2a$. $\square$

$^W$**2.34**   For the problem of proving that, "If $n$ is an integer greater than 2, $a$ and $b$ are the lengths of the legs of a right triangle, and $c$ is the length of the hypotenuse, then $c^n > a^n + b^n$," provide justification for each sentence in the following condensed proof.

> **Proof.** You have that $c^n = c^2 c^{n-2} = (a^2 + b^2)c^{n-2}$. Observing that $c^{n-2} > a^{n-2}$ and $c^{n-2} > b^{n-2}$, it follows that $c^n > a^2(a^{n-2}) + b^2(b^{n-2})$. Consequently, $c^n > a^n + b^n$. $\square$

$^B$**2.35**   Read the following condensed proof that, "If $x$ and $y$ are nonnegative real numbers that satisfy $x + y = 0$, then $x = 0$." What key question did the author ask and how was that question answered?

> **Proof.** From the hypothesis that $x + y = 0$, it follows that $x = -y$. Also, because $y \geq 0$, $-y \leq 0$ and hence $x = -y \leq 0$ and so $x = 0$. $\square$

$^B$**2.36**   Consider the problem of proving that, "If $x$ and $y$ are nonnegative real numbers that satisfy $x + y = 0$, then $x = 0$ and $y = 0$."

a. For the following condensed proof, write an analysis indicating the forward and backward steps and the key questions and answers.

> **Proof.** First, it will be shown that $x \leq 0$, for then, because $x \geq 0$, it must be that $x = 0$. To see that $x \leq 0$, by the hypothesis, $x + y = 0$, so $x = -y$. Also, because $y \geq 0$, it follows that $-y \leq 0$ and hence $x = -y \leq 0$. Finally, because $x = 0$ and $x + y = 0$, it follows that $0 = x + y = 0 + y = y$. $\square$

b. Rewrite the proof of part (a) entirely from the backward process.

**2.37**   Consider the problem of proving that, "If $RST$ is the triangle in Exercise 2.23, then triangle $SUR$ is congruent to triangle $SUT$." For the following condensed proof, write an analysis indicating the forward and backward steps, and the key questions and answers.

> **Proof.** It will be shown that $\overline{RU} = \overline{UT}$, for then you also have $\angle RUS = \angle TUS = 90°$ and $\overline{SU} = \overline{SU}$. To that end, note that

$\overline{RU} = \overline{UT}$ because $SU$ is a perpendicular bisector of $RT$, and so the proof is complete.  □

**2.38**    Consider the following condensed proof that, "If $A$ implies $B$" and "$B$ implies $C$," then "$A$ implies $C$." Write an analysis indicating the forward and backward steps, and the key questions and answers.

> **Proof.** To conclude that "$A$ implies $C$" is true, assume that $A$ is true. By the hypothesis, "$A$ implies $B$" is true, so $B$ must be true. Finally, because "$B$ implies $C$" is true, it must be that $C$ is true, thus completing the proof.  □

$^W$**2.39**    Consider an alphabet consisting of the two letters $s$ and $t$, together with the following rules for creating new "words" from old ones. (You can apply the rules in any order.)

(1)  Double the current word (for example, $sts$ could become $stssts$).

(2)  Erase $tt$ from the current word (for example, $stts$ could become $ss$).

(3)  Replace $sss$ in the current word by $t$ (for example, $stsss$ could become $stt$).

(4)  Add the letter $t$ at the right end of the current word if the last letter is $s$ (for example, $tss$ could become $tsst$).

  a. Use the forward process to derive all possible words that can be obtained in three steps by repeatedly applying the above rules in any order to the initial word $s$.

  b. Apply the backward process one step to the word $tst$. Specifically, list all the words for which an application of one of the above rules would result in $tst$.

  c. Prove that, "If $s$, then $tst$."

$^B$**2.40**    Prove that, if the right triangle $XYZ$ in Figure 2.1 on page 9 is isosceles, then the area of the triangle is $z^2/4$.

$^*$**2.41**    Prove that, if $XYZ$ is a triangle with $\angle X = 30^o$, $\angle Y = 60^o$, and $\angle Z = 90^o$ and $W$ is the midpoint of the hypotenuse, then the line connecting $W$ to $Z$ divides $XYZ$ into an equilateral triangle and an isosceles triangle.

$^W$**2.42**    Prove that the statement in Exercise 2.23 is true.

$^*$**2.43**    Prove that, if the triangle $RST$ in Exercise 2.23 is equilateral and $SU$ is a perpendicular bisector of $RT$, then the area of triangle $RST = \sqrt{3}\left(\overline{RS}\right)^2/4$.

# 3

## On Definitions and Mathematical Terminology

In the previous chapter you learned the forward-backward method and saw the importance of formulating and answering the key question. One of the simplest yet most effective ways of answering a key question is through the use of a definition, as explained in this chapter. You will also learn some of the "vocabulary" of the language of mathematics.

### 3.1 DEFINITIONS

A **definition** in mathematics is an agreement, by all parties concerned, as to the meaning of a particular term. You have already come across a definition in Chapter 1. There, the statement "$A$ implies $B$" was defined—and hence, agreed—to be true in all cases except when $A$ is true and $B$ is false. Nothing says that you must accept this definition as being correct. If you choose not to, then we will be unable to communicate regarding this particular idea.

Definitions are not made randomly. Usually they are motivated by a mathematical concept that occurs repeatedly. You can view a definition as an abbreviation that is agreed on for a particular concept. Take, for example, the notion of "a positive integer greater than one that is not divisible by any positive integer other than one and itself." This type of number is abbreviated (or defined) as a "prime." Surely it is easier to say "prime" than "a positive integer greater than one . . .," especially if the concept comes up frequently. Several other examples of definitions follow:

**Definition 1** *An integer $n$* **divides** *an integer $m$ (written $n|m$) if $m = kn$, for some integer $k$.*

**Definition 2** *An integer $p > 1$ is* **prime** *if the only positive integers that divide $p$ are 1 and $p$.*

**Definition 3** *A triangle is* **isosceles** *if two of its sides have equal length.*

**Definition 4** *Two ordered pairs of real numbers $(x_1, y_1)$ and $(x_2, y_2)$ are* **equal** *if $x_1 = x_2$ and $y_1 = y_2$.*

**Definition 5** *An integer $n$ is* **even** *if and only if the remainder on dividing $n$ by 2 is 0.*

**Definition 6** *An integer $n$ is* **odd** *if and only if $n = 2k+1$ for some integer $k$.*

**Definition 7** *A real number $r$ is* **rational** *if and only if $r$ can be expressed as the ratio of two integers $p$ and $q$ in which the denominator $q$ is not 0.*

**Definition 8** *Two statements $A$ and $B$ are* **equivalent** *if and only if "$A$ implies $B$" and "$B$ implies $A$."*

**Definition 9** *The statement $A$* **AND** *$B$ (written $A \bigwedge B$) is true if and only if $A$ is true and $B$ is true.*

**Definition 10** *The statement $A$* **OR** *$B$ (written $A \bigvee B$) is true in all cases except when $A$ is false and $B$ is false.*

Observe that the words "if and only if" are used in some of the definitions, but often, "if" is used instead of "if and only if." Some terms, such as "set" and "point," are left undefined. One could possibly try to define a set as a collection of objects, but to do so is impractical because the concept of an "object" is too vague; one would then be led to ask for the definition of an "object," and so on, and so on. Such philosophical issues are beyond the scope of this book.

## Using Definitions in Proofs

In the proof of Proposition 1 on page 9, a definition is used to answer a key question. Recall the first one, which is, "How can I show that a triangle is isosceles?" Using Definition 3, to show that a triangle is isosceles, one shows that two of its sides have equal length. Definitions are equally useful in the forward process. For instance, if you know that an integer $n$ is odd, then by Definition 6 you know that $n = 2k + 1$, for some integer $k$. Using definitions to work forward and backward is a common occurrence in proofs.

It is often the case that there seem to be two possible definitions for the same concept. Take, for example, the notion of an even integer introduced in

Definition 5. A second plausible definition for an even integer is "an integer that can be expressed as 2 times some other integer." There can be only one definition for a particular concept, so when more possibilities exist, how do you select the definition and what happens to the other alternatives?

Because a definition is simply something agreed on, any one of the alternatives can be agreed on as the definition. Once the definition is chosen, it is advisable to establish the "equivalence" of the definition and the alternatives. For the case of an even integer, this is accomplished by using Definition 5 and the alternative to create the statements:

**A:** $n$ is an integer whose remainder on dividing by 2 is 0.

**B:** $n$ is an integer that can be expressed as 2 times some integer.

To establish that the definition is equivalent to the alternative, you must show that "$A$ implies $B$" and "$B$ implies $A$" (see Definition 8). Then you would know that, if $A$ is true (that is, $n$ is an integer whose remainder on dividing by 2 is 0), then $B$ is true (that is, $n$ is an integer that can be expressed as 2 times some other integer). Moreover, if $B$ is true then $A$ is true, too.

The statement that $A$ is equivalent to $B$ is often written "$A$ is true if and only if $B$ is true," or, "$A$ if and only if $B$." In mathematical notation, one would write "$A$ iff $B$" or "$A \Leftrightarrow B$." Whenever you need to show that "$A$ if and only if $B$," you must show that "$A$ implies $B$" and "$B$ implies $A$."

It is useful to establish that a definition is equivalent to an alternative. To see why, suppose that in some proof you derive the key question, "How can I show that an integer is even?" As a result of having obtained the equivalence of the two concepts, you now have two possible answers. One is obtained from the definition: show that the remainder on dividing the integer by 2 is 0; the second answer comes from the alternative: show that the integer can be expressed as 2 times some other integer. Similarly, in the forward process, if you know that $n$ is an even integer, then you would have two possible statements that are true as a result of this—the original definition and the alternative. While the ability to answer a key question (or to work forward) in more than one way can be a hindrance, as was the case in the proof of Proposition 1, it can also be advantageous, as is shown now.

**Proposition 2** *If $n$ is an even integer, then $n^2$ is an even integer.*

**Analysis of Proof.**    From the forward-backward method, you are led to the key question, "How can I show that an integer (namely, $n^2$) is even?" Choosing the alternative over the definition, you can answer this question by showing that

**B1:** $n^2$ can be expressed as 2 times some other integer.

The only question is, which integer? The answer is from the forward process.

Because $n$ is an even integer, using the alternative, $n$ can be expressed as 2 times some integer, say $k$; that is,

**A1:** $n = 2k$.

Squaring both sides of $A1$ and rewriting by algebra yields

**A2:** $n^2 = (n)(n) = (2k)(2k) = 4k^2 = 2(2k^2)$.

Thus, it has been shown that $n^2$ is 2 times some other integer, that other integer being $2k^2$ (see $A2$), and this completes the proof. You could also prove this proposition by using Definition 5, but it is harder that way.

**Proof of Proposition 2.** Because $n$ is an even integer, there is an integer $k$ for which $n = 2k$. Consequently, $n^2 = (2k)^2 = 2(2k^2)$, and so $n^2$ is an even integer. $\Box$

A definition is one common method for working forward and for answering key questions. The more statements you can show are equivalent to the definition, the more ammunition you have available for the forward and backward processes; however, a large number of equivalent statements can also make it challenging to know exactly which one to use.

## Notational Issues

Notational difficulties sometimes arise when using definitions in the forward and backward processes. This occurs when the definition uses one set of symbols and notation while the specific problem under consideration uses a second set of symbols and notation. When these two sets of symbols are completely distinct from each other, generally no confusion arises; however, care is needed when these two sets involve **overlapping notation**—that is, when the same symbol is used in both sets.

To illustrate, recall the foregoing Definition 1:

Definition 1. An integer $n$ **divides** an integer $m$ (written $n|m$) if $m = kn$, for some integer $k$.

Suppose the conclusion of the problem you are working with is:

**B:** The integer $p$ divides the integer $q$.

A key question associated with $B$ is, "How can I show that one integer (namely, $p$) divides another integer (namely, $q$)?" Because the symbols $p$ and $q$ in $B$ are distinct from those in the definition, you should have no trouble using the definition to obtain the following answer to the key question:

**B1:** $q = kp$, for some integer $k$.

Observe that $B1$ is obtained by **matching up the notation** in Definition 1 with that in $B$. That is, the symbol $p$ in $B$ is "matched" to the symbol

$n$ in the definition; the symbol $q$ in $B$ is "matched" to the symbol $m$ in the definition. $B1$ is then obtained by replacing $n$ everywhere in the definition with $p$ and $m$ everywhere with $q$.

This process of matching up the notation when using a definition is similar to a process you have seen when working with functions. To illustrate, suppose $f$ is a function of one variable defined by $f(x) = x(x+1)$. To write $f(a+b)$, you "match" $x$ to $a+b$; that is, you replace $x$ everywhere by $a+b$ to obtain $f(a+b) = (a+b)(a+b+1)$.

To see how notational difficulties arise when using definitions, suppose the last statement in the backward process of the problem you are currently working with is:

**B2:** The integer $k$ divides the integer $n$.

Once again, a key question associated with $B2$ is, "How can I show that one integer (namely, $k$) divides another integer (namely, $n$)?" A difficulty in applying Definition 1 to answer this question arises because of overlapping notation—the symbols $k$ and $n$ appear in both $B2$ and in the foregoing Definition 1, but the symbols are used differently in each case. Observe how this overlapping notation makes it challenging to match up the notation in $B2$ with that in the definition.

When overlapping notation occurs, you can avoid notational errors by first rewriting the definition using a new set of symbols that do not overlap with the specific problem under consideration. Then when you apply the definition to the specific problem, the matching up of notation will be clear. For the foregoing example, you could rewrite the definition using the symbols $a$, $b$, and $c$ as follows so that the definition contains no overlapping notation with $B2$:

Definition 1. An integer $a$ **divides** an integer $b$ (written $a|b$) if $b = ca$, for some integer $c$.

Now when you apply the foregoing definition with the new symbols to answer the key question associated with $B2$, "How can I show that one integer (namely, $k$) divides another integer (namely, $n$)?" you should have no trouble matching up the notation correctly (match $k$ to $a$ and $n$ to $b$) to obtain the following answer:

**B3:** $n = ck$, for some integer $c$.

With practice, you will find that there really is no need to rewrite a definition. This is because you will be able to match up notation correctly even when there is overlapping notation. Until you reach that point, however, it is advisable to rewrite the definition. It is critical to learn how to apply a definition correctly in the forward and backward processes because, if you make a mistake, the rest of the proof is incorrect.

## 3.2    USING PREVIOUS KNOWLEDGE

Just as you can use a definition in the forward and backward processes, so you can use **previous knowledge** in the form of a previously proven implication, as is shown now.

### Using Previous Knowledge in the Backward Process

To illustrate how to use previous knowledge in the backward process, suppose you want to prove the following proposition.

**Proposition 3**  *If the right triangle RST with sides of lengths r and s and hypotenuse of length t satisfies $t = \sqrt{2rs}$, then the triangle RST is isosceles (see Figure 3.1).*

**Analysis of Proof.**    The forward-backward method gives rise to the key question, "How can I show that a triangle (namely, $RST$) is isosceles?" One answer is to use Definition 3, but a second answer is provided by the conclusion of Proposition 1, which states that the triangle $XYZ$ is isosceles (see page 9). Perhaps the current triangle $RST$ is also isosceles for the same reason as triangle $XYZ$. To find out, it is necessary to see if $RST$ also satisfies the hypothesis of Proposition 1, as did triangle $XYZ$, for then $RST$ will also satisfy the conclusion, and hence be isosceles.

In verifying the hypothesis of Proposition 1 for the triangle $RST$, it is first necessary to match up the current notation with that of Proposition 1 (just as is done when applying a definition). In this case, the corresponding lengths are $x = r$, $y = s$, and $z = t$. Thus, to check the hypothesis of Proposition 1 for the current problem, you must show that

**B1:** The area of triangle $RST$ equals $t^2/4$,

or equivalently, because the area of triangle $RST$ is $rs/2$, you must show that

**B2:** $rs/2 = t^2/4$.

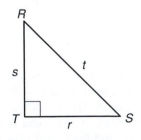

*Fig. 3.1*   The right triangle $RST$.

The fact that $rs/2 = t^2/4$ is established by working forward from the current hypothesis that

> **A:** $t = \sqrt{2rs}.$

To be specific, on squaring both sides of the equality in $A$ and dividing by 4, you obtain the desired conclusion that

> **A1:** $rs/2 = t^2/4.$

Do not forget to observe that the hypothesis of Proposition 1 also requires that the triangle $RST$ be a right triangle, which of course it is, as stated in the current hypothesis.

Notice how much more challenging it would be to match up the notation if the current triangle had been labeled $WXY$ with sides of length $w$ and $x$ and hypotenuse of length $y$. This overlapping notation can (and will) arise and, when it does, you should use the same technique as in the case of definitions; that is, you should rewrite the previous proposition with a set of symbols that do not overlap with the current problem. In the condensed proof that follows, note the complete lack of reference to the matching of notation.

**Proof of Proposition 3.**     By the hypothesis, $t = \sqrt{2rs}$, so $t^2 = 2rs$, or equivalently, $t^2/4 = rs/2$. Thus the area of the right triangle $RST = t^2/4$. As such, the hypothesis, and hence the conclusion, of Proposition 1 are true. Consequently, the triangle $RST$ is isosceles.  $\square$

From the foregoing example you can see that, to use previous knowledge in the backward process to prove that "$A$ implies $B$" is true, you should look for a previously proved proposition of the form "$C$ implies $B$" (that is, a proposition with the same conclusion as the current proposition, except for the notation) and then:

1. Match up the notation of the current proposition with that of the previous proposition.

2. Verify that the hypotheses of the previous proposition are satisfied for the current proposition; that is, prove that "$A$ implies $C$" is true.

## Using Previous Knowledge in the Forward Process

In Proposition 3, previous knowledge was used in the backward process to answer a key question. You can also use previous knowledge to work forward. To understand how, suppose you have already proved that "$A$ implies $C$" is true and that you now want to prove "$A$ implies $B$" is true. Working forward from the hypothesis that $A$ is assumed true, you can use the previous knowledge that "$A$ implies $C$" is true to claim, as a new statement in the forward process, that $C$ is true (see Table 1.1 on page 4). In doing the proof, you would write,

> Because $A$ is assumed to be true and it has already been
> proved that "$A$ implies $C$" is true, it follows that
> **A1:** $C$ is true.

Of course to complete the proof that "$A$ implies $B$" is true, you must still
work forward from $C$ to show that $B$ is true. Indeed, the time to use the
previous knowledge that "$A$ implies $C$" is true is when it is possible to work
forward from $C$ to get to $B$. More specifically, when trying to prove that
"$A$ implies $B$" is true, look for a previously proved proposition of the form
"$A$ implies $C$" (that is, a proposition with the same hypothesis as the current
proposition), for which it is possible to work forward from $C$ to get to $B$.

This technique of working forward using previous knowledge arose in the
proof of Proposition 1 in Chapter 2. Specifically, recall from the hypothesis of
Proposition 1 on page 9 that $XYZ$ is a right triangle, for which it is possible
to use the following previous knowledge:

**Pythagorean Theorem:** If $ABC$ is a right triangle with sides of lengths $a$
and $b$ and hypotenuse of length $c$, then $a^2 + b^2 = c^2$.

After matching up notation ($a = x$, $b = y$, and $c = z$), this previous knowledge
is used in the forward process of the proof of Proposition 1 to state that

**A2:** $x^2 + y^2 = z^2$.

The foregoing statement $A2$ was helpful in reaching the desired conclusion
that triangle $XYZ$ is isosceles and completing the proof of Proposition 1.

In general, to use previous knowledge in the forward process to prove that
"$A$ implies $B$" is true, look for a previously proved proposition of the form
"$A$ implies $C$" (that is, a proposition with the same hypothesis as the current
proposition, except for the notation) and then:

1. Match up the notation of the current proposition with that of the previous proposition.

2. Write, as a new statement in the forward process, the conclusion of the previous proposition using the notation of the current proposition (that is, write a forward statement that $C$ is true).

3. Complete the proof that "$A$ implies $B$" by working forward from $C$ and backward from $B$.

## 3.3  MATHEMATICAL TERMINOLOGY

In dealing with proofs, there are four terms you will often come across in
mathematics: *proposition, theorem, lemma,* and *corollary*. A **proposition** is

*Table 3.1*   Statements Related to "*A* Implies *B*."

| Statement | Alternate Written Form | Name of Statement |
|-----------|------------------------|-------------------|
| "*B* implies *A*" | $B \Rightarrow A$ | converse |
| "*NOT A* implies *NOT B*" | $(\neg A) \Rightarrow (\neg B)$ | inverse |
| "*NOT B* implies *NOT A*" | $(\neg B) \Rightarrow (\neg A)$ | contrapositive |

a true statement of interest that you are trying to prove. Some propositions are subjectively considered to be extremely important, and these are referred to as **theorems**. When the proof of a theorem is long, the proof is often easier to communicate in "pieces." For example, when proving the statement "*A* implies *B*," it may first be convenient to show that "*A* implies *C*," then that "*C* implies *D*," and finally that "*D* implies *B*." Each of these supporting propositions would be presented separately and referred to as a *lemma*. In other words, a **lemma** is a preliminary proposition that is used in the proof of a theorem. Once a theorem is proved, it is often the case that certain propositions follow almost immediately as a result of knowing that the theorem is true. These are called **corollaries**.

Just as there are certain mathematical concepts that are accepted without a formal definition, so certain statements are accepted without a formal proof. These unproved statements are called **axioms**. One example of an axiom is, "the shortest distance between two points is a straight line." A further discussion of axioms is beyond the scope of this book.

Associated with a statement *A* is the statement *NOT A* (sometimes written $\neg A$ or $\sim A$). The statement *NOT A* is true when *A* is false, and vice versa. More is said about the *NOT* of a statement in Chapter 8.

Given two statements *A* and *B*, you have already learned the meaning of the statement "*A* implies *B*." There are other ways of saying the statement "*A* implies *B*," for example:

1. Whenever *A* is true, *B* must also be true.

2. *B* follows from *A*.

3. *B* is a necessary consequence of *A* (meaning that if *A* is true, *B* is necessarily true also.)

4. *A* is sufficient for *B* (meaning that, if you want *B* to be true, it is enough to know that *A* is true.)

5. *A* only if *B*.

Three other statements closely related to "*A* implies *B*" are the **contrapositive statement**, the **converse statement**, and the **inverse statement**,

*Table 3.2*   The Truth Table for "*NOT B* Implies *NOT A*."

| $A$ | $B$ | $NOT\ B$ | $NOT\ A$ | $A \Rightarrow B$ | $NOT\ B \Rightarrow NOT\ A$ |
|-----|-----|----------|----------|-------------------|------------------------------|
| True  | True  | False | False | True  | True  |
| True  | False | True  | False | False | False |
| False | True  | False | True  | True  | True  |
| False | False | True  | True  | True  | True  |

as given in Table 3.1. You can use Table 1.1 to determine when each of these three statements is true. For instance, the contrapositive statement, "*NOT B* implies *NOT A*," is true in all cases except when the statement to the left of the word "implies" (namely, *NOT B*) is true and the statement to the right of the word "implies" (namely, *NOT A*) is false. In other words, the contrapositive statement is true in all cases except when $B$ is false and $A$ is true, which is precisely the same as when the statement "*A* implies *B*" is true (see Table 3.2). That is, the statement "*A* implies *B*" is logically equivalent to the contrapositive statement "NOT *B* implies NOT *A*."

Note from Table 3.2 that the statement "*NOT B* implies *NOT A*" is true under the same conditions as "*A* implies *B*," that is, in all cases except when $A$ is true and $B$ is false. This observation gives rise to a new proof technique known as the contrapositive method that is described in Chapter 10. In the exercises, you are asked to derive truth tables similar to Table 3.2 for the converse and inverse statements.

## Summary

You have now learned how definitions and previous knowledge are used in the forward-backward method. A definition is used in the backward process to answer a key question and in the forward process to derive new statements. To use a definition, you must correctly match the notation of the current proposition to that of the definition. If there is overlapping notation between the two, then you should rewrite the definition using a set of symbols that do not overlap with those of the current proposition.

To use previous knowledge during the backward process to prove that "*A* implies *B*" is true, look for a previously proved proposition of the form "*C* implies *B*" (that is, a proposition with the same conclusion as the current proposition, except for the notation) and then:

1. Match up the notation of the current proposition with that of the previous proposition.

2. Verify that the hypotheses of the previous proposition are satisfied for the current proposition; that is, prove that "*A* implies *C*" is true.

In contrast, to use previous knowledge in the forward process to prove that "*A* implies *B*" is true, look for a previously proved proposition of the form "*A* implies *C*" (that is, a proposition with the same hypothesis as the current proposition, except for the notation) and then:

1. Match up the notation of the current proposition with that of the previous proposition.

2. Write, as a new statement in the forward process, the conclusion of the previous proposition using the notation of the current proposition (that is, write a forward statement that *C* is true).

3. Complete the proof that "*A* implies *B*" by working forward from *C* and backward from *B*.

You have also learned many of the terms used in the language of mathematics: an axiom is a statement that is accepted as being true without having to provide a proof; a proposition is a true statement that you are trying to prove; a theorem is an important proposition; a lemma is a preliminary proposition that is used in the proof of a theorem; and a corollary is a proposition that follows from a theorem.

Now it is time to learn more proof techniques for proving propositions, theorems, corollaries, and lemmas.

### Exercises

**Note:** Solutions to those exercises marked with a *B* are in the back of this book. Solutions to those exercises marked with a *W* are located on the web at http://www.wiley.com/college/solow/.

**Note:** All proofs should contain an analysis of proof and a condensed version. Definitions for all mathematical terms are provided in the glossary at the end of the book.

$^B$**3.1**    For each of the following conclusions, pose a key question. Then use a definition (1) to answer the question abstractly and (2) to apply the answer to the specific problem.

a. If $n$ is an odd integer, then $n^2$ is an odd integer.

b. If $s$ and $t$ are rational numbers with $t \neq 0$, then $s/t$ is rational.

c. Suppose that $a$, $b$, $c$, $d$, $e$, and $f$ are real numbers with the property that $ad - bc \neq 0$. If $(x_1, y_1)$ and $(x_2, y_2)$ are pairs of real numbers satisfying:

$$ax_1 + by_1 = e, \quad cx_1 + dy_1 = f,$$
$$ax_2 + by_2 = e, \quad cx_2 + dy_2 = f,$$

then $(x_1, y_1)$ equals $(x_2, y_2)$.

**3.2**    For each of the following conclusions, pose a key question. Then use a definition (1) to answer the question abstractly and (2) to apply the answer to the specific problem.

  a. If $n$ is an integer for which $n^2$ is even, then $n$ is even.

  b. If $n$ is an integer $> 1$ for which $2^n - 1$ is prime, then $n$ is prime.

  c. If $n - 1$, $n$, and $n + 1$ are three consecutive integers, then 9 divides the sum of their cubes.

**3.3**    For each of the following statements, obtain a new statement in the backward process by using a definition to answer the key question. If necessary, rewrite the definitions so that no overlapping notation occurs.

  a.    The integer $m > 1$ is prime.    b. $p^2$ is even ($p$ is an integer).

  c.    Triangle $ABC$ is equilateral.    d. $\sqrt{n}$ is rational ($n$ is an integer).

$^W$**3.4**    For each of the following hypotheses, use a definition to work forward one step.

  a. If $n$ is an odd integer, then $n^2$ is an odd integer.

  b. If $s$ and $t$ are rational numbers with $t \neq 0$, then $s/t$ is rational.

  c. If the right triangle $XYZ$ of Figure 2.1 on page 9 satisfies the property that $\sin(X) = \cos(X)$, then triangle $XYZ$ is isosceles.

  d. If $a$, $b$, and $c$ are integers for which $a|b$ and $b|c$, then $a|c$.

**3.5**    Use a definition to work forward from each of the following statements.

  a. The hypothesis in Exercise 3.2(b).

  b. For sets $R$, $S$, and $T$, $R = S \cup T$.

  c. For functions $f$ and $g$, the function $f + g$ is convex, where $f + g$ is the function whose value at any point $x$ is $f(x) + g(x)$. (See the definition of a convex function in the glossary.)

  d. For functions $f$ and $g$ and sets $S$ and $T$, the function $f \geq g$ on $S \cap T$. (See the definition of greater-than-or-equal-to functions in the glossary.)

$^B$**3.6**    Write truth tables for the following statements.

  a. The converse of "$A$ implies $B$."

  b. The inverse of "$A$ implies $B$." How are (a) and (b) related?

$^W$**3.7**   Write truth tables for the following statements.

  a.   *A AND B.*          b. *A AND NOT B.*

**3.8**   Write truth tables for the following statements.

  a. *A OR B.*

  b. *(NOT A) OR B.* How is this related to the truth of "*A implies B*"?

$^B$**3.9**   Write the indicated statement for each of the following problems.

  a. The contrapositive of the proposition, "If $n$ is an integer for which $n^2$ is even, then $n$ is even."

  b. The inverse of the proposition, "If $r$ is a real number such that $r^2 = 2$, then $r$ is not rational."

  c. The converse of the proposition, "If the quadrilateral $ABCD$ is a parallelogram with one right angle, then $ABCD$ is a rectangle."

**3.10**   Write the indicated statement for each of the following problems.

  a. The converse of the proposition, "If $a$, $b$, and $c$ are integers for which $a|b$ and $b|c$, then $a|c$."

  b. The contrapositive of the proposition, "If $n > 1$ is an integer for which $2^n - 1$ is prime, then $n$ is prime."

  c. The inverse of the proposition, "If $r$ is a rational number with $r \neq 0$, then $1/r$ is rational."

$^B$**3.11**   Prove that if "$A$ implies $B$" and "$B$ implies $C$," then "$A$ implies $C$."

**3.12**   Suppose you have already proved that "$C$ implies $D$." Using the result in Exercise 3.11, what single implication involving the new statement $E$ would you have to prove in order to show that each of the following statements is true?

  a.   $C$ implies $E$.        b.   $E$ implies $D$.

**3.13**   Use Exercise 3.11 to prove that if "$A$ implies $B$," "$B$ implies $C$," and "$C$ implies $A$," then $A$ is equivalent to $B$ and $A$ is equivalent to $C$.

$^W$**3.14**   Consider a definition in the form of a statement $A$, together with three possible alternative definitions, say $B$, $C$, and $D$. Suppose you want to prove that $A$ is equivalent to each of the three alternatives.

  a. Explain why you can do so by proving that "$A$ implies $B$," "$B$ implies $C$," "$C$ implies $D$," and "$D$ implies $A$".

b. What is the advantage of the approach in part (a) as opposed to proving separately that $A$ is equivalent to each of statements $B$, $C$, and $D$? (Hint: Count the total number of proofs with the two approaches.)

**3.15**    Suppose you want to prove that, "If the right triangle $ABC$ with sides of lengths $a$ and $b$ and hypotenuse of length $c$ has an area of $c^2/4$, then the triangle $ABC$ is isosceles." How can you do so using Proposition 3 on page 30; that is, what would you have to prove to do so?

**3.16**    Suppose you want to prove that, "If the right triangle $ABC$ with sides of lengths $a$ and $b$ and hypotenuse of length $c$ satisfies $c = \sqrt{2ab}$, then angle $A$ is 45°." Write a forward statement that is obtained by using Proposition 3 on page 30.

**3.17**    Use Exercise 2.39(a) to prove each of the following propositions:

a. If $s$, then *ttst*.

b. If *stsss*, then *tst*.

c. What is the difference in the way previous knowledge is used in part (a) and part (b)?

$^B$**3.18**    Explain where and how Proposition 3 on page 30 is used in the following condensed proof that, "If the right triangle $UVW$ with sides of lengths $u$ and $v$ and hypotenuse of length $w$ satisfies $\sin(U) = \sqrt{u/2v}$, then the triangle $UVW$ is isosceles."

> **Proof.** It is given that $\sin(U) = \sqrt{u/2v}$ and from the definition of sine, $\sin(U) = u/w$, thus one has $\sqrt{u/2v} = u/w$. By algebraic manipulations one obtains $w = \sqrt{2uv}$ and hence the right triangle $UVW$ is isosceles.  □

**3.19**    Explain where and how previous knowledge is used in the following condensed proof that, "Two lines $L_1$ and $L_2$ that are tangent to the two endpoints of a diameter $D$ of a circle are parallel."

> **Proof.** The line $L_1$, being tangent to the endpoint of the diameter $D$ of a circle, is perpendicular to $D$. Likewise, $L_2$ is also perpendicular to $D$. As such, $L_1$ and $L_2$ are parallel.  □

$^W$**3.20**    Write an analysis of proof that corresponds to the following condensed proof of the proposition, "If the right triangle $UVW$ with sides of lengths $u$ and $v$ and hypotenuse of length $w$ satisfies $\sin(U) = \sqrt{u/2v}$, then the triangle $UVW$ is isosceles." Fill in the details of any missing steps.

**Proof.** It will be shown that $w = \sqrt{2uv}$, for then, by Proposition 3 on page 30, the right triangle $UVW$ will be isosceles. To that end, by hypothesis, $\sin(U) = \sqrt{u/2v}$ and, from the definition of sine, $\sin(U) = u/w$. Thus, $\sqrt{u/2v} = u/w$. By algebra, you have $w = \sqrt{2uv}$ and so the proof is complete. $\square$

**3.21**  Explain where and how previous knowledge from a proposition in this chapter is used in each of the following condensed proofs that, "If $m$ and $n$ are even integers, then $(m+n)^2$ is an even integer."

  a. **Proof.** Because $m$ and $n$ are even integers, there are integers $p$ and $q$ for which $m = 2p$ and $n = 2q$. But then $m + n = 2p + 2q = 2(p+q)$. This means that $m + n$ is even and, as such, $(m+n)^2$ is also even.

  b. **Proof.** Because $m$ and $n$ are even integers, it follows that $m^2$ and $n^2$ are also even. So, there are integers $p$ and $q$ for which $m^2 = 2p$ and $n^2 = 2q$. Hence, $(m+n)^2 = m^2 + 2mn + n^2 = 2p + 2mn + 2q = 2(p + mn + q)$ and so $(m+n)^2$ is even.

$^B$**3.22**  Prove that if $n$ is an odd integer, then $n^2$ is an odd integer.

**3.23**  Use the proposition in Exercise 3.22 to prove that if $a$ and $b$ are consecutive integers, then $(a+b)^2$ is an odd integer.

**3.24**  Use Proposition 2 on page 27 to prove that if $a$ and $b$ are odd integers, then $(a+b)^2$ is an even integer.

$^*$**3.25**  Suppose you have already proved the proposition that, "If $a$ and $b$ are nonnegative real numbers, then $(a+b)/2 \geq \sqrt{ab}$."

  a. Explain how you could use this proposition to prove that if $a$ and $b$ are real numbers satisfying the property that $b \geq 2|a|$, then $b \geq \sqrt{b^2 - 4a^2}$. Be careful how you match up notation.

  b. Use the foregoing proposition and part (a) to prove that if $a$ and $b$ are real numbers with $a < 0$ and $b \geq 2|a|$, then one of the roots of the equation $ax^2 + bx + a = 0$ is $\leq -b/a$.

$^W$**3.26**  Use the definition of an isosceles triangle to prove that, if the right triangle $UVW$ with sides of lengths $u$ and $v$ and hypotenuse of length $w$ satisfies $\sin(U) = \sqrt{u/2v}$, then the triangle $UVW$ is isosceles.

$^*$**3.27**  Use Proposition 1 on page 9 to prove that, if the right triangle $UVW$ with sides of lengths $u$ and $v$ and hypotenuse of length $w$ satisfies the property that $\sin(U) = \sqrt{u/2v}$, then the triangle $UVW$ is isosceles.

# 4

# *Quantifiers I:*
# *The Construction Method*

In Chapter 3, you saw that a definition is often used to answer a key question and to work forward. The next four chapters provide several other techniques for doing so that arise when $A$ or $B$ have a special form.

Two particular forms of statements appear repeatedly throughout all branches of mathematics. They are always identified by certain keywords. The first one has the words *there is* (*there are*, *there exists*); the second one has *for all* (*for each*, *for every*, *for any*). These two groups of words are referred to collectively as **quantifiers**, and each one gives rise to its own proof technique. The remainder of this chapter deals with the **existential quantifier** "there is." The **universal quantifier** "for all" is discussed in the next chapter.

## 4.1  WORKING WITH THE QUANTIFIER "THERE IS"

The quantifier "there is" arises naturally in many mathematical statements. Recall Definition 7 for a rational number as being a real number that can be expressed as the ratio of two integers in which the denominator is not zero. This definition could be written just as well using the quantifier "there are."

> Definition 7. A real number $r$ is **rational** if and only if there are integers $p$ and $q$ with $q \neq 0$ such that $r = p/q$.

Another such example arises from the alternative definition of an even integer, that being an integer that can be expressed as 2 times some integer.

Using a quantifier to express this statement, one obtains:

> Definition 5. An integer $n$ is **even** if and only if there is an integer $k$ such that $n = 2k$.

It is important to observe that the quantifier "there is" allows for the possibility of more than one such object, as is shown in the next definition.

**Definition 11** *An integer $n$ is a* **square** *if and only if there is an integer $k$ such that $n = k^2$.*

Note that if a nonzero integer $n$ (say, for example, $n = 9$) is a square, then there are two values of $k$ that satisfy $n = k^2$ (in this case, $k = 3$ or $-3$). More is said in Chapter 11 about the issue of uniqueness—that is, the existence of only one such object.

There are many other instances where an existential quantifier is used but, from the foregoing examples, you can see that such statements always have the same basic structure. Each time the quantifier "there is," "there are," or "there exists" appears, the statement will have the following **standard form**:

> There is an "object" with a "certain property" such that "something happens."

The words in quotation marks depend on the particular statement under consideration. You must learn to read, to identify, and to write each of the three components. Consider these examples.

1. There is an integer $x > 2$ such that $x^2 - 5x + 6 = 0$.
   Object:                integer $x$.
   Certain property:      $x > 2$.
   Something happens:     $x^2 - 5x + 6 = 0$.

2. There are real numbers $x$ and $y$ both $> 0$ such that $2x + 3y = 8$ and $5x - y = 3$.
   Object:                real numbers $x$ and $y$.
   Certain property:      $x > 0$, $y > 0$.
   Something happens:     $2x + 3y = 8$ and $5x - y = 3$.

Mathematicians often use the symbol $\exists$ to abbreviate the words "there is" ("there are," and so on) and the symbol $\ni$ for the words "such that" ("for which," and so on). The use of symbols is illustrated in the next example.

3. $\exists$ an angle $t \ni \cos(t) = t$.
   Object:                angle $t$.
   Certain property:      none.
   Something happens:     $\cos(t) = t$.

Observe that the words "such that" (or equivalent words like "for which") always precede the something that happens. Practice is needed to become fluent at reading and writing these statements.

## 4.2   HOW TO USE THE CONSTRUCTION METHOD

When proving that "$A$ implies $B$" is true, suppose you obtain a statement in the forward process that has the quantifier "there is" in the standard form:

> **A:** There is an "object" with a "certain property" such that "something happens."

Thus, you can assume that there is such an object, say, $X$. This object $X$ together with its certain property and the something that happens should help you reach the conclusion that $B$ is true. The technique of working with such an object in the forward process is straightforward and is therefore not given a special name.

In contrast, if you encounter the keywords "there is" during the backward process, then you must show that

> **B:** There is an "object" with a "certain property" such that "something happens."

One way to do so is to use the **construction method**. The idea is to construct (guess, produce, devise an algorithm to produce, and so on) the desired object. The constructed object then becomes a new statement in the forward process. However, you should realize that the construction of the object does not, by itself, constitute the proof. Rather, the proof is completed when you have shown that the object you construct is in fact the correct one; that is, that the object has the certain property and satisfies the something that happens, which becomes the next statement in the backward process.

How you actually construct the desired object is not at all clear. Sometimes it is by trial and error; sometimes an algorithm is designed to produce the desired object—it all depends on the particular problem. In almost all cases, the information in statement $A$ is used to help accomplish the task. Indeed, the appearance of the quantifier "there is" in the backward process strongly suggests turning to the forward process to produce the desired object. The construction method was used subtly in the proof of Proposition 2, but another example will clarify the process.

**Proposition 4** *If $a$, $b$, $c$, $d$, $e$, and $f$ are real numbers such that $ad - bc \neq 0$, then the two equations $ax + by = e$ and $cx + dy = f$ can be solved for the real numbers $x$ and $y$.*

**Analysis of Proof.**   On starting the backward process, you should recognize that statement $B$ has the form discussed above, even though the quantifier "there are" does not appear explicitly. To see that this is so, rewrite statement $B$ as follows to contain the quantifier explicitly:

> **B:** There are real numbers $x$ and $y$ such that $ax + by = e$ and $cx + dy = f$.

Statements containing **hidden quantifiers** occur frequently, and you should watch for them.

Proceeding with the construction method, the first step is to identify the objects, the certain property, and the something that happens. In this case:

| | |
|---|---|
| Objects: | real numbers $x$ and $y$. |
| Certain property: | none. |
| Something happens: | $ax + by = e$ and $cx + dy = f$. |

The next step is to construct these real numbers. Turn to the forward process to do so. If you are able to "guess" that $x = (de - bf)/(ad - bc)$ and $y = (af - ce)/(ad - bc)$, then you are fortunate. (Observe that, by guessing these values for $x$ and $y$, you have used the information in $A$ because the denominators are not 0.) However, recall that constructing these objects does not constitute the proof—you must still show that these objects satisfy the certain property and the something that happens. In this case, that means you must show that, for the foregoing values of $x$ and $y$, $ax + by = e$ and $cx + dy = f$. In doing this proof, you might write the following:

Noting that $ad - bc \neq 0$, construct
**A1:** $x = (de - bf)/(ad - bc)$    and    $y = (af - ce)/(ad - bc)$.
It must be shown that, for these values of $x$ and $y$,
**B1:** $ax + by = e$    and    $cx + dy = f$.

The remainder of this proof consists of working forward from $A1$ using algebra to show that $B1$ is true. The details are straightforward and are omitted.

While this "guess-and-check" approach is perfectly acceptable for producing the desired $x$ and $y$, it is not informative as to how these particular values are obtained. A more instructive proof is desirable. For example, these values for $x$ and $y$ are obtained by working backward from the following two equations that you want $x$ and $y$ to satisfy:

$$ax + by = e \tag{4.1}$$

$$cx + dy = f. \tag{4.2}$$

On multiplying (4.1) by $d$ and (4.2) by $b$ and then subtracting (4.2) from (4.1), you obtain:

$$(ad - bc)x = de - bf. \tag{4.3}$$

You can then use the information in $A$ to divide (4.3) through by $ad - bc$ because, from the hypothesis, this number is not 0, thus obtaining:

$$x = (de - bf)/(ad - bc).$$

A similar process is used to obtain $y = (af - ce)/(ad - bc)$. In general, it is advisable to work backward from the desired properties to construct the object.

Even with all this added explanation as to how the objects are constructed, the mere construction of the objects does not constitute the proof. You must still show that, for these values of $x$ and $y$, $ax + by = e$ and $cx + dy = f$.

**Proof of Proposition 4.**    On multiplying the equation $ax + by = e$ by $d$, and the equation $cx + dy = f$ by $b$, and then subtracting the two equations one obtains $(ad - bc)x = (de - bf)$. From the hypothesis, $ad - bc \neq 0$, and so dividing by $ad - bc$ yields $x = (de - bf)/(ad - bc)$. A similar argument shows that $y = (af - ce)/(ad - bc)$. It is not hard to check that, for these values of $x$ and $y$, $ax + by = e$ and $cx + dy = f$.    □

In the foregoing condensed proof, observe that the author glosses over the fact that the constructed objects do satisfy the needed properties. When writing such proofs, be sure to include those details. When reading such proofs, you must fill in those details yourself, as shown next.

## 4.3    READING A PROOF

The process of reading and understanding a proof is demonstrated with the following proposition.

**Proposition 5** *If $m < n$ are consecutive integers and $m$ is even, then 4 divides $m^2 + n^2 - 1$.*

**Proof of Proposition 5.**    (For reference purposes, each sentence of the proof is written on a separate line.)

> **S1:**    Let $n = m + 1$.
> **S2:**    Then $m^2 + n^2 - 1 = m^2 + (m + 1)^2 - 1 = 2m(m + 1)$.
> **S3:**    Because $m$ is even, there is an integer $k$ such that $m = 2k$.
> **S4:**    Letting $p = k(m + 1)$, it follows that

$$m^2 + n^2 - 1 = 2m(m + 1) = 4k(m + 1) = 4p,$$

and so 4 divides $m^2 + n^2 - 1$.

The proof is now complete.    □

**Analysis of Proof.**    An interpretation of statements $S1$ through $S4$ follows.

**Interpretation of S1:** *Let $n = m + 1$.*
The author is working forward from the hypothesis that $m$ and $n$ are consecutive integers.

**Interpretation of S2:** *Then $m^2 + n^2 - 1 = m^2 + (m + 1)^2 - 1 = 2m(m + 1)$.*
The author is working forward from $m^2 + n^2 - 1$ by substituting $n = m + 1$ from $S1$ and then using algebra.

**Interpretation of S3:** *Because m is even, there is an integer k such that* $m = 2k$.

The author is working forward from the hypothesis that $m$ is an even integer by using the definition.

**Interpretation of S4:** *Letting* $p = k(m + 1)$*, it follows that* ....

Up to this point, it is unclear why the author derives the statements $S1$, $S2$, and $S3$. The answer lies in $S4$. Specifically, the author is working backward from the conclusion $B$ and asks the key question, "How can I show that an integer (namely, 4) divides another integer (namely, $m^2 + n^2 - 1$)?" The author then uses Definition 1 on page 26 to answer the question, whereby it must be shown that

**B1:** There is an integer $p$ such that $m^2 + n^2 - 1 = 4p$.

Recognizing the keywords "there is" in $B1$, the author uses the construction method to construct the integer $p$; namely, $p = k(m + 1)$ (which is the reason for statements $S1$, $S2$, and $S3$). Finally, as required by the construction method, the author verifies that this value of $p = k(m + 1)$ is correct by showing that $m^2 + n^2 - 1 = 4p$.

The following points about reading the foregoing condensed proof of Proposition 5 are worth noting.

- No mention is made of the techniques being used (the forward-backward method, the key question and answer, and the construction method).

- Several steps are condensed into the single sentence $S4$.

- Although $S1$, $S2$, and $S3$ are true, it is not clear where the author is heading with these statements. When you are not sure of why the author is doing something, ask yourself what technique you would use to do the proof. For example, if you work backward yourself from $B$ and recognize that the quantifier "there is" arises, then you will realize that $S1$, $S2$, and $S3$ are part of the construction method.

### Summary

The construction method is a technique for dealing with statements in the backward process that have the quantifier "there is" in the standard form:

> There is an "object" with a "certain property" such that "something happens."

To use the construction method:

1. Identify the object, the certain property, and the something that happens in the backward statement containing the quantifier "there is."

2. Turn to the forward process and use the hypothesis together with your creative ability to construct the desired object. To do so, you might also find it useful to work backward from the properties you want the object to satisfy. (The actual construction of the object becomes the new statement in the forward process.)

3. Establish that the object you construct does satisfy the certain property and the something that happens. (The statement that the constructed object satisfies the desired properties becomes the new statement in the backward process.)

## Exercises

**Note:** Solutions to those exercises marked with a $B$ are in the back of this book. Solutions to those exercises marked with a $W$ are located on the web at http://www.wiley.com/college/solow/.

**Note:** All proofs should contain an analysis of proof and a condensed version. Definitions for all mathematical terms are provided in the glossary at the end of the book.

$^B$**4.1**   For each of the following there-is statements, identify the object, the certain property, and the something that happens.

   a. At a party of $n$ $(\geq 2)$ people, at least two of the people have the same number of friends.

   b. The function $f(x)$ has an integer root.

   c. There is a point $(x, y)$ with $x \geq 0$ and $y \geq 0$ that lies on the two lines $y = m_1 x + b_1$ and $y = m_2 x + b_2$.

   d. Given an angle $t$, one can find an angle $t'$ between 0 and $\pi$ whose tangent is larger than that of $t$.

   e. For the two integers $a$ and $b$, at least one of which is not zero and whose greatest common divisor is $c$, there are integers $m$ and $n$ such that $am + bn = c$.

**4.2**   Rewrite each of the following statements so that the quantifier "there is" appears in standard form. Then identify the object, the certain property, and the something that happens.

   a. Some element of the set $S$ is $\geq 0$.

   b. The set $T \neq \emptyset$ (the empty set).

   c. $x^2 - kx + 2 = 0$, for some positive integer $k$.

**4.3**   After constructing each of the following objects, what would you have to show about the object to complete the construction method?

   a. There is a positive integer $n$ for which $n! > 3^n$.

   b. The integer $n > 1$ can be divided by some integer $p$ with $1 < p < n$.

   c. For a given integer $n > 0$ and real numbers $a_0, \ldots, a_n$, the polynomial $a_0 + a_1 x^1 + \cdots + a_n x^n$ has $n$ (possibly complex) roots.

**4.4**   Suppose each of the statements in Exercise 4.1 [except for part (a)] is the conclusion of a proposition. Explain how you would apply the construction method to do the proof. (You need not actually do the proof.)

**4.5**   Would you use the construction method to prove each of the following propositions? Why or why not? Explain.

   a. The polynomial $x^{71} - 4x^{44} + 11x - 3$ has a real root.

   b. If $a$ and $b$ are integers with $a \neq 0$ for which $a$ does not divide $b$, then there is no positive integer $x$ such that $ax^2 + bx + b - a = 0$.

   c. If $ABCD$ is a square whose sides have length $s$, then you can inscribe in $ABCD$ a circle whose area is at least $3s^2/4$.

$^B$**4.6**   Explain why and how the construction method is used in the proof of Proposition 2 on page 27.

$^B$**4.7**   Use the construction method to prove that there is an integer $x$ such that $x^2 - 5x/2 + 3/2 = 0$. Is the constructed object unique?

$^W$**4.8**   Use the construction method to prove that there is a real number $x$ such that $x^2 - 5x/2 + 3/2 = 0$. Is the constructed object unique?

*__4.9__   Use the construction method to produce the desired object in each of the following statements. (You may have to use trial-and-error to produce the desired object.)

   a. The smallest positive integer $n$ such that $n! > 3^n$.

   b. The first positive integer $n$ such that you have more than double your money after $n$ years of investing at 5% interest compounded annually. (Recall that if \$P are invested at $r\%$ interest compounded annually for $n$ years, then you will have $\$\left(1 + \frac{r}{100}\right)^n P$ at the end of $n$ years.)

   c. An angle $x$ between 0 and $\pi/4$ such that the first three decimal digits of $\cos(x)$ and $x$ are the same.

$^B$**4.10**   Write an analysis of proof that corresponds to the following condensed proof that, "If $s$ and $t$ are rational numbers, then $s + t$ is a rational number." Indicate in which sentence the desired object is constructed.

**Proof.** It will be shown that there are integers $p$ and $q$ with $q \neq 0$ such that $s + t = p/q$. Now because $s$ and $t$ are rational, there are integers $a, b, c, d$ with $b, d \neq 0$ such that $s = a/b$ and $t = c/d$. Now set $p = ad + bc$ and $q = bd$. Then $q \neq 0$ and

$$s + t = \frac{a}{b} + \frac{c}{d} = \frac{ad + bc}{bd} = \frac{p}{q}.$$

The proof is now complete.  □

**4.11**    Write an analysis of proof that corresponds to the following condensed proof that, "If $a$, $b$, and $c$ are integers with $a \neq 0$ and for which $b^2 - 4ac$ is a square (see the definition on page 42), then the equation $ax^2 + bx + c = 0$ has a rational root." Indicate in which sentence the desired objects are constructed.

**Proof.** Because $b^2 - 4ac$ is a square, there is an integer $k$ such that $b^2 - 4ac = k^2$. Hence $\sqrt{b^2 - 4ac} = |k|$ is an integer. Thus, $x = \frac{-b + |k|}{2a}$ is a rational root of $ax^2 + bx + c = 0$.  □

**4.12**    Identify the error in the following condensed proof that, "If $r$ is a positive real number with $r \neq 1$, then there is an integer $n$ such that $2^{\frac{1}{n}} < r$."

**Proof.** Let $n$ be any integer with $n > 1/\log_2(r)$. It then follows that $\frac{1}{n} < \log_2(r)$ and hence that $2^{\frac{1}{n}} < 2^{\log_2(r)} = r$. The proof is now complete.  □

*\***4.13**    Answer the given questions pertaining to the following condensed proof that, "If $m < n$ are positive integers, then there is a rational number $r$ with $\frac{1}{n} < r < \frac{1}{m}$."

**Proof.** Let $q$ be a positive integer for which $q \left( \frac{1}{m} - \frac{1}{n} \right) > 1$. Then let $p$ be any integer for which $\frac{q}{n} < p < \frac{q}{m}$. On defining $r = \frac{p}{q}$, it follows that $\frac{1}{n} < r < \frac{1}{m}$, completing the proof.  □

a. Why is the author defining $q$ and $p$ in the first two sentences?

b. Is there a positive integer for which $q \left( \frac{1}{m} - \frac{1}{n} \right) > 1$, as the author claims in the first sentence? Why?

c. Is there an integer $p$ for which $\frac{q}{n} < p < \frac{q}{m}$, as the author claims in the second sentence? Why?

d. Is the author correct in stating that $\frac{1}{n} < r < \frac{1}{m}$ in the last sentence? Why?

e. Where did the author show that $q \neq 0$?

$^B$**4.14**    Explain what is wrong with the following condensed proof that, "If $a$, $b$, and $c$ are real numbers for which the function $ax^2 + bx + c$ has a rational root, then the function $cx^2 + bx + a$ has a rational root."

> **Proof.** Because the function $ax^2 + bx + c$ has a rational root, there are integers $p$ and $q$ with $q \neq 0$ such that $a(p/q)^2 + b(p/q) + c = 0$. Multiplying through by $q^2$ and then dividing by $p^2$ shows that $x = q/p$ is a rational root of $cx^2 + bx + a$.  $\Box$

$^W$**4.15**    Do you agree with the following condensed proof that, "If $R$, $S$, and $T$ are sets with $R \cap S \neq \emptyset$ and $S \cap T \neq \emptyset$, then $R \cap T \neq \emptyset$." Why or why not? Explain. (Note: The symbol $\emptyset$ stands for the empty set.)

> **Proof.** It is shown that there is an element in $R \cap T$. Now, because $R \cap S \neq \emptyset$, by definition, there is an element $x \in R \cap S$. Likewise, because $S \cap T \neq \emptyset$, there is an element $x \in S \cap T$. It then follows that $x \in R$ and $x \in T$, so $x \in R \cap T$.  $\Box$

**4.16**    Explain what is wrong with the following condensed proof that, "If $m < n$ are consecutive integers and $m$ is even, then $4 \mid (m^2 + n^2 - 1)$."

> **Proof.** Suppose that $n = m + 1$. The proof follows by noting that, for $k = m(m + 1)$, $m^2 + n^2 - 1 = 2k$.  $\Box$

*$^*$**4.17**    Explain what is wrong with the following condensed proof that, "If $m$ and $p$ are positive integers with $m < p$, then there is a rational number $r$ with $\frac{1}{p} < r < \frac{1}{m}$."

> **Proof.** Let $n$ be any integer with $0 < m < n < p$. Now $n \neq 0$ and so $r = \frac{1}{n}$ is a rational number for which $\frac{1}{p} < r < \frac{1}{m}$.  $\Box$

$^B$**4.18**    Prove that if $a$, $b$, and $c$ are integers for which $a|b$ and $b|c$, then $a|c$.

**4.19**    Prove that if $a$ and $b$ are integers with $a \neq 0$ and $x$ is a positive integer such that $ax^2 + bx + b - a = 0$, then $a|b$.

**4.20**    Prove that if $a$, $b$, and $c$ are integers with $a|(b + c)$ and $a|b$, then $a|c$.

**4.21**    Prove that if $s$ and $t$ are rational and $t \neq 0$, then $s/t$ is rational.

*$^*$**4.22**    Consider proving that, "If $a$, $b$, $c$, $d$, and $e$ are real numbers for which a certain property $P$ holds, then the intersection of the two sets $S = \{(x, y) : y = ax^2 + bx + c\}$ and $T = \{(x, y) : y = dx + e\}$ is nonempty; that is, $S \cap T \neq \emptyset$." Find a property $P$ on $a$, $b$, $c$, $d$, and $e$ that allows you to prove this proposition. Then prove the proposition.

*$^*$**4.23**    Consider proving that, "If $i = \sqrt{-1}$ and $a + bi$ is a complex number for which the real numbers $a$ and $b$ satisfy a certain property $P$, then there is a complex number $c + di$ such that $(a + bi)(c + di) = 1$." Find a property $P$ that allows you to prove this proposition. Then prove the proposition.

$$5$$

# *Quantifiers II:*
# *The Choose Method*

This chapter develops a technique for dealing with statements in the backward process that contain the quantifier "for all." Such statements arise quite naturally in many mathematical areas, one of which is set theory, as you will now see.

## 5.1  WORKING WITH THE QUANTIFIER "FOR ALL"

A **set** is a collection of items. For example, you can think of the numbers 1, 4, and 7 as a collection of items, and hence they form a set. Each of the individual items is called a **member** or **element of the set** and each member of the set is said to *be in* or *belong to* the set. The set is often denoted by enclosing the list of its members, separated by commas, in braces. Thus, the set consisting of the numbers 1, 4, and 7 is written as follows:

$$\{1, 4, 7\}.$$

To indicate that the number 4 belongs to this set, mathematicians write:

$$4 \in \{1, 4, 7\},$$

where the symbol $\in$ stands for the words "is a member of." Similarly, to indicate that 2 is not a member of $\{1, 4, 7\}$, one would write:

$$2 \notin \{1, 4, 7\}.$$

While it is desirable to make a list of all the elements in a set, sometimes it is impractical to do so because the list is too long. For example, imagine having to write down every integer between 1 and 100,000. When a set has an infinite number of elements (such as the set of real numbers that are greater than or equal to 0) it is impossible to make a complete list, even if you want to. Fortunately there is a way to describe such "large" sets through the use of what is known as **set-builder notation**, which involves using a verbal and mathematical description for the members of the set. With set-builder notation, the set of all real numbers that are greater than or equal to 0 is written as follows:

$$S = \{\text{real numbers } x : x \geq 0\},$$

where the ":" stands for the words "such that." Everything following the ":" is referred to as the **defining property of the set**. The question that one always has to be able to answer is, "How do I know if a particular item belongs to the set?" To answer such a question, you need only check if the item $y$ satisfies the defining property. If so, then $y$ is an element of the set; otherwise, $y$ is not in the set. For the foregoing set $S$, to see if the real number 3 belongs to $S$, simply replace $x$ everywhere by 3 and see if the defining property is true. In this case, 3 does belong to $S$ because $3 \geq 0$.

Sometimes part of the defining property appears to the left of the ":" as well as to the right, so, when trying to determine if a particular item belongs to such a set, be sure to verify this portion of the defining property, too. For example, if $T = \{\text{real numbers } x \geq 0 : x^2 - x + 2 \geq 0\}$, then $-1$ does not belong to $T$ even though $-1$ satisfies the defining property to the right of the ":". The reason is that $-1$ does not satisfy the defining property to the left of the ":" because $-1$ is not $\geq 0$.

From the point of view of doing proofs, the defining property plays the same role as a definition—the defining property is used to answer the key question, "How can I show that an item belongs to a particular set?" One answer is to check that the item satisfies the defining property.

While discussing sets, observe that it can happen that no item satisfies the defining property. Consider, for example,

$$\{\text{real numbers } x \geq 0 : x^2 + 3x + 2 = 0\}.$$

The only real numbers for which $x^2 + 3x + 2 = 0$ are $-1$ and $-2$. Neither of these satisfies the defining property to the left of the ":". Such a set is said to be **empty**, meaning that the set has no members. The special symbol $\emptyset$ is used to denote the empty set.

To motivate the use of the quantifier "for all," observe that it is usually possible to write a set in more than one way—for example, the two sets

$$\begin{aligned} S &= \{\text{real numbers } x : x^2 - 3x + 2 \geq 0\}, \\ T &= \{\text{real numbers } x : 1 \leq x \leq 2\}, \end{aligned}$$

where $1 \leq x \leq 2$ means that $1 \leq x$ and $x \leq 2$. Surely for two sets $S$ and $T$ to be the same, each element of $S$ should appear in $T$ and vice versa. Using the quantifier "for all," a definition can now be made.

**Definition 12** *A set S is a* **subset** *of a set T (written $S \subseteq T$ or sometimes $S \subset T$) if and only if for each element $x \in S$, $x \in T$.*

**Definition 13** *Two sets S and T are* **equal** *(written $S = T$) if and only if S is a subset of T and T is a subset of S.*

Like any definition, these are used to answer a key question. Definition 12 answers the question, "How can I show that a set (say, $S$) is a subset of another set (say, $T$)?" by requiring you to show that for each element $x \in S$, $x \in T$. How you do so is explained in Section 5.2. Definition 13 answers the key question, "How can I show that two sets (say, $S$ and $T$) are equal?" by requiring you to show that $S$ is a subset of $T$ and $T$ is a subset of $S$.

In addition to set theory, there are many other instances where the quantifier "for all" is used, but, from the foregoing example, you can see that such statements appear to have the same consistent structure. When the quantifier "for all," "for each," "for every," or "for any" appears, the statement will have the following **standard form** (which is similar to the one in Chapter 4):

For every "object" with a "certain property," "something happens."

The words in quotation marks depend on the particular statement under consideration, and you must learn to read, to write, and to identify these three components. Consider these examples.

1. For every angle $t$, $\sin^2(t) + \cos^2(t) = 1$.
   Object:                angle $t$.
   Certain property:      none.
   Something happens:     $\sin^2(t) + \cos^2(t) = 1$.

Mathematicians often use the symbol $\forall$ to abbreviate the words "for all" ("for each," and so on). The use of symbols is illustrated in the next example.

2. $\forall$ real numbers $y > 0$, $\exists$ a real number $x \ni 2^x = y$.
   Object:                real numbers $y$.
   Certain property:      $y > 0$.
   Something happens:     $\exists$ a real number $x \ni 2^x = y$.

Observe that a comma always precedes the something that happens.

Sometimes the quantifier is hidden; for example, the statement "the cosine of an angle strictly between 0 and $\pi/4$ is larger than the sine of the angle" could be phrased equally well as "for every angle $t$ with $0 < t < \pi/4$, $\cos(t) > \sin(t)$." Also, some authors write the quantifier after the something that happens; for example: "$2^n > n^2$, for all integers $n \geq 5$." Practice is needed to become fluent at reading and writing these statements, regardless of how they are presented.

## 5.2    USING THE CHOOSE METHOD

During the backward process, if you come across a statement having the quantifier "for all" in the standard form:

**B:**   For all "objects" with a "certain property," "something happens,"

then one approach to showing that the statement is true is to make the following list of all of the objects having the certain property:

**Objects with the Certain Property**

$X_1$

$X_2$

$X_3$

$\vdots$

Then, for each object on the list, you would need to show that the something happens. When the list consists of only a few objects, this might be a reasonable way to proceed. However, when the list is long or infinite, this approach is not practical. You have already dealt with this type of obstacle in set theory, where the problem is overcome by using set-builder notation to describe the set. Here, the **choose method** allows you to circumvent the difficulty.

To understand the idea of the choose method consider again the foregoing list of objects, each with the certain property. Suppose you are able to prove that, for object $X_1$, the something happens. Further, imagine that your proof is such that when you replace $X_1$ everywhere with $X_2$, the resulting proof correctly establishes that the something happens for $X_2$. In this case, you do not need to write a separate proof to establish that the something happens for $X_2$. You would simply say, "To see that the something happens for $X_2$, repeat the proof, replacing $X_1$ everywhere with $X_2$."

Extending this idea to the remaining objects on the list, the goal of the choose method is to construct a "model proof" for establishing that the something happens for a general object $X$ that has the certain property, in such a way that you could, in theory, repeat the proof for each and every object on the list (see Figure 5.1). If you had such a model proof, then you would not need to check the whole (possibly infinite) list of objects because you would know that you could always do so by a simple substitution in the model proof. In other words, rather than actually proving that the something happens for every object having the certain property, the choose method provides the *capability* of doing so through the use of a model proof. The way in which this model proof is designed is now described.

To understand how the choose method is used, recall that the model proof must establish that the something happens for a general object having the certain property. As such, suppose that you have one of these objects, say, $X$, but remember, you do not know precisely which one. All you do know

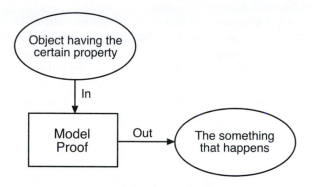

*Fig. 5.1*  A model proof for the choose method.

is that the particular object $X$ does have the certain property. You must somehow use that property to reach the conclusion that, for this object $X$, the something happens. This is most easily accomplished by working forward from the certain property and backward from the something that happens. In other words, with the choose method,

- You choose an object with the certain property, which then becomes a new statement in the forward process.

- You must show that, for the chosen object, the something happens, which then becomes the new statement in the backward process.

If successful, then you have the capability of repeating the foregoing model proof for any object having the certain property.

### An Example of Using the Choose Method

To illustrate how the choose method is used, suppose that, in some proof, you need to show that

**B:** For all real numbers $x$ with $x^2 - 3x + 2 \leq 0$, $1 \leq x \leq 2$.

The first step is to identify, in this for-all statement, the objects (real numbers $x$), the certain property ($x^2 - 3x + 2 \leq 0$), and the something that happens ($1 \leq x \leq 2$). To apply the choose method, you choose a real number that has the certain property. In this case, you might write the following:

**A1:** Let $y$ be a real number with $y^2 - 3y + 2 \leq 0$.

Then, by working forward from $y^2 - 3y + 2 \leq 0$, you must reach the conclusion that for $y$ the something happens; that is, you must show that

**B1:** $1 \leq y \leq 2$.

Here, the symbol $y$ is used to distinguish the chosen object from the general object $x$ in $B$. Notice in $A1$ and in $B1$ that the certain property and the something that happens are written specifically for the chosen object, not for the general one. This is done by replacing the general object $(x)$ everywhere in $B$ with the chosen object $(y)$. In many condensed proofs, the same symbol is used for both the general object and the chosen one. In such cases, be careful to interpret the symbol correctly. Consider the following example.

**Proposition 6** *If $S$ and $T$ are the two sets defined by*

$$
\begin{aligned}
S &= \{\text{real numbers } x : x^2 - 3x + 2 \le 0\} \\
T &= \{\text{real numbers } x : 1 \le x \le 2\},
\end{aligned}
$$

*then $S = T$.*

**Analysis of Proof.**    When doing a proof, learn to choose a technique consciously, based on the form of the statements under consideration. In this proposition, the hypothesis $A$ and conclusion $B$ do not contain keywords (such as "there is" or "for all"). In the absence of keywords, the forward-backward method is a reasonable technique to use. Doing so in this case gives rise to the key question, "How can I show that two sets (namely, $S$ and $T$) are equal?" Definition 13 provides the answer that you must show that

**B1:** $S$ is a subset of $T$ and $T$ is a subset of $S$.

So first try to establish that

**B2:** $S$ is a subset of $T$,

and, afterward, that

**B3:** $T$ is a subset of $S$.

To show that $S$ is a subset of $T$ (see $B2$), you obtain the key question, "How can I show that a set (namely, $S$) is a subset of another set (namely, $T$)?" Using Definition 12 leads to the answer that you must show that

**B4:** For all elements $x \in S$, $x \in T$.

This new statement, $B4$, contains the quantifier "for all," thus indicating that you should proceed by the choose method. To do so, first identify, in $B4$, the objects (elements $x$), the certain property (being in $S$), and the something that happens ($x \in T$).

To apply the choose method to $B4$, you must now choose an object having the certain property and then show that, for this chosen object, the something happens. In this case that means you should choose

**A1:** An element $x \in S$.

Using the fact that $x \in S$ (that is, that $x$ satisfies the defining property of $S$) together with the information in $A$, you must show that the something happens in $B4$; that is,

**B5:** $x \in T$.

Note that you do not want to pick one specific element in $S$, say, $3/2$. Also, note the double use of the symbol $x$ for both the general object in $B4$ and the chosen object in $A1$.

Working backward from $B5$, you should ask the key question, "How can I show that an element (namely, $x$) belongs to a set (namely, $T$)?" One answer is to show that $x$ satisfies the defining property of $T$; that is,

**B6:** $1 \le x \le 2$.

Turning now to the forward process, you can make use of the information in $A$ to show that $1 \le x \le 2$ because you have assumed that $A$ is true. However, additional information is available. Recall that, during the backward process, you used the choose method, at which time you chose $x \in S$ (see $A1$). Now is the time to use this fact. Specifically, because $x \in S$, from the defining property of the set $S$, you have that

**A2:** $x^2 - 3x + 2 \le 0$.

Then, by factoring, you obtain

**A3:** $(x - 2)(x - 1) \le 0$.

The only way that the product of $x - 2$ and $x - 1$ can be $\le 0$ is for one of the terms to be $\le 0$ and the other $\ge 0$. In other words,

**A4:** Either $x - 2 \ge 0$ and $x - 1 \le 0$, or else $x - 2 \le 0$ and $x - 1 \ge 0$.

The first situation can never happen because, if it did, $x \ge 2$ and $x \le 1$, which is impossible. Thus the second condition must happen; that is,

**A5:** $x \le 2$ and $x \ge 1$.

But this is precisely the last statement obtained in the backward process ($B6$), and hence it has been shown successfully that $S$ is a subset of $T$. Do not forget that you still have to show that $T$ is a subset of $S$ ($B3$) in order to complete the proof that $S = T$. This part is done in Section 5.3.

**Proof of Proposition 6.**    To show that $S = T$, it is shown that $S \subseteq T$ and $T \subseteq S$. To see that $S \subseteq T$, let $x \in S$. (The use of the word "let" in condensed proofs frequently indicates that the choose method is being invoked.) Consequently, $x^2 - 3x + 2 \le 0$ and so $(x - 2)(x - 1) \le 0$. This means that either $x - 2 \ge 0$ and $x - 1 \le 0$ or else $x - 2 \le 0$ and $x - 1 \ge 0$. The former cannot happen because, if it did, $x \geq 2$ and $x \leq 1$. Hence it must be that $x \leq 2$ and $x \geq 1$, which means that $x \in T$. The proof that $T$ is a subset of $S$ is given subsequently in Proposition 7.  □

On a final note, whenever you apply the choose method to choose an object with the certain property, you must first be sure that, indeed, there is at least one such object, for if there are none, how can you choose such an object? To illustrate, suppose you are trying to prove that

**B:** For all real numbers $x \geq 0$ with $x^2 + 3x + 2 = 0$, $x^2 \geq 4$.

According to the choose method, you should choose a real number $x \geq 0$ with the property that $x^2 + 3x + 2 = 0$. However, there are no such numbers (because the only values for $x$ that satisfy the equation are $x = -1$ and $x = -2$, and neither of these is $\geq 0$). In this case, you cannot apply the choose method; however, there is no need to do so. The reason is that, when there is no object with the certain property, the associated for-all statement is automatically true. To understand why, recall again that, when using the choose method to prove that

**S1:** For every object with a certain property, something happens,

you choose

**A:** An object $X$ with the certain property,

for which you must then show that

**B:** $X$ satisfies the something that happens.

Observe that this approach is exactly how you would proceed if you were using the forward-backward method to prove that

**S2:**   If $X$ is an object with the certain property, then
        $X$ satisfies the something that happens.

Specifically, to show that the statement $S2$ is true with the forward-backward method, you work forward from the hypothesis $A$ given above and backward from the conclusion $B$ given above. In summary, the foregoing statements $S1$ and $S2$ are equivalent.

Now you can see why, if there is no object with the certain property, $S1$ is true. This is because, if there is no object with the certain property, then the hypothesis in $S2$ is false and so, from Table 1.1 on page 4, the implication in $S2$ is true. Because $S1$ is equivalent to $S2$, $S1$ is also true. This means that, whenever you need to prove a statement in the form $S1$, you can prove $S2$ instead, and vice versa.

## 5.3 READING A PROOF

The process of reading and understanding a proof is now demonstrated.

**Proposition 7** *If $S$ and $T$ are the two sets defined by*

$$
\begin{aligned}
S &= \{real\ numbers\ x : x^2 - 3x + 2 \leq 0\} \\
T &= \{real\ numbers\ x : 1 \leq x \leq 2\},
\end{aligned}
$$

*then $T \subseteq S$.*

**Proof of Proposition 7.** (For reference purposes, each sentence of the proof is written on a separate line.)

**S1:** To show that $T \subseteq S$, let $t \in T$.
**S2:** It must be shown that $t \in S$.
**S3:** Because $t \in T$, $1 \leq t \leq 2$, so $t - 1 \geq 0$ and $t - 2 \leq 0$.
**S4:** Thus, $t \in S$ because $t^2 - 3t + 2 = (t - 1)(t - 2) \leq 0$.

The proof is now complete. □

**Analysis of Proof.** An interpretation of statements $S1$ through $S4$ follows.

**Interpretation of S1:** *To show that $T \subseteq S$, let $t \in T$.*
  The author has worked backward from the conclusion $B$ and asked the key question, "How can I show that a set (namely, $T$) is a subset of another set (namely, $S$)?" Applying Definition 12 means it must be shown that

  **B1:** For all elements $x \in T$, $x \in S$.

The author then recognizes the keywords "for all" in $B1$ and uses the choose method to choose an object with the certain property, as indicated by the words "... let $t \in T$."

**Interpretation of S2:** *It must be shown that $t \in S$.*
  According to the choose method, it is necessary to show that, for the chosen object, the something happens. This is exactly what the author is saying must be done for the chosen object $t$.

**Interpretation of S3:** *Because $t \in T$, $1 \leq t \leq 2$, so $t - 1 \geq 0$ and $t - 2 \leq 0$.*
  The author is working forward from the fact that $t \in T$, using the defining property of $T$. Presumably this is being done to show that the something happens for the chosen object; that is, that $t \in S$.

**Interpretation of S4.** *Thus, $t \in S$ because $t^2 - 3t + 2 = (t - 1)(t - 2) \leq 0$.*
  The author is now claiming that $t \in S$ by showing that $t$ satisfies the defining property of $S$. In essence, the author has asked the key question, "How can I show that an element (namely, $t$) belongs to a set (namely, $S$)?"

and has answered the question by using the defining property of the set. Having shown that $t \in S$, the choose method, and hence the proof, is now complete.

The following points about reading the condensed proof of Proposition 7 are in order.

- No mention is made of the techniques being used (the forward-backward and choose method, and the key questions and answers). However, in this case, the word "let" indicates that the choose method is used.

- The techniques used in the proof vary as the form of the statement currently under consideration varies. For example, the author starts with the forward-backward method and then changes to the choose method when a backward statement contains the quantifier "for all."

- Several steps are condensed into the single sentence $S1$.

## Summary

Use the choose method when the last statement in the backward process contains the quantifier "for all" in the standard form:

For all "objects" with a "certain property," "something happens."

To use the choose method, proceed as follows to create a model proof that could, in theory, be repeated for every object with the certain property.

1. Identify the object, the certain property, and the something that happens in the for-all statement.

2. Verify that there is at least one object with the certain property. If there is no such object, then the for-all statement is true and you are done.

3. Choose an object that has the certain property. (Write the fact that the chosen object has the certain property as a new statement in the forward process.)

4. Show that, for this chosen object, the something happens. (Write this objective as the next statement in the backward process.)

Step 4 is accomplished by the forward-backward method. That is, work forward from the fact that the chosen object in Step 3 has the certain property and backward from the fact that this chosen object must be shown to satisfy the something that happens. In so doing, you can use the assumption that the hypothesis $A$, or any other statement in the forward process, is true.

## Exercises

**Note:** Solutions to those exercises marked with a $B$ are in the back of this book. Solutions to those exercises marked with a $W$ are located on the web at http://www.wiley.com/college/solow/.

**Note:** All proofs should contain an analysis of proof and a condensed version. Definitions for all mathematical terms are provided in the glossary at the end of the book.

$^B$**5.1**   For each of the following definitions, identify the objects, the certain property, and the something that happens in the for-all statements.

    a. The real number $x^*$ is a **maximizer** of the function $f$ if and only if for every real number $x$, $f(x) \leq f(x^*)$.

    b. Suppose that $f$ and $g$ are functions of one variable. Then $g \geq f$ **on the set** $S$ of real numbers if and only if for every element $x \in S$, $g(x) \geq f(x)$.

    c. A real number $u$ is an **upper bound** for a set $S$ of real numbers if and only if for all elements $x \in S$, $x \leq u$.

**5.2**   For each of the following definitions, identify the objects, the certain property, and the something that happens in the for-all statements.

    a. A function $f$ of one real variable is **strictly increasing** if and only if for all real numbers $x$ and $y$ with $x < y$, $f(x) < f(y)$.

    b. The set $C$ of real numbers is a **convex set** if and only if for all elements $x, y \in C$, and for every real number $t$ with $0 \leq t \leq 1$, $tx + (1-t)y \in C$.

    c. The function $f$ of one real variable is a **convex function** if and only if for all real numbers $x$ and $y$ and for all real numbers $t$ with $0 \leq t \leq 1$, it follows that $f(tx + (1-t)y) \leq tf(x) + (1-t)f(y)$.

$^W$**5.3**   Reword the following statements in standard form using the appropriate symbols $\forall$, $\exists$, $\ni$, as necessary.

    a. Some mountain is taller than every other mountain.

    b. If $t$ is an angle, then $\sin(2t) = 2\sin(t)\cos(t)$.

    c. The square root of the product of any two nonnegative real numbers $p$ and $q$ is not less than their sum divided by 2.

    d. If $x$ and $y$ are real numbers such that $x < y$, then there is a rational number $r$ such that $x < r < y$.

**5.4**    Reword each of the following for-all statements as an equivalent statement in the form, "If ... then ...".

a. For every prime number $p$, $p + 7$ is composite.

b. For all sets $A$, $B$, and $C$ with the property that $A \subseteq B$ and $B \subseteq C$, it follows that $A \subseteq C$.

c. For all integers $p$ and $q$ with $q \neq 0$, $p/q$ is rational.

$^B$**5.5**    For each of the parts in Exercise 5.1, describe how you would apply the choose method to show that the for-all statement is true. Use a different symbol to distinguish the chosen object from the general object. For instance, for Exercise 5.1(a), to show that $x^*$ is the maximizer of the function $f$, you would choose

**A1:**  a real number, say, $x'$,

for which it must then be shown that

**B1:**  $f(x') \leq f(x^*)$.

**5.6**    For each of the parts in Exercise 5.2, describe how you would apply the choose method to show that the for-all statement is true. Use a different symbol to distinguish the chosen object from the general object.

**5.7**    Could you use the choose method to prove each of the following statements? Why or why not? Explain. ($S$ is a set of real numbers.)

a. There is an integer $n \geq 4$ such that $n! \geq 2^n$.

b. For all integers $n \geq 4$, $n! \geq 2^n$.

c. If for all elements $x \in S$, $|x| < 20$, then there is an element $s \in S$ such that $s < 5$.

d. If there is an element $x \in S$ such that $x < 5$, then for all elements $x \in S$, $|x| < 20$.

e. For all real numbers $a$, $b$, and $c$, if $4ac \leq b^2$, then $ax^2 + bx + c$ has real roots.

$^B$**5.8**    Suppose you are trying to prove that, "If $R$, $S$, and $T$ are sets for which $R \subseteq S$ and $S \subseteq T$, then $R \subseteq T$." Write an appropriate key question and answer to create a new statement, $B1$, in the backward process. Then indicate how the choose method would be applied to $B1$ by writing a new statement $A1$ in the forward process that is a result of choosing the appropriate object and a new statement $B2$ in the backward process indicating what you would have to show about your chosen object. (Do not complete the proof.)

<sup>W</sup>**5.9**    Repeat Exercise 5.8 when you are trying to prove that, "If the real number $u$ is an upper bound for a set $S$ of real numbers and the real number $v \geq u$, then $v$ is an upper bound for $S$." [See the definition in Exercise 5.1(c)].

**5.10**    Repeat Exercise 5.8 when you are trying to prove that, "If the function $f(x) = x^3$, then $f$ is strictly increasing." [See the definition in Exercise 5.2(a)].

**5.11**    Repeat Exercise 5.8 when you are trying to prove that, "If $f$ is a convex function and $y$ is a given real number, then {real numbers $x : f(x) \leq y$} is a convex set." [See the definition in Exercise 5.2(b)].

**5.12**    Repeat Exercise 5.8 when you are trying to prove that, "If $f$ and $g$ are convex functions, then the function $f + g$ is a convex function." [See the definition in Exercise 5.2(c)].

<sup>B</sup>**5.13**    Suppose you are trying to prove that, "If $S$ and $T$ are the sets defined by $S = \{(x, y) : x^2 + y^2 \leq 16\}$ and $T = \{(x, y) : 3x^2 + 2y^2 \leq 125\}$, then for every element $(x, y) \in S$, $(x, y) \in T$." Which of the following constitutes a correct application of the choose method? For those that are incorrect, explain what is wrong.

a.    Choose
    $A1$ : real numbers $x'$ and $y'$.
It must be shown that
    $B1 : (x', y') \in T$.

b.    Choose
    $A1$ : real numbers $x'$ and $y'$ with $(x', y') \in S$.
It must be shown that
    $B1 : (x', y') \in T$.

c.    Choose
    $A1$ : real numbers $x'$ and $y'$ with $(x', y') \in T$.
It must be shown that
    $B1 : (x', y') \in S$.

d.    Choose
    $A1$ : real numbers $x$ and $y$, say 1 and 2, with
        $x^2 + y^2 = 5 \leq 16$ and therefore $(x, y) \in S$.
It must be shown that
    $B1 : 3x^2 + 2y^2 \leq 125$ and therefore that $(x, y) \in T$.

e.    Choose
    $A1$ : real numbers $x$ and $y$ with $(x, y) \in S$.
It must be shown that
    $B1 : (x, y) \in T$.

**5.14**    Suppose you are trying to prove that, "If $p(x) = a_0 + a_1 x^1 + \cdots + a_n x^n$ is a polynomial of degree $n > 1$ such that $a_i > 0$ for all integers $i = 0, \ldots, n$, then for all real numbers $x$ and $y$ with $0 < x < y$, $p(x) < p(y)$." Which of the following constitutes a correct application of the choose method? For those that are incorrect, explain what is wrong.

a.    Choose
      $A1$ : an integer $i$ between 0 and $n$.
      It must be shown that
      $B1 : a_i > 0$.

b.    Choose
      $A1$ : real numbers $x$ and $y$ with $p(x) < p(y)$.
      It must be shown that
      $B1 : 0 < x < y$.

c.    Choose
      $A1$ : real numbers $x$ and $y$ with $x < y$.
      It must be shown that
      $B1 : p(x) < p(y)$.

d.    Choose
      $A1$ : real numbers, say $x = 3$ and $y = 5$, with $0 < x < y$.
      It must be shown that
      $B1 : p(3) < p(5)$.

e.    Choose
      $A1$ : real numbers $x$ and $y$ with $0 < x < y$.
      It must be shown that
      $B1 : p(x) < p(y)$.

$^B$**5.15**    For the proposition and condensed proof given below, explain where (that is, in which sentence), why, and how the choose method is used. Identify any mistakes you encounter. [Refer to the definition in Exercise 5.1(a).]

> **Proposition.** If $a$, $b$, and $c$ are real numbers for which $a < 0$, then $x^* = -b/(2a)$ is a maximizer of $f(x) = ax^2 + bx + c$.

> **Proof.** Let $x$ be a real number. If $x^* \geq x$, then $x^* - x \geq 0$ and $a(x^* + x) + b \geq 0$. So, $(x^* - x)[a(x^* + x) + b] \geq 0$. On multiplying the term $x^* - x$ through, rearranging terms, and adding $c$ to both sides, one obtains that $a(x^*)^2 + bx^* + c \geq ax^2 + bx + c$. A similar argument applies when $x^* < x$.  $\square$

**5.16**    For the proposition and condensed proof given below, explain where (that is, in which sentence), why, and how the choose method is used. Identify any mistakes you encounter.

**Proposition.** If

$$
\begin{aligned}
R &= \{\text{real numbers } x : x^2 - x \geq 0\}, \\
S &= \{\text{real numbers } x : -(x-1)(x-3) \leq 0\}, \quad \text{and} \\
T &= \{\text{real numbers } x : x \leq 1 \text{ or } x \geq 3\},
\end{aligned}
$$

then $R \cap S \subseteq T$.

**Proof.** To reach the conclusion, it will be shown that for all $x \in R \cap S$, $x \in T$. To that end, let $x \in R \cap S$. Because $x \in R$, $x^2 - x = x(x-1) \geq 0$. Likewise, because $x \in S$, $-(x-1)(x-3) \leq 0$. Combining these two yields that, in all cases, $x \leq 1$ or $x \geq 3$ and so $x \in T$. □

$^B$**5.17**  Write an analysis that corresponds to the condensed proof given below. Indicate which techniques are used and how they are applied. Fill in the details of any missing steps where appropriate.

> **Proposition.** If $m$ and $b$ are real numbers with $m > 0$, and $f$ is the function defined by $f(x) = mx + b$, then for all real numbers $x$ and $y$ with $x < y$, $f(x) < f(y)$.
>
> **Proof.** Let $x$ and $y$ be real numbers with $x < y$. Then because $m > 0$, $mx < my$. On adding $b$ to both sides, it follows that $f(x) < f(y)$, and so the proof is complete. □

$^W$**5.18**  Write an analysis that corresponds to the condensed proof given below. Indicate which techniques are used and how they are applied. Fill in the details of any missing steps where appropriate.

> **Proposition.** If $S = \{\text{real numbers } x : x(x-3) \leq 0\}$ and $T = \{\text{real numbers } x : x \geq 3\}$, then every element of $T$ is an upper bound for $S$. [See the definition in Exercise 5.1(c).]
>
> **Proof.** Let $t$ be an element of $T$. It will be shown that $t$ is an upper bound for $S$. To that end, let $x \in S$. Consequently, $x(x-3) \leq 0$. Therefore, either $x \leq 0$ and $x - 3 \geq 0$ or else $x \geq 0$ and $x - 3 \leq 0$. The former cannot happen, so it must be that $x \geq 0$ and $x - 3 \leq 0$. But then $x \leq 3$, and because $t$ belongs to $T$, $t \geq 3$, and so $x \leq t$. □

**5.19**  Write an analysis that corresponds to the condensed proof given below. Indicate which techniques are used and how they are applied. Fill in the details of any missing steps where appropriate.

> **Proposition.** If $p$ is a positive integer, then for all nonzero integers $q$ and $r$ having the same sign and for which $q < r$, $p/q > p/r$.

**Proof.** Because $q$ and $r$ have the same sign, $p/(qr) > 0$. Now $r > q$, so multiplying both sides of $r > q$ by $p/(qr)$ results in $rp/(qr) > qp/(qr)$. It now follows that $p/q > p/r$, and so the proof is complete.  $\square$

*5.20    Answer the given questions pertaining to the following condensed proof of the proposition: If $x > 0$ is a real number for which $\frac{2-x^2}{2x+1} > 0$ and $S = \left\{ \frac{1}{n} : n \text{ is a positive integer with } \frac{1}{n} < \frac{2-x^2}{2x+1} \right\}$, then $\sqrt{2} - x$ is an upper bound for $S$. [See the definition in Exercise 5.1(c).]

**Proof.** Let $\frac{1}{n} \in S$. Then you have,

$$\left( x + \frac{1}{n} \right)^2 = x^2 + \frac{1}{n}\left( 2x + \frac{1}{n} \right) \le x^2 + \frac{1}{n}(2x+1) < x^2 + 2 - x^2 = 2.$$

On taking the positive square root of both sides, it follows that $x + \frac{1}{n} \le \sqrt{2}$ and so $\frac{1}{n} \le \sqrt{2} - x$, completing the proof.  $\square$

a. What key question did the author ask and what was the answer?

b. Where and how did the author use the choose method?

c. In the first sentence, the author states that $\frac{1}{n} \in S$. How does the author know that there is such an element in $S$? Explain.

d. In the first sentence, the author states that $\frac{1}{n} \in S$. As a result of the defining property of $S$, this means that $n \ge 1$ and $\frac{1}{n} < \frac{2-x^2}{2x+1}$. Where does the author use this information?

$^B$**5.21**    Prove that, if $f(x) = (x-1)^2$ and $g(x) = x+1$, then $g \ge f$ on the set $S = \{\text{real numbers } x : 0 \le x \le 3\}$. [See the definition in Exercise 5.1(b).]

$^W$**5.22**    Prove that, if $a$ and $b$ are real numbers, then the set $C = \{\text{real numbers } x : ax \le b\}$ is a convex set. [See the definition in Exercise 5.2(b).]

**5.23**    Prove that, if $a$, $b$, and $c$ are real numbers with $a \ge 0$, then the function $f(x) = ax^2 + bx + c$ satisfies the property that for all real numbers $x$ and $y$, $f(x) \ge f(y) + (2ay + b)(x - y)$.

**5.24**    Prove that, for all real numbers $a$ and $b$, at least one of which is not $0$, $a^2 + ab + b^2 > 0$. (Hint: Use the fact that $a^2 + b^2 > (a^2 + b^2)/2$.)

**5.25**    Use the result in Exercise 5.24 to prove that the function $f(x) = x^3$ is strictly increasing. [See the definition in Exercise 5.2(a).]

*5.26    Prove that, if $m$ and $b$ are real numbers and $f$ is the function defined by $f(x) = mx + b$, then $f$ is convex. [See the definition in Exercise 5.2(c).]

# 6

## *Quantifiers III: Specialization*

The previous two chapters illustrate how to proceed when a quantifier appears in the statement $B$. A method is introduced in this chapter for working forward from a statement that contains the universal quantifier "for all."

### 6.1  HOW TO USE SPECIALIZATION

When the statement $A$ contains the quantifier "for all" in the standard form:

**A:** For all "objects" with a "certain property,"
"something happens,"

one typical method emerges for working forward from $A$—**specialization**. In general terms, specialization works as follows. As a result of assuming $A$ is true, you know that, for all objects with the certain property, something happens. If, at some point, you were to come across one of these objects that does have the certain property, then you can use the information in $A$ by being able to conclude that, for this particular object, the something does indeed happen. That fact should help you to conclude that $B$ is true. In other words, you will have specialized the statement $A$ to one particular object having the certain property.

To illustrate the idea of specialization in a more tangible way, suppose you know that

**A:** All cars with 4 cylinders get good gas mileage.

In this statement, you can identify the following three items:

| | |
|---|---|
| Objects: | cars. |
| Certain property: | with 4 cylinders. |
| Something happens: | get good gas mileage. |

Suppose that you are interested in buying a car that gets good gas mileage, so your objective is

**B:** To buy a car that gets good gas mileage.

You can work forward from the information in $A$ by specialization to establish $B$ as follows. Suppose you are in a dealer's lot one day and you see a particular car that you like. Talking with the salesperson, suppose you verify that the car has 4 cylinders. Recalling that foregoing statement $A$ is assumed to be true, you can use this information to conclude that

**A1:** This particular car gets good gas mileage.

In other words, you have specialized the for-all statement in $A$ to one particular object.

If you analyze this example in detail, you can identify the following steps associated with applying specialization to a forward statement of the form:

**A:** For all "objects" with a "certain property," "something happens."

## Steps for Using Specialization

1. Identify, in the for-all statement, the object, the certain property, and the something that happens.

2. Look for one particular object to apply specialization to. (This object often arises in the backward process.)

3. Verify that this particular object does have the certain property specified in the for-all statement.

4. Conclude, by writing a new statement in the forward process, that the something happens for this particular object.

The following example demonstrates the proper use of specialization in doing mathematical proofs.

**Definition 14** *A real number $u$ is an* **upper bound** *for a set of real numbers $T$ if and only if for all elements $t \in T$, $t \le u$.*

**Proposition 8** *If $R$ is a subset of a set $S$ of real numbers and $u$ is an upper bound for $S$, then $u$ is an upper bound for $R$.*

**Analysis of Proof.**     The forward-backward method gives rise to the key question, "How can I show that a real number (namely, $u$) is an upper bound for a set of real numbers (namely, $R$)?" Definition 14 is used to answer the question. Thus it must be shown that

**B1:** For all elements $r \in R$, $r \leq u$.

The appearance of the quantifier "for all" in the backward process suggests proceeding with the choose method, whereby one chooses

**A1:** An element, say $r$, in $R$,

for which it must be shown that

**B2:** $r \leq u$.

(Here, the symbol $r$ is used both for the chosen object in $A1$ and the general object in the for-all statement in $B1$, though they have different meanings.)

Turning now to the forward process, you will see how specialization is used to reach the conclusion that $r \leq u$ in $B2$. From the hypothesis that $R$ is a subset of $S$, and by Definition 12 on page 53, you know that

**A2:** For each element $x \in R$, $x \in S$.

Recognizing the keywords "for each" in the forward process, you should consider using specialization. According to the discussion preceding Proposition 8, the first step in doing so is to identify, in $A2$, the object (element $x$), the certain property (being in the set $R$), and the something that happens ($x \in S$). Next, you must look for one particular object with which to specialize. In the backward process you chose the particular element $r$ (see $A1$). The third step of specialization requires you to verify that the particular object ($r$) has the certain property in $A2$; namely, being in the set $R$. In fact, the particular object $r$ does have this property because $r$ is chosen to be in the set $R$ (see $A1$). The final step of specialization is to conclude, by writing a new statement in the forward process, that the something in $A2$ happens for the particular object $r$. In this case, specialization of $A2$ allows you to conclude that

**A3:** $r \in S$.

The proof is not yet complete because the last statement in the backward process ($B2$) has not yet been reached in the forward process. To do so, continue to work forward. For example, from the hypothesis you know that $u$ is an upper bound for $S$. By Definition 14 this means that

**A4:** For every element $s \in S$, $s \leq u$.

Once again, the appearance of the quantifier "for every" in the forward process suggests using specialization. Accordingly, identify, in $A4$, the object (element $s$), the certain property (being in the set $S$), and the something that happens

$(s \leq u)$. Now look for one particular object with which to apply specialization. The same element $r$ chosen in $A1$ serves the purpose. Next, verify that this particular $r$ satisfies the certain property in $A4$; namely, being in the set $S$. Indeed, $r \in S$, as stated in $A3$. Finally, conclude that, for this particular object $r$, the something in $A4$ happens, so

**A5:** $r \leq u$.

The proof is now complete because $A5$ is the last statement obtained in the backward process (see $B2$).

In the condensed proof that follows, note the lack of reference to the forward-backward, choose, and specialization methods.

**Proof of Proposition 8.**    To show that $u$ is an upper bound for $R$, let $r \in R$. (The word "let" here indicates that the choose method is used.) By hypothesis, $R \subseteq S$ and so $r \in S$. (Here is where specialization is used.) Furthermore, by hypothesis, $u$ is an upper bound for $S$, thus, every element in $S$ is $\leq u$. In particular, $r \in S$, so $r \leq u$. (Again specialization is used.)    □

When using specialization, be careful to keep your notation and symbols in order. Doing so involves a correct "matching up of notation," similar to what you learned in Chapter 3 when using definitions. To illustrate, suppose you are going to apply specialization to a statement of the form:

**A:** For all objects $X$ with a certain property,
      something happens.

After finding a particular object, say $Y$, with which to specialize, it is necessary to verify that $Y$ satisfies the certain property in $A$. To do so, replace $X$ with $Y$ everywhere in the certain property in $A$ and see if the resulting condition is true. Similarly, when concluding that the particular object $Y$ satisfies the something that happens in $A$, again replace $X$ everywhere with $Y$ in the something that happens to obtain the correct statement in the forward process. (This is done when writing statements $A3$ and $A5$ in the foregoing analysis of the proof of Proposition 8.) Be careful of overlapping notation, for example, when the particular object you have identified has precisely the same symbol as the one in the for-all statement you are specializing.

## 6.2    READING A PROOF

The process of reading and understanding a proof that uses specialization is demonstrated with the following proposition.

**Definition 15** *A real number $u$ is a* **least upper bound** *for a set $S$ of real numbers if and only if (1) $u$ is an upper bound for $S$ and (2) for every upper bound $v$ for $S$, $u \leq v$.*

**Proposition 9** *If $v^*$ and $w^*$ are least upper bounds for a set $T$, then $v^* = w^*$.*

**Proof of Proposition 9.**    (For reference purposes, each sentence of the proof is written on a separate line.)

**S1:**  From the hypothesis, both $v^*$ and $w^*$ are upper bounds for $T$.

**S2:**  Because $v^*$ is a least upper bound for $T$, $v^* \leq u$, for any upper bound $u$ for $T$.

**S3:**  In particular, $w^*$ is an upper bound for $T$, so $v^* \leq w^*$.

**S4:**  Similarly, $w^*$ is a least upper bound for $T$ and, because $v^*$ is an upper bound for $T$, $w^* \leq v^*$.

**S5:**  It now follows that $v^* = w^*$.

The proof is now complete.  □

**Analysis of Proof.**    An interpretation of statements $S1$ through $S5$ follows.

**Interpretation of S1:** *From the hypothesis, both $v^*$ and $w^*$ are upper bounds for $T$.*

The author is working forward from the hypothesis using part (1) of the definition of a least upper bound to claim that

**A1:** $v^*$ and $w^*$ are upper bounds for $T$.

When reading a proof, it is advisable to determine where the author is heading. To do so, work backward from $B$ yourself. In this case, you are led to the key question, "How can I show that two real numbers (namely, $v^*$ and $w^*$) are equal?" Read forward in the proof to see how the author answers this question. From $S3$ and $S4$, the answer in this case is to show that

**B1:** $v^* \leq w^*$ and $w^* \leq v^*$.

**Interpretation of S2:** *Because $v^*$ is a least upper bound for $T$, $v^* \leq u$, for any upper bound $u$ for $T$.*

The author is continuing to work forward by stating part (2) of the definition of a least upper bound applied to $v^*$; that is,

**A2:** For every upper bound $u$ for $T$, $v^* \leq u$.

**Interpretation of S3:** *In particular, $w^*$ is an upper bound for $T$, so $v^* \leq w^*$.*

It is here that specialization is applied to the for-all statement in $A2$, indicated by the words "in particular." Specifically, $A2$ is specialized to the value $u = w^*$, which is an upper bound for $T$ (see $A1$). The result of specialization, as the author claims is $S3$, is

**A3:** $v^* \leq w^*$.

**Interpretation of S4:** *Similarly, $w^*$ is a least upper bound for $T$ and, because $v^*$ is an upper bound for $T$, $w^* \leq v^*$.*

The author is using the same analysis as in $S3$ but, this time, applied to the least upper bound $w^*$ and the upper bound $v^*$ for $T$. The result of this specialization is that

**A4:** $w^* \leq v^*$.

**Interpretation of S5:** *It now follows that $v^* = w^*$, and so the proof is complete.*

The author is working forward by combining $v^* \leq w^*$ from $A3$ and $w^* \leq v^*$ from $A4$ to claim correctly that $v^* = w^*$. Finally, the author states that the proof is complete, which is true because the conclusion $B$ has been established.

## Summary

You now have various techniques for dealing with quantifiers that can appear in either $A$ or $B$. As always, let the form of the statement guide you. When $B$ contains the quantifier "there is," the construction method is used to produce the desired object. The choose method is associated with the quantifier "for all" in the backward process. Finally, if the quantifier "for all" appears in the forward process, use specialization. To do so, follow these steps:

1. Identify, in the for-all statement, the object, the certain property, and the something that happens.

2. Look for one particular object to apply specialization to. (This object often arises from the backward process, especially when the choose method is used.)

3. Verify that this particular object does have the certain property specified in the for-all statement.

4. Conclude, by writing a new statement in the forward process, that the something happens for this one particular object.

It is common to confuse the choose method with the specialization method. Use the choose method when you encounter the keywords "for all" in the backward process; use specialization when the keywords "for all" arise in the forward process. Another way to say this is to use the choose method when you want to *show that* "for all objects with a certain property, something happens"; use specialization when you *know that* "for all objects with a certain property, something happens."

All of the statements thus far have contained only one quantifier. In the next chapter you will learn what to do when statements contain more than one quantifier.

## Exercises

**Note:** Solutions to those exercises marked with a $B$ are in the back of this book. Solutions to those exercises marked with a $W$ are located on the web at http://www.wiley.com/college/solow/.

**Note:** All proofs should contain an analysis of proof and a condensed version. Definitions for all mathematical terms are provided in the glossary at the end of the book.

$^W$**6.1**  Suppose you want to specialize the statement that "for every object with a certain property, something happens" to the particular object $Y$. Explain why you need to show that $Y$ has the certain property before you can do so.

$^B$**6.2**  Suppose you are working backward and want to show that, for a particular object $Y$ with a certain property $P$, $S$ happens. When and how can you reach this conclusion by working forward from the statement that, for every object $X$ with a certain property $Q$, $T$ happens? State your answer in terms of the objects, the certain properties, and the somethings that happen.

$^B$**6.3**  For each definition in Exercise 5.1 on page 61, explain how you would work forward from the associated for-all statement. For example, to work forward from the for-all statement in Exercise 5.1(b), (1) look for a specific element, say $y$, with which to apply specialization, (2) show that $y \in S$, and (3) conclude that $g(y) \geq f(y)$ as a new statement in the forward process.

**6.4**  For each definition in Exercise 5.2 on page 61, explain how you would work forward from the associated for-all statement.

**6.5**  Would you use specialization to prove each of the following statements? Why or why not? Explain.

a. If $a$, $b$, and $c$ are real numbers for which there is a real number $x \neq 0$ such that $ax^2 + bx + c = 0$, then $cx^2 + bx + a$ has a rational root.

b. If $a$ and $b$ are real numbers with $a < 0$ and $y = -b/(2a)$, then for all real numbers $x$, $ax^2 + bx \leq ay^2 + by$.

c. If $a \neq 0$ and $b$ are real numbers and $y = -b/(2a)$ satisfies the property that, for all real numbers $x$, $ax^2 + bx \leq ay^2 + by$, then $a < 0$.

d. If $f$ is a real-valued function such that, for all real numbers $x, y$, and $t$ with $0 \leq t \leq 1$, $f(tx + (1 - t)y) \leq tf(x) + (1 - t)f(y)$, then the set $C = \{$real numbers $x : f(x) \leq 0\}$ is a convex set.

**6.6**  For each of the following for-all statements, what properties must the given object satisfy so that you can apply specialization? Given that the object does satisfy those properties, what can you conclude about the object?

a. Statement:         For all prime numbers $p$, $p + 7$ is composite.
   Given object:      An integer $m$.

b. Statement:         For all elements $x > 0$ in a set $S$ of real numbers, $x$ is a root of the polynomial $p(x)$.
   Given object:      A real number $y$.

c. Statement:         Every triangle $ABC$ with sides of length $a = \overline{BC}$, $b = \overline{CA}$, and $c = \overline{AB}$ satisfies $c^2 = a^2 + b^2 - 2ab\cos(C)$.
   Given object:      The isosceles right triangle $ABC$ whose legs $a = \overline{BC}$ and $b = \overline{CA}$ are both equal to $m$.

d. Statement:         For all pairs of equilateral triangles $ABC$ and $DEF$, if one side of triangle $ABC$ is parallel to one side of triangle $DEF$, then the other two sides of triangle $ABC$ are parallel to the corresponding sides of triangle $DEF$.
   Given object:      The triangle $CDE$ whose side $DE$ is parallel to side $DA$ of triangle $FDA$.

$^{B}$6.7    To what specific object could you specialize each of the following for-all statements so that the result of specialization leads to the desired conclusion? Verify that the object to which you are applying specialization satisfies the certain property in the for-all statement so that you can apply specialization.

a. For-all statement:     For all angles $\alpha$ and $\beta$, $\sin(\alpha + \beta) = \sin(\alpha)\cos(\beta) + \cos(\alpha)\sin(\beta)$.
   Desired conclusion:    For a particular angle $X$, $\sin(2X) = 2\sin(X)\cos(X)$.

b. For-all statement:     For any sets $S$ and $T$, $(S \cup T)^c = S^c \cap T^c$ (where $X^c$ is the complement of the set $X$).
   Desired conclusion:    For two sets $A$ and $B$, $(A \cap B)^c = A^c \cup B^c$.

$^{*}$6.8    To what specific object could you specialize each of the following for-all statements so that the result of specialization leads to the desired conclusion? Verify that the object to which you are applying specialization satisfies the certain property in the for-all statement so that you can apply specialization.

a. For-all statement:     $f$ is a function of one variable such that, for all real numbers $x$, $y$, and $t$ with $0 \le t \le 1$, $f(tx + (1 - t)y) \le tf(x) + (1 - t)f(y)$.
   Desired conclusion:    the function $f$ satisfies $f(1/2) \le (f(0) + f(1))/2$.

b. For-all statement:     For all real numbers $c$ and $d$ for which $c^2 \ge d^2$, $\sqrt{c^2 - d^2} \le c$.
   Desired conclusion:    For the real numbers $a$, $b \ge 0$, $\sqrt{ab} \le (a+b)/2$.

*6.9   Answer the given questions about the following proof that, "If $f$ is a convex function and $y$ is a real number, then $C = \{$real numbers $x : f(x) \le y\}$ is a convex set (see the definitions in Exercise 5.2(b) and (c) on page 61).

> **Proof.** Let $a, b \in C$, and $t$ be a real number with $0 \le t \le 1$. Because $f$ is convex, $f(ta + (1 - t)b) \le tf(a) + (1 - t)f(b)$. Furthermore, $f(a) \le y$ and $f(b) \le y$ and hence it follows that $tf(a)+(1-t)f(b) \le ty+(1-t)y = y$. Thus, $f(ta+(1-t)b) \le y$ and so $ta + (1 - t)b \in C$. □

a. Explain what the author is doing in the first sentence of the proof. What techniques have been used?

b. Where and how is specialization used?

c. Why is the author justified in saying that $f(a) \le y$ and $f(b) \le y$?

$^B$6.10   What, if anything, is wrong with the following proof of the statement, "If $R$ is a nonempty subset of a set $S$ of real numbers and $R$ is convex (see the definition in Exercise 5.2(b) on page 61), then $S$ is convex?"

> **Proof.** To show that $S$ is a convex set, let $x, y \in S$, and $t$ be a real number with $0 \le t \le 1$. Because $R$ is a convex set, by definition, for any two elements $u$ and $v$ in $R$, and for any real number $s$ with $0 \le s \le 1$, $su + (1 - s)v \in R$. In particular, for the specific elements $x$ and $y$, and for the real number $t$, it follows that $tx + (1 - t)y \in R$. Because $R \subseteq S$, it follows that $tx + (1 - t)y \in S$ and so $S$ is convex. □

*6.11   What, if anything, is wrong with the following proof of the statement, "If $a$, $b$, and $c$ are real numbers with $a < 0$ and $x^*$ is a maximizer of the function $f(x) = ax^2 + bx + c$ (see the definition in Exercise 5.1(a) on page 61), then for every real number $\epsilon > 0$, $\epsilon \le (2ax^* + b)/a?$"

> **Proof.** Let $\epsilon > 0$. It will be shown that $\epsilon \le (2ax^* + b)/a$. Because $x^*$ is a maximum of $f$, by definition, for every real number $x$, $f(x^*) \ge f(x)$. Thus, for $x = x^* - \epsilon$, you have that
>
> $$\begin{aligned} a(x^*)^2 + bx^* + c &\ge a(x^* - \epsilon)^2 + b(x^* - \epsilon) + c \\ &= a(x^*)^2 + bx^* + c - (b + 2ax^*)\epsilon + a\epsilon^2. \end{aligned}$$
>
> Subtracting $a(x^*)^2 + bx^* + c$ from both sides and dividing by $\epsilon > 0$, it follows that
>
> $$a\epsilon - (b + 2ax^*) \le 0.$$
>
> The result that $\epsilon \le (2ax^* + b)/a$ follows by adding $b + 2ax^*$ to both sides of the foregoing inequality and dividing by $a$. □

**\*6.12**    Identify two errors in the following proof of the statement, "If $f$ is a convex function (see the definition in Exercise 5.2(c) on page 61), then for all real numbers $a_1, a_2, a_3$ with $a_1 + a_2 + a_3 = 1$ and $1 - a_3 > 0$, it follows that $f(a_1x_1 + a_2x_2 + a_3x_3) \le a_1f(x_1) + a_2f(x_2) + a_3f(x_3)$."

> **Proof.** Let $a_1, a_2, a_3$ be real numbers with $a_1 + a_2 + a_3 = 1$ and $1 - a_3 > 0$. Now, because $f$ is a convex function, by definition, for all real numbers $x$, $y$, and $t$ with $0 \le t \le 1$, $f(tx + (1-t)y) \le tf(x) + (1-t)f(y)$. In particular, for $t = 1 - a_3 > 0$, $x = \frac{a_1}{1-a_3}x_1 + \frac{a_2}{1-a_3}x_2$ and $y = x_3$, you have
>
> $$f(tx + (1-t)y) = f\left((1-a_3)\left[\frac{a_1}{1-a_3}x_1 + \frac{a_2}{1-a_3}x_2\right] + a_3x_3\right)$$
>
> $$\le (1-a_3)f\left(\frac{a_1}{1-a_3}x_1 + \frac{a_2}{1-a_3}x_2\right) + a_3f(x_3).$$
>
> Now let $t = \frac{a_1}{1-a_3}$ and note that, because $a_1 + a_2 + a_3 = 1$, $1 - t = 1 - \frac{a_1}{1-a_3} = \frac{a_2}{1-a_3}$. Thus, from the convexity of $f$ with $x = x_1$ and $y = x_2$, you have
>
> $$f\left(\frac{a_1}{1-a_3}x_1 + \frac{a_2}{1-a_3}x_2\right) \le \frac{a_1}{1-a_3}f(x_1) + \frac{a_2}{1-a_3}f(x_2).$$
>
> Multiplying both sides of the foregoing inequality by $1-a_3 > 0$ and combining the result with the previous inequality yields the desired conclusion that $f(a_1x_1 + a_2x_2 + a_3x_3) \le a_1f(x_1) + a_2f(x_2) + a_3f(x_3)$. $\square$

$^B$**6.13**    For sets $R$, $S$, and $T$, prove that, if $R \subseteq S$ and $S \subseteq T$, then $R \subseteq T$.

**6.14**    Prove that, if $a$ and $b$ are real numbers such that for every integer $n > 0$, $a \le b + \frac{1}{n}$, then for all real numbers $\epsilon > 0$, $a \le b + \epsilon$.

**6.15**    Prove that, if for all real numbers $x$ and $y$, $|x + y| \le |x| + |y|$, then for all real numbers $x, y$, and $z$, $|x - z| \le |x - y| + |y - z|$. (Be careful of overlapping notation.)

$^W$**6.16**    For functions $f, g$, and $h$, prove that, if $f \ge g$ on a set $S$ of real numbers (see the definition in Exercise 5.1(b) on page 61) and $g \ge h$ on $S$, then $f \ge h$ on $S$.

**6.17**    Prove that, if the real numbers $u$ and $v$ are upper bounds, respectively, for the set $S$ of real numbers and the set $-S = \{-x : x \in S\}$ and $v \ge u$, then for every element $x \in S$, $|x| \le v$.

**6.18**    For real numbers $u$ and $v$, prove that, if $u$ is an upper bound for a set $S$ of real numbers and $u \le v$, then $v$ is an upper bound for $S$.

$^B$**6.19**    Prove that, if $S$ and $T$ are convex sets (see the definition in Exercise 5.2(b) on page 61), then $S \cap T$ is a convex set.

$^W$**6.20**    Prove that, if $f$ is a convex function (see the definition in Exercise 5.2(c) on page 61), then for all real numbers $s \geq 0$, the function $sf$ is convex [where the value of the function $sf$ at any point $x$ is $sf(x)$].

**6.21**    For functions $f$ and $g$ of one variable, prove that, if $g \geq f$ on the set of real numbers and $x^*$ is a maximizer of $g$ (see the definitions in Exercise 5.1(a) and (b) on page 61), then for every real number $x$, $f(x) \leq g(x^*)$ .

$^*$**6.22**    Suppose that $a$, $b$, and $c$ are real numbers with $a \neq 0$. Prove that, if $x^* = -b/(2a)$ is a maximizer of the function $f(x) = ax^2 + bx + c$ (see Exercise 5.1(a) on page 61), then $a < 0$. (Hint: Specialize $x$ to $x^* + \epsilon$, where $\epsilon > 0$.)

$^B$**6.23**    Write an analysis of proof that corresponds to the condensed proof given below. Indicate which techniques are used and how they are applied. Fill in the details of any missing steps, where appropriate.

> **Proposition.** If $R$ is subset of a set $S$ of real numbers and if $f$ and $g$ are functions for which $g \geq f$ on $S$ (see the definition in Exercise 5.1(b) on page 61), then $g \geq f$ on $R$.
>
> **Proof.** To show that $g \geq f$ on $R$, let $x \in R$. Because $R$ is a subset of $S$, every element $r \in R$ is in $S$. In particular, $x \in R$, so $x \in S$. Also, because $g \geq f$ on $S$, it follows that, for every element $s \in S$, $g(s) \geq f(s)$. In particular, $x \in S$, so $g(x) \geq f(x)$.  □

**6.24**    Write an analysis of proof that corresponds to the condensed proof given below. Indicate which techniques are used and how they are applied. Fill in the details of any missing steps, where appropriate.

> **Proposition.** If $f$ and $g$ are convex functions (see the definition in Exercise 5.2(c) on page 61), then $f + g$ is a convex function.
>
> **Proof.** To see that $f + g$ is convex, let $x$, $y$, and $t$ be real numbers with $0 \leq t \leq 1$. Then, because $f$ is convex, it follows that
>
> $$f(tx + (1-t)y) \leq tf(x) + (1-t)f(y) \quad \text{(i)}$$
>
> Likewise, because $g$ is convex,
>
> $$g(tx + (1-t)y) \leq tg(x) + (1-t)g(y) \quad \text{(ii)}$$
>
> Adding (i) and (ii) yields that
>
> $$f(tx+(1-t)y)+g(tx+(1-t)y) \leq t[f(x)+g(x)]+(1-t)[f(y)+g(y)].$$
>
> Thus, $f + g$ is convex and the proof is complete.  □

*6.25    Write an analysis of proof that corresponds to the condensed proof given below. Indicate which techniques are used and how they are applied. Fill in the details of any missing steps, where appropriate.

> **Proposition.** If $f$ is a convex function (see the definition in Exercise 5.2(c) on page 61), then for all real numbers $x_1, x_2, x_3$, $a_1, a_2$, and $a_3$ with $a_1, a_2, a_3 \geq 0$ and $a_1 + a_2 + a_3 = 1$, it follows that $f(a_1 x_1 + a_2 x_2 + a_3 x_3) \leq a_1 f(x_1) + a_2 f(x_2) + a_3 f(x_3)$.
>
> **Proof.** Let $a_1, a_2, a_3 \geq 0$ be real numbers with $a_1 + a_2 + a_3 = 1$. If $a_3 = 1$, then $a_1 = a_2 = 0$ and so the conclusion is true. Thus, it can be assumed that $1 - a_3 > 0$. Now because $f$ is a convex function, by definition, for all real numbers $x$, $y$, and $t$ with $0 \leq t \leq 1$, $f(tx + (1-t)y) \leq tf(x) + (1-t)f(y)$. In particular, for $t = 1 - a_3 > 0$, $x = \frac{a_1}{1-a_3} x_1 + \frac{a_2}{1-a_3} x_2$, and $y = x_3$, it follows that
>
> $$f(tx + (1-t)y) = f\left((1-a_3)\left[\frac{a_1}{1-a_3}x_1 + \frac{a_2}{1-a_3}x_2\right] + a_3 x_3\right)$$
>
> $$\leq (1-a_3)f\left(\frac{a_1}{1-a_3}x_1 + \frac{a_2}{1-a_3}x_2\right) + a_3 f(x_3)$$
>
> Now let $t = \frac{a_1}{1-a_3} \geq 0$ and note that, because $a_1 + a_2 + a_3 = 1$, $1 - t = 1 - \frac{a_1}{1-a_3} = \frac{a_2}{1-a_3} \geq 0$. Thus, from the convexity of $f$ with $x = x_1$ and $y = x_2$, you have
>
> $$f\left(\frac{a_1}{1-a_3}x_1 + \frac{a_2}{1-a_3}x_2\right) \leq \frac{a_1}{1-a_3}f(x_1) + \frac{a_2}{1-a_3}f(x_2).$$
>
> Multiplying both sides of the foregoing inequality by $1 - a_3 > 0$ and combining the result with the previous inequality yields the desired conclusion that $f(a_1 x_1 + a_2 x_2 + a_3 x_3) \leq a_1 f(x_1) + a_2 f(x_2) + a_3 f(x_3)$.  $\square$

# 7

# *Quantifiers IV: Nested Quantifiers*

The statements in the previous three chapters contain only one quantifier–either "there is" or "for all." You will now learn what to do when statements contain more than one quantifier–that is, when there are **nested quantifiers**.

## 7.1 UNDERSTANDING STATEMENTS WITH NESTED QUANTIFIERS

The use of nested quantifiers is illustrated in the following statement containing both "for all" and "there is":

**S1:** For all real numbers $x$ with $0 \leq x \leq 1$, there is a real number $y$ with $-1 \leq y \leq 1$ such that $x + y^2 = 1$.

When reading, writing, or processing such statements, always work from left to right. For the foregoing statement $S1$, the first quantifier encountered from the left is "for all." For that quantifier, identify the following components:

| | |
|---|---|
| Object: | real number $x$. |
| Certain property: | $0 \leq x \leq 1$. |
| Something happens: | there is a real number $y$ with $-1 \leq y \leq 1$ such that $x + y^2 = 1$. |

The something that happens in this case contains the nested quantifier "there is," for which you can then identify the following components:

Object:                   real number $y$.
Certain property:         $-1 \le y \le 1$.
Something happens:        $x + y^2 = 1$.

As another example of nested quantifiers, suppose that $T$ is a set of real numbers and consider the statement:

**S2:** There is a real number $M > 0$ such that, for all elements
$x \in T$, $x < M$.

In $S2$, the quantifier "there is" is the first one encountered when reading from the left. Associated with that quantifier are the following three components:

Object:                   real number $M$.
Certain property:         $M > 0$.
Something happens:        for all elements $x \in T$, $x < M$.

The something that happens in this case contains the nested quantifier "for all," for which you can identify the following components:

Object:                   element $x$.
Certain property:         $x \in T$.
Something happens:        $x < M$.

It is important to realize that the order in which the quantifiers appear is critical to the meaning of the statement. For example, compare the foregoing statement $S2$ with the following statement:

**S3:** For all real numbers $M > 0$, there is an element $x \in T$
such that $x < M$.

$S3$ states that, for each positive real number $M$, you can find a (possibly different) element $x \in T$ for which $x < M$. Note that the element $x$ may depend on the value of $M$; that is, if the value of $M$ is changed, the value of $x$ may change. In contrast, $S2$ says that there is a positive real number $M$ such that, no matter which element $x$ you choose in $T$, $x < M$. It should be clear that $S2$ and $S3$ are not the same.

Statements can have any number of quantifiers, as illustrated in the following example containing three nested quantifiers (in which $f$ is a function of one real variable):

**S4:** For every real number $\epsilon > 0$, there is a real number $\delta > 0$
such that, for all real numbers $x$ and $y$ with $|x - y| < \delta$,
$|f(x) - f(y)| < \epsilon$.

Applying the principle of working from left to right, associated with the first quantifier "for all" in $S4$ you should identify:

| | |
|---|---|
| Object: | real number $\epsilon$. |
| Certain property: | $\epsilon > 0$. |
| Something happens: | there is a real number $\delta > 0$ such that, for all real numbers $x$ and $y$ with $|x - y| < \delta$, $|f(x) - f(y)| < \epsilon$. |

The something that happens in this case contains the nested quantifiers "there is" and "for all." Working from left to right once again, identify the three components associated with the quantifier "there is":

| | |
|---|---|
| Object: | real number $\delta$. |
| Certain property: | $\delta > 0$. |
| Something happens: | for all real numbers $x$ and $y$ with $|x-y| < \delta$, $|f(x) - f(y)| < \epsilon$. |

The three components of the last nested quantifier "for all" in the foregoing something that happens are:

| | |
|---|---|
| Objects: | real numbers $x$ and $y$. |
| Certain property: | $|x - y| < \delta$. |
| Something happens: | $|f(x) - f(y)| < \epsilon$. |

When writing statements, it is important that all symbols be defined beforehand or by using an appropriate quantifier, for otherwise, a **syntax error** occurs. To illustrate, suppose that $a$ and $b$ are integers and you know $a$ divides $b$. Using Definition 1, you might then write the following statement:

**A1:** $b = ca$.

The foregoing statement $A1$ contains a syntax error because, while the symbols $a$ and $b$ are known to be integers, the symbol $c$ is undefined. The correct way to write the foregoing statement is:

**A1:** There is an integer $c$ such that $b = ca$.

In summary, when writing statements, make sure that all symbols are properly defined using appropriate quantifiers, when necessary.

## 7.2   USING PROOF TECHNIQUES WITH NESTED QUANTIFIERS

When a statement in the forward or backward process contains nested quantifiers, apply appropriate techniques (construction, choose, and specialization) based on the order in which the quantifiers appear from left to right in the statement. To illustrate, suppose you want to show that the foregoing statement $S2$ is true; that is, you want to show that

**B:** There is a real number $M > 0$ such that, for all elements $x \in T$, $x < M$.

Because $B$ is in the backward process and the first quantifier encountered from the left is "there is," the construction method is the correct proof technique to use first. Thus, you should turn to the forward process in an attempt to construct a real number $M > 0$. Suppose you have done so. According to the construction method, you must show that the value of $M$ you constructed satisfies the something that happens in $B$; that is, you must show that

**B1:** For all elements $x \in T$, $x < M$.

When trying to show that $B1$ is true, you should apply the choose method because of the appearance of the quantifier "for all" in the backward process. In this case, you would choose

**A1:** An element $x \in T$,

for which it must be shown that

**B2:** $x < M$.

Applying appropriate techniques to nested quantifiers is illustrated again in the proof of the following proposition.

**Definition 16** *A function $f$ from the set of real numbers to the set of real numbers is* **onto** *(or* **surjective**) *if and only if for every real number $y$ there is a real number $x$ such that $f(x) = y$.*

**Proposition 10** *If $m$ and $b$ are real numbers with $m \neq 0$, then the function $f(x) = mx + b$ is onto.*

**Analysis of Proof.** The forward-backward method is used to begin the proof because the hypothesis $A$ and the conclusion $B$ do not contain keywords (such as "for all" or "there is"). Working backward, you are led to the key question, "How can I show that a function (namely, $f(x) = mx + b$) is onto?" Using Definition 16 with $f(x) = mx + b$, you must show that

**B1:** For every real number $y$, there is a real number $x$ such that $mx + b = y$.

The statement $B1$ contains nested quantifiers. In deciding which proof technique to apply next, observe that the quantifier "for all" is the first one encountered from the left. Thus, use the choose method (because the keywords "for all" appear in the backward process). Accordingly, you should choose

**A1:** A real number $y$,

for which it must be shown that

**B2:** There is a real number $x$ such that $mx + b = y$.

Recognizing the keywords "there is" in the backward statement $B2$, you should now proceed with the construction method. To that end, turn to the forward process in an attempt to construct the desired real number $x$.

Looking at the fact that you want $mx + b = y$ and observing that $m \neq 0$ by the hypothesis, you might produce this statement:

**A2:** Construct the real number $x = (y - b)/m$.

Recall that with the construction method you must show that the object you construct satisfies the certain property and the something that happens (in $B2$). Thus you must show that

**B3:** $mx + b = y$.

But, from $A2$, the proof is complete because

**A3:** $mx + b = m[(y - b)/m] + b = (y - b) + b = y$,

In the condensed proof that follows, observe that the names of the techniques are omitted. Note also that several steps in the foregoing analysis-of-proof are combined into a single statement.

**Proof of Proposition 10.**  To show that $f$ is onto, let $y$ be a real number. (The word "let" here indicates that the choose method is used.) By hypothesis, $m \neq 0$, so let $x = (y - b)/m$. (Here, the word "let" indicates that the construction method is used.) You can now see that $f(x) = mx + b = y$.  $\square$

## 7.3   READING A PROOF

The process of reading a proof is demonstrated with the following proposition.

**Proposition 11** *If $a$, $b$, and $c$ are real numbers with $a < 0$, then there is a real number $y$ such that for every real number $x$, $ax^2 + bx + c \leq y$.*

**Proof of Proposition 11.**    (For reference purposes, each sentence of the proof is written on a separate line.)

**S1:**   Let $y = \frac{4ac - b^2}{4a}$.
**S2:**   For a real number $x$, you have that

$$ax^2 + bx + c = a\left(x + \frac{b}{2a}\right)^2 + \frac{4ac - b^2}{4a}.$$

**S3:**   Now, because $a < 0$, $ax^2 + bx + c \leq \frac{4ac - b^2}{4a} = y$.

The proof is now complete.  $\square$

**Analysis of Proof.**   An interpretation of statements $S1$ through $S3$ follows.

**Interpretation of S1:** *Let* $y = \frac{4ac-b^2}{4a}$.
The author recognizes the first quantifier "there is" in the conclusion $B$ and is therefore using the construction method to construct the value

**A1:** $y = \frac{4ac-b^2}{4a}$.

**Interpretation of S2:** *For a real number* $x$, *you have that*

$$ax^2 + bx + c = a\left(x + \frac{b}{2a}\right)^2 + \frac{4ac - b^2}{4a}.$$

The author is following the construction method and is showing that the value of $y$ in $A1$ satisfies the something that happens associated with the quantifier "there is" in $B$; namely, that

**B1:** For all real numbers $x$, $ax^2 + bx + c \leq y$.

Recognizing the quantifier "for all" in $B1$, the author uses the choose method, as indicated by the words, "For a real number $x$...." In other words, the author chooses

**A2:** A real number $x$,

for which it must be shown that

**B2:** $ax^2 + bx + c \leq y$.

In showing that $B2$ is true, the author first establishes the following statement in $S2$, omitting several algebraic steps:

**A3:** It follows that:

$$\begin{aligned} ax^2 + bx + c &= a\left(x^2 + \frac{b}{a}x + \frac{b^2}{4a^2}\right) + c - \frac{b^2}{4a} \\ &= a\left(x + \frac{b}{2a}\right)^2 + \frac{4ac-b^2}{4a}. \end{aligned}$$

**Interpretation of S3:** *Now, because* $a < 0$, $ax^2 + bx + c \leq \frac{4ac-b^2}{4a} = y$.
The author is working forward from the first term on the right side of $A3$. Specifically, the author notes that $a\left(x + \frac{b}{2a}\right)^2 \leq 0$ because $a < 0$ by hypothesis and $\left(x + \frac{b}{2a}\right)^2 \geq 0$ and therefore

**A4:** $ax^2 + bx + c \leq \frac{4ac-b^2}{4a}$.

Finally, the author uses the fact that $\frac{4ac-b^2}{4a} = y$, from $A1$. This completes both the choose and the construction methods, and so the proof is finished.

## Summary

When statements contain nested quantifiers, follow these steps:

1. For each quantifier encountered from left to right, identify the object, the certain property, and the something that happens.

2. Apply the appropriate construction, choose, and specialization methods based on the order of the quantifiers as they appear from left to right.

In Chapters 9 and 10 you will learn two new techniques for showing that "$A$ implies $B$." Those techniques require the material in Chapter 8.

## Exercises

**Note:** Solutions to those exercises marked with a $B$ are in the back of this book. Solutions to those exercises marked with a $W$ are located on the web at http://www.wiley.com/college/solow/.

**Note:** All proofs should contain an analysis of proof and a condensed version. Definitions for all mathematical terms are provided in the glossary at the end of the book.

$^B$**7.1**   Identify the object, the certain property, and the something that happens for each of the quantifiers as they appear from left to right in each of the following definitions.

a. The function $f$ of one real variable is **bounded above** if and only if there is a real number $y$ such that, for every real number $x$, $f(x) \leq y$.

b. A set of real numbers $S$ is **bounded** if and only if there is a real number $M > 0$ such that, $\forall$ element $x \in S$, $|x| < M$.

c. The function $f$ of one real variable is **continuous at the point** $x$ if and only if, for every real number $\epsilon > 0$, there is a real number $\delta > 0$ such that, for all real numbers $y$ with $|x - y| < \delta$, $|f(x) - f(y)| < \epsilon$.

d. Suppose that $x_1, x_2, \ldots$ and $x$ are real numbers. The sequence $x_1, x_2, \ldots$ **converges to** $x$ if and only if $\forall$ real numbers $\epsilon > 0$, $\exists$ an integer $j \geq 1$ $\ni \forall$ integer $k$ with $k > j$, $|x_k - x| < \epsilon$.

**7.2**   Rewrite each the following statements introducing appropriate notation and using nested quantifiers.

a. For a set $S$ of real numbers, no matter which element is chosen in the set, you can find another element in the set that is strictly larger.

b. A given function of one real variable has the property that the absolute value of all values of the function is always less than some number.

$^{B}$**7.3**    Are each of the following pairs of statements the same; that is, are these two statements both true? Why or why not?

a.   $S1$ :   For all real numbers $x$ with $0 \leq x \leq 1$, for all real
numbers $y$ with $0 \leq y \leq 2$, $2x^2 + y^2 \leq 6$.
$S2$ :   For all real numbers $y$ with $0 \leq y \leq 2$, for all real
numbers $x$ with $0 \leq x \leq 1$, $2x^2 + y^2 \leq 6$.

b.   $S1$ :   For all real numbers $x$ with $0 \leq x \leq 1$, for all real
numbers $y$ with $0 \leq y \leq 2x$, $2x^2 + y^2 \leq 6$.
$S2$ :   For all real numbers $y$ with $0 \leq y \leq 1$, for all real
numbers $x$ with $0 \leq x \leq 2y$, $2x^2 + y^2 \leq 6$.

c.   Based on your answer to parts (a) and (b), when is the statement, "For all objects $X$ with a certain property $P$ and for all objects $Y$ with a certain property $Q$, something happens," the same as the statement, "For all objects $Y$ with the property $Q$ and for all objects $X$ with the property $P$, something happens"?

**7.4**    Are each of the following pairs of statements the same; that is, are these two statements both true? Why or why not?

a.   $S1$ :   There is a real number $x \geq 2$ such that there is a
real number $y \geq 1$ such that $x^2 + 2y^2 < 9$.
$S2$ :   There is a real number $y \geq 1$ such that there is a
real number $x \geq 2$ such that $x^2 + 2y^2 < 9$.

b.   $S1$ :   There is a real number $0 \leq x \leq 1$ such that there is
a real number $0 \leq y \leq 2x$ such that $2x^2 + y^2 > 6$.
$S2$ :   There is a real number $0 \leq y \leq 1$ such that there is
a real number $0 \leq x \leq 2y$ such that $2x^2 + y^2 > 6$.

c.   Based on your answers to parts (a) and (b), when is the statement, "There is an object $X$ with a certain property $P$ such that there is an object $Y$ with a certain property $Q$ such that something happens," the same as the statement, "There is an object $Y$ with the property $Q$ such that there is an object $X$ with the property $P$ such that something happens"?

$^{B}$**7.5**    Explain how to work backward from each of the following statements. Which proof techniques would you use and in which order? How would you apply those techniques?

a.   For all objects $X$ with a certain property $P$, there is an object $Y$ with a certain property $Q$ such that something happens.

b.   There is an object $X$ with a certain property $P$ such that, for all objects $Y$ with a certain property $Q$, something happens.

$^B$**7.6**    Explain how to work forward from each of the following statements. Which proof techniques would you use and in which order? How would you apply those techniques?

  a. For all objects $X$ with a certain property $P$, there is an object $Y$ with a certain property $Q$ such that something happens.

  b. There is an object $X$ with a certain property $P$ such that, for all objects $Y$ with a certain property $Q$, something happens.

**7.7**    For each of the following statements in the backward process, indicate which techniques you would use (construction and choose) and in which order. Also, explain how you would apply the techniques to the particular problem; that is, what would you construct, what would you choose, and so on?

  a. There is a real number $M > 0$ such that, for all elements $x$ in the set $T$ of real numbers, $|x| \leq M$.

  b. For all real numbers $M > 0$, there is an element $x$ in the set $T$ of real numbers such that $|x| > M$.

  c. $\forall$ real numbers $\epsilon > 0$, $\exists$ a real number $\delta > 0$ $\ni$  $\forall$ real numbers $x$ and $y$ with $|x - y| < \delta$, $|f(x) - f(y)| < \epsilon$ (where $f$ is a function of one real variable).

**7.8**    When working forward from each of the statements in the previous exercise, indicate which techniques you would use and in which order. Explain how you would apply those techniques to the particular problem.

**7.9**    For each of the following propositions, ask and answer an appropriate key question. Then indicate which proof techniques you would use next and in which order. Explain how you would apply those techniques to the problem.

  a. If $S$ is a subset of a set $T$ of real numbers and $T$ is bounded, then $S$ is bounded. (See the definition in Exercise 7.1(b).)

  b. If the functions $f$ and $g$ are onto (see Definition 16 on page 82), then the function $f \circ g$ is onto, where $(f \circ g)(x) = f(g(x))$.

  c. If $f$ and $g$ are functions of one real variable for which $g \geq f$ on the set of real numbers and $g$ is bounded above, then $f$ is bounded above. (See the definitions in Exercise 5.1(b) and Exercise 7.1(a).)

**7.10**    For each of the propositions in the previous exercise, use a definition to work forward from one statement in the hypothesis to create a new statement with nested quantifiers. Then indicate which proof techniques you would use next and in which order. Explain how you would apply those techniques to the problem.

$^B$**7.11**    Prove that, for all real numbers $\epsilon > 0$ and $a > 0$, there is an integer $n > 0$ such that $\frac{a}{n} < \epsilon$.

$^B$**7.12**    Answer the given questions about the following proof that, "For all real numbers $x$ and $y$ with $x < y$, there is a rational number $r$ such that $x < r < y$."

> **Proof.** Let $x$ and $y$ be real numbers with $x < y$. Then let $\epsilon = y - x > 0$ and so, from the proposition in Exercise 7.11, there is an integer $n > 0$ such that $n\epsilon > 2$. Now let $m$ be an integer with $nx < m < ny$. Then the desired rational number is $r = \frac{m}{n}$ and so the proof is complete. $\Box$

a. What proof technique is the author using in the first sentence of the proof? Why did the author choose to use that technique? Explain.

b. What is the author doing in the second sentence?

c. What proof technique is the author using in the last sentence of the proof? Why did the author choose to use that technique? Explain.

d. Is the author justified in claiming that the proof is complete in the last sentence of the proof? Why or why not?

$^*$**7.13**    Answer the given questions about the following proof that, "For every prime number $n$, there is a prime number $p$ such that $p > n$."

> **Proof.** Let $n$ be a prime number and let $p$ be any prime number that divides $n! + 1$. To see that $p > n$, observe first that every integer $k$ with $2 \le k \le n$ divides $n!$. Thus, every such integer does not divide $n! + 1$. In particular, if $2 \le p \le n$, then it would follow that $p$ would not divide $n! + 1$. Because $p$ does divide $n! + 1$, $p$ cannot be any of the integers $2, \ldots, n$ and so $p > n$, completing the proof. $\Box$

a. What techniques are used in the first sentence of the proof. Why did the author use those techniques? What statement remains to be shown to complete the proof and in which sentence is this done?

b. Justify the statement that "every integer $k$ with $2 \le k \le n$ divides $n!$" in the second sentence by providing a proof of this statement.

c. Justify the statement, "Thus, every such integer does not divide $n! + 1$" in the third sentence by considering $\frac{n!+1}{k}$.

d. What technique is the author using in the fourth sentence of the proof?

$^*$**7.14**    Answer the given questions about the following proof that, "If $f$ is a function of one real variable for which there is a real number $m$ such that, for all real numbers $x$ and $y$, $f(y) \geq f(x) + m(y - x)$, then the function $f$ is convex [see the definition in Exercise 5.2(c)]."

> **Proof.** Let $x$, $y$, and $t$ be real numbers with $0 \leq t \leq 1$, for which it will be shown that $f(tx + (1-t)y) \leq tf(x) + (1-t)f(y)$. To that end, let $z = tx + (1 - t)y$. Now, from the hypothesis, it follows that
>
> $$\begin{aligned} f(x) &\geq f(z) + m(x - z) &\quad\text{(a)} \\ f(y) &\geq f(z) + m(y - z). &\quad\text{(b)} \end{aligned}$$
>
> Multiplying inequality (a) through by $t$ and inequality (b) through by $1 - t$ and then adding the two inequalities yields:
>
> $$tf(x) + (1 - t)f(y) \geq f(z) + m(tx + (1 - t)y - z) = f(z). \quad\text{(c)}$$
>
> The proof is now complete. $\square$

a. What technique does the author use in the first sentence of the proof and how is that technique applied? Why is this technique used?

b. What is the author doing in the second sentence?

c. What proof technique is the author using in the third sentence of the proof to obtain the inequalities (a) and (b)? Explain.

d. In the first sentence, the author states that $0 \leq t \leq 1$. Where in the proof is this fact used?

e. Justify the equality $f(z) + m(tx + (1 - t)y - z) = f(z)$ in (c).

f. Is the author correct in stating that the proof is complete? Why or why not?

**7.15**  Identify and then correct the error in the following condensed proof of the proposition, "If $a$ is a positive real number, then for all real numbers $b$ and $m$, there is a real number $x > 0$ such that $ax^2 + bx \geq mx$."

> **Proof.** Let $b$ and $m$ be real numbers. Now it is necessary to find a real number $x$ that satisfies $ax^2 + bx \geq mx$, or equivalently, $x(ax + b - m) \geq 0$. Now, noting that $a > 0$ and $x > 0$, it is seen that any value of $x \geq \frac{m-b}{a}$ will suffice, and so the proof is complete. $\square$

*7.16  Identify the error in the following condensed proof of the statement, "If $v$ is an upper bound for a set $S$ of real numbers, then, for every real number $\epsilon > 0$, $v - \epsilon$ is an upper bound for $S$."

**Proof.** Let $\epsilon > 0$. To see that $v - \epsilon$ is an upper bound for $S$, let $x \in S$. Now, because $v$ is an upper bound for $S$, $x \leq v$ and so $x - \epsilon \leq v - \epsilon$. This means that $v - \epsilon$ is an upper bound for $S$ and so the proof is complete. $\square$

$^B$**7.17**  Prove that, for every real number $x > 2$, there is a real number $y < 0$ such that $x = 2y/(1 + y)$.

**7.18**  Let $f$ and $g$ be functions of one variable. Prove that, if $f$ and $g$ are onto (see Definition 16), then the function $f \circ g$ is onto, where $(f \circ g)(x) = f(g(x))$.

$^*$**7.19**  Prove that, if $S = \{$real numbers $x > 0 : x^2 < 2\}$, then for every real number $\epsilon > 0$, there is an element $x \in S$ such that $x^2 > 2 - \epsilon$.

$^W$**7.20**  Prove that the function $f(x) = -x^2 + 2x$ is bounded above; that is, there is a real number $y$ such that, for all real numbers $x$, $f(x) \leq y$.

$^*$**7.21**  Prove that, if

$$
\begin{aligned}
S &= \{(x, y) : x^2 + y^2 \leq 1\} \quad \text{and} \\
T &= \{(x, y) : (x - 3)^2 + (y - 4)^2 \leq 1\},
\end{aligned}
$$

then there are real numbers $a$ and $b$ such that, for every $(x, y) \in S$, $y \leq ax + b$, and for every $(x, y) \in T$, $y \geq ax + b$. (Hint: Draw a picture of $S$ and $T$ and find an appropriate line $y = ax + b$.)

$^W$**7.22**  Prove that the set $S = \{1 - \frac{1}{2}, 1 - \frac{1}{3}, 1 - \frac{1}{4}, \ldots\}$ satisfies the property that, for every $\epsilon > 0$, there is an $x \in S$ such that $x > 1 - \epsilon$. (Hint: You can write $S$ as $\{$real numbers $x :$ there is an integer $n \geq 2$ such that $x = 1 - \frac{1}{n}\}$.)

$^*$**7.23**  Write an analysis of proof that corresponds to the condensed proof given below. Indicate which techniques are used and how they are applied. Fill in the details of any missing steps, where appropriate.

**Definition.** A real-valued function $f$ of one real variable is **linear** if and only if there are real numbers $m$ and $b$ such that, for all real numbers $x$, $f(x) = mx + b$.

**Proposition.** If $f$ and $g$ are linear functions, then $f + g$ is a linear function [where for any real number $x$, $(f + g)(x) = f(x) + g(x)$].

**Proof.** Because $f$ and $g$ are linear functions, there are real numbers $m_1, m_2, b_1, b_2$ such that, for every real number $x$, $f(x) = m_1 x + b_1$ and $g(x) = m_2 x + b_2$. Let $m = m_1 + m_2$ and $b = b_1 + b_2$. Then, for a given real number $x$, you have

$$
\begin{aligned}
(f + g)(x) &= (m_1 x + b_1) + (m_2 x + b_2) \\
&= (m_1 + m_2)x + (b_1 + b_2) \\
&= mx + b.
\end{aligned}
$$

But this means that $f + g$ is linear, completing the proof. $\square$

# 8

---

# *Nots of Nots*
# *Lead to Knots*

As you will see, the proof techniques in Chapters 9 and 10 require that you be able to write the negation of a statement $A$, hereafter written as $NOT\ A$. This chapter provides rules for writing the $NOT$ of various statements that contain keywords, such as quantifiers.

## 8.1  WRITING THE *NOT* OF STATEMENTS HAVING *NOT*, *AND*, AND *OR*

In some instances, the $NOT$ of a statement is easy to find. For example, if $A$ is the statement, "the real number $x > 0$," then the $NOT$ of $A$ is, "it is not the case that the real number $x > 0$," or equivalently, "the real number $x$ is not $> 0$." You can eliminate the word "not" altogether by incorporating this word into the statement to obtain "the real number $x \leq 0$."

When taking the $NOT$ of a statement that already contains the word "no" or "not," the result is that the $NOT$ cancels the existing "not." For example, if $A$ is the statement,

> **A:** There is no integer $x$ such that $x^2 + x - 11 = 0$,

then the $NOT$ of this statement is,

> **NOT A:** There is an integer $x$ such that $x^2 + x - 11 = 0$.

Special rules apply when writing the *NOT* of a statement that contains the words *AND* or *OR*. Specifically,

| | | |
|---|---|---|
| *NOT* [*A AND B*] | becomes | [*NOT A*] *OR* [*NOT B*], |
| *NOT* [*A OR B*] | becomes | [*NOT A*] *AND* [*NOT B*]. |

For example,

| | | |
|---|---|---|
| *NOT* [($x \geq 3$) *AND* ($y < 2$)] | becomes | [($x < 3$) *OR* ($y \geq 2$)], |
| *NOT* [($x \geq 3$) *OR* ($y < 2$)] | becomes | [($x < 3$) *AND* ($y \geq 2$)]. |

## 8.2   WRITING THE *NOT* OF A STATEMENT WITH QUANTIFIERS

A more challenging situation arises when the statements contain quantifiers. For instance, suppose that the statement $B$ contains the quantifier "for all" in the standard form:

**B:** For all "objects" with a "certain property," "something happens."

Then the *NOT* of this statement is:

**NOT B:** It is not the case that for all "objects" with the "certain property," "something happens,"

which really means that

**NOT B:** There is an object with the certain property for which the something does not happen.

Similarly, if the statement $B$ contains the quantifier "there is" in the standard form:

**B:** There is an "object" with a "certain property" such that "something happens,"

then the *NOT* of this statement is:

**NOT B:** It is not the case that there is an "object" with the "certain property" such that "something happens,"

or, in other words,

**NOT B:** For all objects with the certain property, the something does not happen.

Notice that, when a statement contains a quantifier, the *NOT* of that statement contains the opposite quantifier; that is, "for all" becomes "there is" and "there is" becomes "for all." In general, there are three steps to finding the *NOT* of a statement containing one or more quantifiers:

### Steps for Finding the *NOT* of a Statement with Quantifiers

**Step 1.**   Put the word *NOT* in front of the entire statement.

**Step 2.**   If the word *NOT* appears to the left of a quantifier, then move the word *NOT* to the right of the quantifier and place the *NOT* just before the something that happens. As you do so, change the quantifier to its opposite—"for all" becomes "there is" and "there is" becomes "for all." Repeat this step so long as the word *NOT* appears to the left of a quantifier.

**Step 3.**   When all of the quantifiers appear to the left of the *NOT*, eliminate the *NOT* by incorporating the *NOT* into the statement that appears immediately to its right.

These steps are demonstrated with the following examples.

1. For every real number $x \geq 2$, $x^2 + x - 6 \geq 0$.
   Step 1.    *NOT* [for every real number $x \geq 2$, $x^2 + x - 6 \geq 0$.]
   Step 2.    There is a real number $x \geq 2$ such that
               *NOT* $[x^2 + x - 6 \geq 0]$.
   Step 3.    There is a real number $x \geq 2$ such that $x^2 + x - 6 < 0$.

Note in Step 2 that, when the *NOT* is passed from left to right, the quantifier changes but the certain property (namely, $x \geq 2$) does not. Also, because the quantifier "for every" changes to "there exists," it is necessary to replace the comma by the words "such that." If the quantifier "there exists" is changed to "for all," then the words "such that" are replaced with a comma, as illustrated in the next example.

2. There is a real number $x \geq 2$ such that $x^2 + x - 6 \geq 0$.
   Step 1.    *NOT* [there is a real number $x \geq 2$ such that
               $x^2 + x - 6 \geq 0$.]
   Step 2.    For all real numbers $x \geq 2$, *NOT* $[x^2 + x - 6 \geq 0]$.
   Step 3.    For all real numbers $x \geq 2$, $x^2 + x - 6 < 0$.

If the statement you are taking the *NOT* of contains nested quantifiers (see Chapter 7), then Step 2 is performed on each quantifier, in turn, as it appears from left to right. Step 2 is repeated until all quantifiers appear to the left of the *NOT*, as demonstrated in the next two examples.

3. For every real number $x$ between $-1$ and $1$, there is a real number $y$ between $-1$ and $1$ such that $x^2 + y^2 \leq 1$.

Step 1.    $NOT$ [for every real number $x$ between $-1$ and $1$, there is a real number $y$ between $-1$ and $1$ such that $x^2 + y^2 \leq 1$.]

Step 2.    There is a real number $x$ between $-1$ and $1$ such that $NOT$ [there is a real number $y$ between $-1$ and $1$ such that $x^2 + y^2 \leq 1$].

Step 2.    There is a real number $x$ between $-1$ and $1$ such that, for all real numbers $y$ between $-1$ and $1$, $NOT$ [$x^2 + y^2 \leq 1$].

Step 3.    There is a real number $x$ between $-1$ and $1$ such that, for all real numbers $y$ between $-1$ and $1$, $x^2 + y^2 > 1$.

4. There is a real number $M > 0$ such that, for all elements $x$ in a set $S$ of real numbers, $|x| < M$.

Step 1.    $NOT$ [there is a real number $M > 0$ such that, for all elements $x$ in a set $S$ of real numbers, $|x| < M$.]

Step 2.    For all real numbers $M > 0$, $NOT$ [for all elements $x$ in a set $S$ of real numbers, $|x| < M$.]

Step 2.    For all real numbers $M > 0$, there is an element $x$ in a set $S$ of real numbers such that $NOT$ [ $|x| < M$ ].

Step 3.    For all real numbers $M > 0$, there is an element $x$ in a set $S$ of real numbers such that $|x| \geq M$.

## 8.3   COUNTEREXAMPLES

All of the statements you have seen so far have been true and you have been able to prove them. However, the truth of some mathematical statements is not always so clear. Consider, for instance, the following statement:

**B:** For every integer $n \geq 2$, $n^2 \geq 2^n$.

You can easily verify that $B$ is true for $n = 2, 3$, and $4$. However, if you try to prove this statement by using the choose method (or any other method, for that matter), you will not succeed. The reason is that the foregoing statement is not true, and so no proof technique will work. In general, how can you show that a particular statement is not true? One answer is to prove that the negation of that statement is true. For example, to show that the foregoing statement $B$ is not true, use the rules you have just learned in Section 8.2 to write the following negation:

**NOT B:** There is an integer $n \geq 2$ such that $n^2 < 2^n$.

To prove that *NOT B* is true, you should now use the construction method because of the keywords "there is" in *NOT B*. In this case, you can construct $n = 5$, which works because $n = 5 \geq 2$ and $n^2 < 2^n$ as $5^2 = 25 < 32 = 2^5$.

In summary, to show that a statement $B$ is not true, try to prove that *NOT B* is true. When the statement $B$ contains the keywords "for all" in the standard form:

> **B:** For all objects with a certain property, something happens,

the statement *NOT B* contains the keywords "there is" in the form:

> **NOT B:** There is an object with the certain property such that the something does not happen.

To show that *NOT B* is true, you can then use the construction method to produce (often by trial-and-error) an object with the certain property and for which the something does not happen. This single object for which statement $B$ is not true is referred to as a **counterexample** to statement $B$. Thus, $n = 5$ is a counterexample to the foregoing statement $B$.

Another example of producing a counterexample arises when you are trying to prove that a statement of the form "$A$ implies $B$" is not true. Consider the following example (recalling that an integer $m$ divides an integer $n$, written $m|n$, if and only if there is an integer $k$ such that $n = km$):

> **S:** If $a$, $b$, and $c$ are integers for which $a|(bc)$, then $a|b$ and $a|c$.

After unsuccessfully attempting to prove statement $S$, you might begin to suspect that $S$ is not true. To prove that $S$ is not true, recall from the truth table on page 4 that an implication of the form "$A$ implies $B$" is not true when $A$ is true and $B$ is false. Thus, to show that the foregoing statement $S$ is not true, you must show that

> **A:** $a$, $b$ and $c$ are integers for which $a|(bc)$ and

> **NOT B:** *NOT* ($a|b$ and $a|c$); that is, using the rules in Section 8.1, either $a$ does not divide $b$ or $a$ does not divide $c$.

In other words, to show that $S$ is not true, you must produce integers $a$, $b$, and $c$ for which $a|(bc)$ and either $a$ does not divide $b$ or else $a$ does not divide $c$. For instance, $a = 2$, $b = 4$, and $c = 5$ satisfy $a|(bc)$ because $2|20$ and yet $a = 2$ does not divide $c = 5$. Thus, the values $a = 2$, $b = 4$, and $c = 5$ provide a counterexample to statement $S$.

In summary, to show that a statement of the form "$A$ implies $B$" is not true, you must show that $A$ is true and $B$ is false. In this case, the counterexample consists of the values for the appropriate items that make $A$ true and $B$ false.

## Summary

The following list summarizes the rules for writing the *NOT* of statements that have a special form.

1. *NOT* [*NOT A*]    becomes    *A*.

2. *NOT* [*A AND B*]    becomes    [(*NOT A*) *OR* (*NOT B*)].

3. *NOT* [*A OR B*]    becomes    [(*NOT A*) *AND* (*NOT B*)].

4. *NOT* [there is an object with a certain property such that something happens] becomes "for all objects with the certain property, the something does not happen."

5. *NOT* [for all objects with a certain property, something happens] becomes "there is an object with the certain property such that the something does not happen."

In the event that a statement contains nested quantifiers, the word *NOT* is processed through each quantifier, from left to right.

## Exercises

**Note:** Solutions to those exercises marked with a *B* are in the back of this book. Solutions to those exercises marked with a *W* are located on the web at http://www.wiley.com/college/solow/.

**Note:** All proofs should contain an analysis of proof and a condensed version. Definitions for all mathematical terms are provided in the glossary at the end of the book.

$^B$**8.1**    Write the *NOT* of each of the definitions in Exercise 5.1 on page 61; that is, write the *NOT* of the statement that constitutes the definition of the term being defined. For instance, the *NOT* of the definition in Exercise 5.1(a) is, "The real number $x^*$ is not a maximum of the function $f$ if there is a real number $x$ such that $f(x) > f(x^*)$."

**8.2**    Write the *NOT* of each of the definitions in Exercise 5.2 on page 61; that is, write the *NOT* of the statement that constitutes the definition of the term being defined.

**8.3**    Write the *NOT* of each of the following definitions so that the word "not" does not appear in the statement after the words "if and only if."

a. A positive integer $p > 1$ is **prime** if and only if there is no integer $n$ with $2 \leq n < p$ such that $n$ divides $p$.

b. A sequence $x_1, x_2, \ldots$ of real numbers is **increasing** if and only if for every integer $k = 1, 2, \ldots$, $x_k < x_{k+1}$.

c. A sequence $x_1, x_2, \ldots$ of real numbers is **decreasing** if and only if for every integer $k = 1, 2, \ldots, x_k > x_{k+1}$.

d. A sequence $x_1, x_2, \ldots$ of real numbers is **strictly monotone** if and only if the sequence is increasing or decreasing. (Use the word "and" in the negation.)

e. An integer $d$ is the **greatest common divisor** of the integers $a$ and $b$ if and only if (i) $d|a$ and $d|b$ and (ii) whenever $c$ is an integer for which $c|a$ and $c|b$, it follows that $c|d$.

f. The real number $f'(\bar{x})$ is the **derivative** of the function $f$ at the point $\bar{x}$ if and only if $\forall$ real number $\epsilon > 0$, $\exists$ a real number $\delta > 0$ such that $\forall$ real number $x$ with $0 < |x - \bar{x}| < \delta$, $\left| \frac{f(x) - f(\bar{x})}{x - \bar{x}} - f'(\bar{x}) \right| < \epsilon$.

$^B$**8.4**    Reword the following statements so that the word "not" appears explicitly. For example, reword the statement "$x > 0$" to read "$x$ is not $\leq 0$."

a. For each element $x$ in the set $S$, $x$ is in $T$.

b. There is an angle $t$ between 0 and $\pi/2$ such that $\sin(t) = \cos(t)$.

c. For every object with a certain property, something happens.

d. There is an object with a certain property such that something happens.

**8.5**    Write the *NOT* of the conclusion in each of the following implications. State your answer in such a way that the words "no" and "not" do not appear.

a. If $a$ is a positive real number, then there is a real number $x$ such that $x = a^{-x}$.

b. $A$ implies *NOT B*.

c. $A$ implies ($B$ implies $C$).

d. If $u$ is a least upper bound for a set $S$ of real numbers, then $\forall$ real number $\epsilon > 0$, $\exists$ an element $x \in S$ such that $x > u - \epsilon$.

$^B$**8.6**    For each of the following statements $A$ implies $B$, what statement(s) will you work forward from and what statement(s) will you work backward from if you want to prove that *NOT B* implies *NOT A*?

a. ($C$ *AND D*) implies $B$.          b. ($C$ *OR D*) implies $B$.

c. $A$ implies ($C$ *AND D*).          d. $A$ implies ($C$ *OR D*).

**8.7**    For each of the following statements $A$ implies $B$, what statement(s) will you work forward from and what statement(s) will you work backward from if you want to prove that $NOT\ B$ implies $NOT\ A$?

    a. If $k$ is an integer that divides an integer $n$, then $k$ does not divide $n+1$.

    b. If $n$ is an even integer and $m$ is an odd integer, then either $mn$ is divisible by 4 or $n$ is not divisible by 4.

    c. Suppose that $m$ and $n$ are integers. If either $mn$ is divisible by 4 or $n$ is not divisible by 4, then $n$ is an even integer and $m$ is an odd integer.

*\*8.8*    For each of the following statements $A$ implies $B$, what proof technique(s) will you use and in what order to prove $NOT\ B$ implies $NOT\ A$?

    a. Let $a$, $b$, and $c$ be real numbers. If $a > 0$, then there does not exist a real number $M$ such that, for every real number $x$, $ax^2 + bx + c \leq M$.

    b. Let $a$, $b$, and $c$ be real numbers. If there is a real number $M$ such that, for all real numbers $x$, $ax^2 + bx + c \leq M$, then $a \leq 0$.

    c. Let $p$ be a positive integer. If there is no integer $m$ with $1 < m \leq \sqrt{p}$ such that $m|p$, then $p$ is prime.

    d. Let $a$ be a real number. If $a > 0$, then there is a real number $x$ such that $x = a^{-x}$.

    e. Let $m$ and $n$ be positive integers. If there is an integer $k$ with $1 < k < m$ such that $k$ does not divide $n$, then $n \neq m!$.

    f. Let $S$ and $T$ be sets of real numbers with $S \subset T$. If for every real number $M > 0$, there is an element $x \in S$ such that $|x| \geq M$, then for every real number $N > 0$, there is an element $y \in T$ such that $|y| \geq N$.

$^B$**8.9**    Provide a counterexample to show that each of the following statements is not true.

    a. For every real number $x$, $x^2 \leq x$.

    b. For every integer $n \geq 1$, $n^2 \geq n!$, where $n! = n(n-1)\cdots 1$.

    c. If $a$, $b$, and $c$ are integers for which $a|(b+c)$, then $a|b$ and $a|c$.

*\*8.10*    Provide a counterexample to show that each of the following statements is not true.

    a. For every real number $x > 0$, $\sqrt{x} \leq x$.

    b. For every positive integer $n$, $n^2 + n + 41$ is prime.

    c. If $p$ is a positive integer that is not prime, then for every integer $m$ with $1 < m \leq \sqrt{p}$, $m$ does not divide $p$.

# 9

## *The Contradiction Method*

With all the techniques you have learned so far, you may still find yourself unable to complete a proof for one reason or another. This chapter presents a new technique that often provides a successful alternative. This method is used when the conclusion of the proposition contains appropriate keywords.

### 9.1  WHY THE NEED FOR ANOTHER PROOF TECHNIQUE?

As powerful as the forward-backward method is, it may not always lead to a successful proof, as shown in the next example.

**Proposition 12** *If $n$ is an integer and $n^2$ is even, then $n$ is even.*

**Analysis of Proof.**    The forward-backward method gives rise to the key question, "How can I show that an integer (namely, $n$) is even?" One answer is to show that

   **B1:** There is an integer $k$ such that $n = 2k$.

The appearance of the quantifier "there is" in the backward process suggests proceeding with the construction method, and so the forward process is used in an attempt to produce the desired integer $k$.

   Working forward from the hypothesis that $n^2$ is even, you can state that

   **A1:** There is an integer, say $m$, such that $n^2 = 2m$.

The objective is to produce an integer $k$ for which $n = 2k$, so it is natural to take the positive square root of both sides of the equality in A1 to obtain

**A2:**  $n = \sqrt{2m}$,

but how can you rewrite $\sqrt{2m}$ to look like $2k$? It would seem that the forward-backward method has failed.

**Proof of Proposition 12.**   The technique you are about to learn leads to a simple proof of this proposition, which is left as an exercise.   $\square$

Fortunately, there are several other techniques that you might want to try before you give up. In this chapter, the **contradiction method** is described, together with an indication of how and when it should be used.

## 9.2   HOW AND WHEN TO USE THE CONTRADICTION METHOD

With the contradiction method, you begin by assuming that $A$ is true, just as you do in the forward-backward method. However, to reach the desired conclusion that $B$ is true, you proceed by asking yourself the simple question, "Why can't $B$ be false?" After all, if $B$ is supposed to be true, then there must be some reason why $B$ cannot be false. The objective of the contradiction method is to discover that reason.

In other words, the idea of a proof by contradiction is to assume that $A$ is true and $B$ is false, and see why this cannot happen. So what does it mean to "see why this cannot happen?" Suppose, for example, that, as a result of assuming that $A$ is true and $B$ is false (hereafter written as *NOT B*), you were somehow able to reach the conclusion that $1 = 0$! Would that not convince you that it is impossible for $A$ to be true and $B$ to be false simultaneously? Thus, in a proof by contradiction, you assume that $A$ is true and *NOT B* is true, using the techniques in Chapter 8 to write the statement *NOT B*. You must use this information to reach a contradiction to something that you absolutely know is true.

Another way of viewing the contradiction method is to recall from Table 1.1 on page 4 that the statement "$A$ implies $B$" is true in all cases except when $A$ is true and $B$ is false. With a proof by contradiction, you rule out this one unfavorable case by assuming that it actually does happen, and then reaching a contradiction. At this point, several natural questions arise:

1. What contradiction should you be looking for?

2. Exactly how do you use the assumption that $A$ is true and $B$ is false to reach the contradiction?

3. Why and when should you use this approach instead of the forward-backward method?

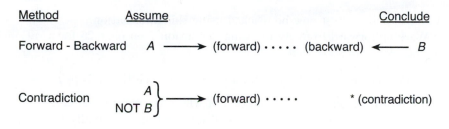

*Fig. 9.1* Comparing the forward-backward and contradiction methods.

The first question is by far the hardest to answer because there are no specific guidelines. Each problem gives rise to its own contradiction. It takes creativity, insight, persistence, and sometimes luck to produce a contradiction.

As to the second question, the most common approach to finding a contradiction is to work forward from the assumptions that $A$ and $NOT\ B$ are true, as is illustrated in a moment.

The foregoing discussion also indicates why you might wish to use contradiction instead of the forward-backward method. With the forward-backward method you assume only that $A$ is true, while in the contradiction method you assume that both $A$ and $NOT\ B$ are true. Thus, you have two statements from which to reason forward instead of just one (see Figure 9.1). On the other hand, you do not know what the contradiction will be and therefore cannot work backward.

As a general rule, use contradiction when the statement $NOT\ B$ gives you useful information. There are at least two recognizable instances when this happens. Recall the statement $B$ associated with Proposition 12, "$n$ is an even integer." Because an integer is either odd or even, when you assume that $B$ is not true—that is, that $n$ is not an even integer—it must be the case that $n$ is odd, resulting in some useful information. In general, when the statement $B$ is one of two possible alternatives, the contradiction method is likely to be effective because, by assuming $NOT\ B$, you will know that the other case must happen, and that should help you to reach a contradiction.

A second instance when contradiction is likely to be successful is when the statement $B$ contains the keyword "no" or "not". This is because, as you learned in Chapter 8, assuming $NOT\ B$ eliminates the word "no" or "not," which can result in a useful statement to work forward from, as shown now.

**Proposition 13** *If $r$ is a real number such that $r^2 = 2$, then $r$ is irrational.*

**Analysis of Proof.** It is important to note that you can rewrite the conclusion of Proposition 13 to read "$r$ is not rational." In this form, the appearance of the keyword "not" now suggests using the contradiction method, whereby you assume that $A$ and $NOT\ B$ are both true—in this case,

$$\textbf{A: } r^2 = 2, \quad \text{and}$$
$$\textbf{A1 (NOT B): } r \text{ is a rational number.}$$

A contradiction must now be reached using this information.

Working forward from $A1$ by using Definition 7 on page 26 for a rational number, you can state that

   **A2:**  There are integers $p$ and $q$ with $q \neq 0$ such that $r = p/q$.

There is still the unanswered question of where the contradiction arises, and this takes a lot of creativity. A crucial observation here really helps—it is possible to assume that

   **A3:**  $p$ and $q$ have no common divisor (that is, there is no integer
              that divides both $p$ and $q$).

The reason for this is that, if $p$ and $q$ did have a common divisor, you could divide this integer out of both $p$ and $q$.

Now you can reach a contradiction by showing that 2 is a common divisor of $p$ and $q$. This is done by working forward to show that $p$ and $q$ are even, and hence 2 divides them both.

Working forward by squaring both sides of the equality in $A2$, it follows that

   **A4:**  $r^2 = p^2/q^2$.

But from $A$ you also know that $r^2 = 2$, so

   **A5:**  $2 = p^2/q^2$.

The rest of the forward process is mostly rewriting $A5$ via algebraic manipulations to reach the desired contradiction that both $p$ and $q$ are even integers. Those steps and their justifications are provided in the following table.

| **Statement** | **Reason** |
|---|---|
| **A6:** $2q^2 = p^2$. | Multiply both sides of $A5$ by $q^2$. |
| **A7:** $p^2$ is even. | From $A6$ because $p^2$ is 2 times some integer; namely, $q^2$. |
| **A8:** $p$ is even. | From Proposition 12. |
| **A9:** $p = 2k$, for some integer $k$. | Definition of an even integer. |
| **A10:** $2q^2 = (2k)^2 = 4k^2$. | Substitute $p = 2k$ from $A9$ in $A6$. |
| **A11:** $q^2 = 2k^2$. | Divide $A10$ through by 2. |
| **A12:** $q^2$ is even. | From $A11$ because $q^2$ is 2 times some integer; namely, $k^2$. |
| **A13:** $q$ is even. | From Proposition 12. |

So, both $p$ and $q$ are even (see $A8$ and $A13$), and this contradicts $A3$, thus completing the proof.

**Proof of Proposition 13.**     Assume, to the contrary, that $r$ is a rational number of the form $p/q$ (where $p$ and $q$ are integers with $q \neq 0$) and that $r^2 = 2$. Furthermore, it can be assumed that $p$ and $q$ have no common divisor for, if they did, this number could be canceled from both the numerator $p$ and the denominator $q$. Because $r^2 = 2$ and $r = p/q$, it follows that $2 = p^2/q^2$, or equivalently, $2q^2 = p^2$. Noting that $2q^2$ is even, $p^2$, and hence $p$, are even. Thus, there is an integer $k$ such that $p = 2k$. On substituting this value for $p$, one obtains $2q^2 = p^2 = (2k)^2 = 4k^2$, or equivalently, $q^2 = 2k^2$. From this it then follows that $q^2$, and hence $q$, are even. Thus it has been shown that both $p$ and $q$ are even and have the common divisor 2. This contradiction establishes the claim.   □

This proof, discovered in ancient times by a follower of Pythagoras, epitomizes the use of contradiction. Try to prove the statement by some other method.


## 9.3   ADDITIONAL USES FOR THE CONTRADICTION METHOD

There are several other valuable uses for the contradiction method. Recall that, when the statement $B$ contains the quantifier "there is," the construction method is recommended in spite of the difficulty of actually having to produce the desired object. The contradiction method opens up a whole new approach. Instead of trying to show that there is an object with the certain property such that the something happens, why not proceed from the assumption that there is no such object? Now your job is to use this information to reach some kind of contradiction. How and where the contradiction arises is not at all clear, but finding a contradiction might be easier than producing or constructing the object. Consider the following example.

Suppose you wish to show that at a party of 367 people there are at least two people whose birthdays fall on the same day of the year. If the construction method is used, then you would actually have to go to the party and find two such people. Using the contradiction method saves you the time and trouble of having to do so. With the contradiction method, you can assume that no two people's birthdays fall on the same day of the year, or equivalently, that everyone's birthday falls on a different day of the year.

To reach a contradiction, assign numbers to the people in such a way that the person with the earliest birthday of the year receives the number 1, the person with the next earliest birthday receives the number 2, and so on. Recall that each person's birthday is assumed to fall on a different day. Thus, the birthday of the person whose number is 2 must occur at least one day later than the person whose number is 1, and so on. Consequently, the birthday of the person whose number is 367 must occur at least 366 days after the person whose number is 1. But a year has at most 366 days, and so this is impossible—you have therefore reached a contradiction.

This example illustrates a subtle but significant difference between a proof using the construction method and one that uses contradiction. If the construction method is successful, then you have produced the desired object, or at least indicated how the object might be produced, perhaps with the aid of a computer. On the other hand, if you establish the same result by contradiction, then you know that the object exists but have not physically constructed the object. For this reason, it is often the case that proofs done by contradiction are quite a bit shorter and easier than those done by construction because you do not have to create the desired object. You only have to show that the object's nonexistence is impossible. This difference has led to some great philosophical debates in mathematics. Moreover, an active area of current research consists of finding constructive proofs where previously only proofs by contradiction were known.

## 9.4   READING A PROOF

The process of reading and understanding a proof is demonstrated with the following proposition.

**Proposition 14** *If $m$ and $n$ are odd integers, then the equation $x^2 + 2mx + 2n = 0$ has no rational root.*

**Proof of Proposition 14.**   (For reference purposes, each sentence of the proof is written on a separate line.)

> **S1:**   Assume that the equation $x^2 + 2mx + 2n = 0$ has a rational root, say $x = p/q$, where $q \neq 0$ and one of $p$ and $q$ is odd.
>
> **S2:**   It now follows that $q$ is odd, for $p^2 = -2mqp - 2q^2 n$, which means that $p^2$, and hence $p$, are even.
>
> **S3:**   Note first that, if $m'$ and $n'$ are odd, then $y^2 + 2m'y + 2n' = 0$ has no root that is odd, for if $y$ is such a root, then $y^2$ is odd and also $y^2 = -2m'y - 2n'$ is even, which cannot happen.
>
> **S4:**   Also, $y^2 + 2m'y + 2n' = 0$ has no root that is even, for if $y$ is such a root, then $y = 2k$, for some integer $k$.
>
> **S5:**   But then $4k^2 + 4m'k + 2n' = 0$; that is, $2k^2 + 2m'k + n' = 0$, which cannot happen because $2k^2 + 2m'k$ is even and $n'$ is odd.
>
> **S6:**   However, the fact that $x = p/q$ satisfies $x^2 + 2mx + 2n = 0$ means that $y = p$ satisfies $y^2 + 2m'y + 2n' = 0$, where $m' = mq$ is odd and $n' = nq^2$ is also odd, which cannot happen.

This contradiction establishes the claim.   $\square$

**Analysis of Proof.**    An interpretation of statements $S1$ through $S6$ follows.

**Interpretation of S1:** *Assume that the equation $x^2 + 2mx + 2n = 0$ has a rational root, say, $x = p/q$, where one of $p$ and $q$ is odd and $q \neq 0$.*
  The author is assuming that the conclusion is not true; that is:

  **A1 (NOT B):** There is a rational root $x = p/q$ to the equation,

thus indicating that this is a proof by contradiction. Note that the author has assumed further that

  **A2:** One of $p$ and $q$ is odd.

This is justified because, if both are even, then you can repeatedly divide both $p$ and $q$ by 2 until one of them is odd.
  Having recognized that this is a proof by contradiction, did you read ahead to identify the contradiction? If not, then you should do so now.

**Interpretation of S2:** *It now follows that $q$ is odd, for otherwise $p^2 = -2mqp - 2q^2n$, which would mean that $p^2$, and hence $p$, are even.*
  The author is working forward from $A1$ to claim that

  **A3:** $q$ is odd.

The author reaches this conclusion by contradiction. That is, by assuming that $q$ is even, the author claims that $p$ is also even, which contradicts $A2$. Indeed this is correct because the author has substituted $x = p/q$ into $x^2 + 2mx + 2n = 0$, multiplied through by $q^2$, and solved for $p^2$ to obtain $p^2 = -2mqp - 2q^2n$. Now you can see that the right side is even and so $p^2$ is even. Finally, Proposition 12 ensures that $p$ is also even.

**Interpretation of S3:** *Note first that, if $m'$ and $n'$ are odd, then $y^2 + 2m'y + 2n' = 0$ has no root that is odd, for if $y$ is such a root, then $y^2$ is odd and also $y^2 = -2m'y - 2n'$ is even, which cannot happen.*
  This statement is directed toward reaching the final contradiction. The author is establishing that

  **A4:** If $m'$ and $n'$ are odd integers, then $y^2 + 2m'y + 2n' = 0$
      has no root that is an odd integer.

This also is shown by contradiction, which is indicated when the author assumes that $y$ is an odd integer root. The author then reaches the contradiction that $y^2$ is both odd (because odd times odd is odd) and even (because, from $A4$, $y^2 = -2m'y - 2n'$, which is even).

**Interpretation of S4:** *Also, $y^2 + 2m'y + 2n' = 0$ has no root that is even, for if $y$ is such a root, then $y = 2k$, for some integer $k$.*
  Here, the author is establishing that

> **A5:** If $m'$ and $n'$ are odd, then $y^2 + 2m'y + 2n' = 0$ has no root
> that is even.

This is also shown by contradiction, which is indicated when the author assumes that $y$ is an even integer root. By definition, this means that $y = 2k$, for some integer $k$.

**Interpretation of S5:** *But then $4k^2 + 4m'k + 2n' = 0$; that is, $2k^2 + 2m'k + n' = 0$, which cannot happen because $2k^2 + 2m'k$ is even and $n'$ is odd.*
The author is continuing the contradiction argument started in $S4$. In particular, the author substitutes $y = 2k$ into $y^2 + 2m'y + 2n' = 0$ and divides the result by 2 to obtain $2k^2 + 2m'k + n' = 0$. The author then claims correctly that a contradiction is reached because it is impossible for the even integer $2k^2 + 2m'k$ plus the odd integer $n'$ to be 0.

**Interpretation of S6:** *However, the fact that $x = p/q$ satisfies $x^2 + 2mx + 2n = 0$ means that $y = p$ satisfies $y^2 + 2m'y + 2n' = 0$, where $m' = mq$ is odd and $n' = nq^2$ is also odd, which cannot happen.*
The author is working forward from $A1$ by substituting $x = p/q$ into $x^2 + 2mx + 2n = 0$, multiplying through by $q^2$, and then realizing that

> **A6:** $y = p$ satisfies the equation $y^2 + 2m'y + 2n' = 0$, where
> $m' = mq$ and $n' = nq^2$.

The author further claims correctly that both $m' = mq$ and $n' = nq^2$ are odd [because $m$ and $n$ are odd (see the hypothesis), and $q$ is also odd (see $A3$)].
Finally, the author claims to have reached the desired contradiction. Indeed, $A4$ and $A5$ together mean that the equation $y^2 + 2m'y + 2n' = 0$, where $m'$ and $n'$ are odd, has no integer solution. Yet in $A6$, the author shows that $y = p$ is an integer solution to such an equation. Thus the proof is complete.

### Summary

The contradiction method is a useful technique when the statement $B$, or the last statement in the backward process, contains the keyword "no" or "not." To use this method, follow these steps:

1. Assume that $A$ is true and $B$ is not true (that is, assume that $A$ and *NOT B* are true).

2. Work forward from $A$ and *NOT B* to reach a contradiction.

One of the disadvantages of this method is that you do not know exactly what the contradiction is going to be and so you cannot work backward. The next chapter describes another proof technique in which you attempt to reach a specific contradiction. As such, you will have a "guiding light" because you will know what contradiction you are looking for.

## Exercises

**Note:** Solutions to those exercises marked with a $B$ are in the back of this book. Solutions to those exercises marked with a $W$ are located on the web at http://www.wiley.com/college/solow/.

**Note:** All proofs should contain an analysis of proof and a condensed version. Definitions for all mathematical terms are provided in the glossary at the end of the book.

$^B$**9.1**    When applying the contradiction method to the following propositions, what should you assume?

    a. If $l$, $m$, and $n$ are three consecutive integers, then 24 does not divide $l^2 + m^2 + n^2 + 1$.

    b. For every integer $n > 2$, $x^n + y^n = z^n$ has no integer solution for $x$, $y$, and $z$.

    c. If $f$ and $g$ are two functions such that (1) $g \geq f$ and (2) $f$ is unbounded above, then $g$ is unbounded above.

**9.2**    When applying the contradiction method to the following propositions, what should you assume?

    a. If $n$ is an integer with $n > 2$, then there do not exist positive integers $x$, $y$, and $z$ such that $x^n + y^n = z^n$.

    b. If $a$ is a positive real number, then for all real numbers $b$, $c$, and $M$ with $M > 0$, there is a real number $x$ such that $ax^2 + bx + c > M$.

    c. If the matrix $M$ is not singular, then the rows of $M$ are not linearly dependent.

**9.3**    Consider applying contradiction to show that, if $a$ and $b$ are integers and $b$ is odd, then $\pm 1$ are not roots of $ax^2 + bx + a$.

    a. What statement(s) would you work backward from?

    b. At the end of the proof, a mathematics student said, "... and because I have been able to show that $b$ is even, the proof is complete." Do you agree with the student? Why or why not? Explain.

$^B$**9.4**    Reword each of the following statements so that the word "not" appears.

    a. There are an infinite number of primes.

    b. The only positive integers that divide the positive integer $p$ are 1 and $p$.

c. The two different lines $l$ and $l'$ in a plane $P$ are parallel.

**9.5**     Reword each of the following statements so that the word "not" does not appear.

a. The real number $ad - bc$ is not equal to 0.

b. The triangle $ABC$ is not equilateral.

c. The polynomial $a_0 + a_1x + \cdots + a_nx^n$ has no real root.

[B]**9.6**     When trying to prove each of the following statements, which techniques would you use and in which order? Specifically, state what you would assume and what you would try to conclude. (Throughout, $S$ and $T$ are sets of real numbers and all of the variables refer to real numbers.)

a. $\exists s \in S \ni s \in T$.

b. $\forall s \in S$, $\nexists t \in T$ such that $s > t$.

c. $\nexists M > 0$ such that $\forall x \in S$, $|x| < M$.

*__9.7__     When trying to prove each of the following statements, which techniques would you use and in which order? Specifically, state what you would assume and what you would try to conclude. (Throughout, $S$ is a given set of real numbers and all of the variables refer to real numbers.)

a. For every real number $\epsilon > 0$, there is an element $x \in S$ such that $x > u - \epsilon$ (where $u$ is a given real number).

b. There is a real number $y > 0$ such that, for every element $x \in S$, $f(x) < y$ (where $f$ is a function of one real variable).

c. For every line $l$ in the plane that is parallel to, but different from, $l'$, there is no point on $l$ that is also in $S$ (here, $S$ is some given set of points in the plane and $l'$ is a given line in the plane).

[B]**9.8**     Prove, by contradiction, that, if $n$ is an integer and $n^2$ is even, then $n$ is even.

**9.9**     Prove, by contradiction, that there do not exist positive real numbers $x$ and $y$ with $x \neq y$ such that $x^3 - y^3 = 0$.

[B]**9.10**     Prove, by contradiction, that no chord of a circle is longer than a diameter. (Hint: Draw an appropriate inscribed right triangle.)

**9.11**     Suppose that $a$, $b$, and $c$ are real numbers with $c \neq 0$. Prove, by contradiction, that, if $cx^2 + bx + a$ has no rational root, then $ax^2 + bx + c$ has no rational root.

$^W$**9.12**    Prove, by contradiction, that, if $n - 1$, $n$, and $n + 1$ are consecutive positive integers, then the cube of the largest cannot be equal to the sum of the cubes of the other two.

**9.13**    Prove, by contradiction, that, if $p$ and $q$ are integers with $p \neq q$ and $p$ is prime and divides $q$, then $q$ is not prime.

$^B$**9.14**    Prove, by contradiction, that, at a party of $n \geq 2$ people, there are at least two people who have the same number of friends at the party.

**9.15**    Prove, by contradiction, that there are at least two people on the planet who were born on the same second of the same hour of the same day of the same year in the twentieth century. (You can assume that there were at least 4 billion people born in that century.)

**9.16**    Prove, by contradiction, that, if $n$ is a positive integer such that $n^3 - n - 6 = 0$, then, for every positive integer $m$ with $m \neq n$, $m^3 - m - 6 \neq 0$.

$^W$**9.17**    Prove, by contradiction, that, if $x$ and $y$ are real numbers such that $x \geq 0$, $y \geq 0$, and $x + y = 0$, then $x = 0$ and $y = 0$.

**9.18**    For a set $S$ whose elements come from a set $U$, the complement of $S$ is the set $S^c = \{x \in U : x \notin S\}$. Prove that, for a set $S$, $(S^c)^c = S$.

*\***9.19**    Answer the given questions about the following proof.

> **Definition.** A function $f$ of one variable is **one-to-one** if and only if, for all real numbers $x$ and $y$ with $x \neq y$, $f(x) \neq f(y)$.
>
> **Proposition.** If $a > 0$ is a real number, then $f(x) = a^x$ is a one-to-one function.
>
> **Proof.** Let $x$ and $y$ be real numbers with $x \neq y$ for which it will be shown that $a^x \neq a^y$. So assume that $a^x = a^y$. It then follows that $\log(a^x) = \log(a^y)$, that is, $x \log(a) = y \log(a)$. But this means that $x = y$ and so the proof is complete. $\square$

   a. Explain what the author is doing in the first sentence. What techniques are used?

   b. What proof technique is the author using in the second sentence? Why did the author choose that technique?

   c. Where does the author use the hypothesis that $a > 0$?

   d. Is the author justified in claiming that the proof is complete in the last sentence? Why or why not? Explain.

*\***9.20**    Prove that, if $a$, $b$, and $c$ are real numbers with $a \neq 0$, then the function $f(x) = ax^2 + bx + c$ is not one-to-one (see the definition in Exercise 9.19).

**9.21**    Find the errors in the following proof that, "If $a \neq 0$ is a real number, then the function $f(x) = x^a$ is one-to-one (see the definition in Exercise 9.19)."

> **Proof.** Let $x$ and $y$ be real numbers with $x \neq y$ for which it will be shown that $x^a \neq y^a$. So assume that $x^a = y^a$. It then follows that $\log(x^a) = \log(y^a)$, that is, $a\log(x) = a\log(y)$. But this means that $x = y$ and so the proof is complete.  □

$^B$**9.22**    Identify the contradiction in the following proof.

> **Proposition.** If $a$ and $b$ are integers and $b$ is odd, then $\pm1$ are not roots of $ax^4 + bx^2 + a$.

> **Proof.**    Assume, to the contrary, that $+1$ or $-1$ is a root of $ax^4 + bx^2 + a$. Then $a(\pm1)^4 + b(\pm1)^2 + a = 0$; that is, $b + 2a = 0$. Thus, $b = -2a$, which cannot happen, and so the proof is complete.  □

$^*$**9.23**    Identify and justify the contradiction in the following proof.

> **Proposition.** If $n$ is a positive integer, then, for every integer $k$ with $1 < k \leq n$, $k$ does not divide $n! + 1$.

> **Proof.**    If there is an integer $k$ with $1 < k \leq n$ such that $k$ divides $n! + 1$, then there is an integer $c$ such that $n! + 1 = ck$. This means that $1 \cdots k \cdots (n-1)n = ck - 1$. But then the integer $k$ divides the left side but not the right side, and so the proof is complete.  □

$^W$**9.24**    Prove, by contradiction, that there are an infinite number of primes. (Hint: Assume that $n$ is the largest prime. Then consider any prime number $p$ that divides $n! + 1$. Now look at the proposition in Exercise 9.23.)

**9.25**    Identify the contradiction in the following proof.

> **Proposition.** If $a$ and $b$ are integers with $a \neq 0$ and the number of rational roots of $ax^4 + bx^2 + a$ is odd, then $b$ is even.

> **Proof.** Assume, to the contrary, that $b$ is odd. Then, by the proposition in Exercise 9.22, $\pm1$ are not roots of $ax^4 + bx^2 + a$. Now consider a rational root $p/q$ of $ax^4 + bx^2 + a$. Note that $p \neq 0$ for otherwise $a = 0$. It is easy to verify that $q/p$ is also a rational root of $ax^4 + bx^2 + a$. Thus, rational roots come in pairs. Because the number of rational roots of $ax^4 + bx^2 + a$ is odd, it must be that one of these roots is repeated, so, $p/q = q/p$. But then $p = \pm q$, which cannot happen, and so the proof is complete.  □

$^B$**9.26**    Write an analysis of proof that corresponds to the condensed proof given below. Indicate which techniques are used and how they are applied. Fill in the details of any missing steps where appropriate.

> **Proposition.** If $a$, $b$, $c$, $x$, $y$, and $z$ are real numbers with $b \neq 0$ such that (1) $az - 2by + cx = 0$ and (2) $ac - b^2 > 0$, then it must be that $xz - y^2 \leq 0$.

> **Proof.** Assume that $xz - y^2 > 0$. From this and (2) it follows that $(ac)(xz) > b^2 y^2$. Rewriting (1) and squaring both sides, one obtains $(az + cx)^2 = 4b^2 y^2 < 4(ac)(xz)$. Rewriting, one has that $(az - cx)^2 < 0$, which cannot happen.  □

$^W$**9.27**    Write an analysis of proof that corresponds to the condensed proof given below. Indicate which techniques are used and how they are applied. Fill in the details of any missing steps where appropriate.

> **Proposition.** The polynomial $x^4 + 2x^2 + 2x + 2$ cannot be expressed as the product of the two polynomials $x^2 + ax + b$ and $x^2 + cx + d$ in which $a$, $b$, $c$, and $d$ are integers.

> **Proof.** Suppose that
>
> $$\begin{aligned} x^4 + 2x^2 + 2x + 2 &= (x^2 + ax + b)(x^2 + cx + d) \\ &= x^4 + (a + c)x^3 + (b + ac + d)x^2 + \\ &\quad (bc + ad)x + bd. \end{aligned}$$
>
> It would then follow that the integers $a$, $b$, $c$, and $d$ satisfy
>
> 1. $a + c = 0$.
> 2. $b + ac + d = 2$.
> 3. $bc + ad = 2$.
> 4. $bd = 2$.
>
> The only way (4) can happen is if one of the factors $b$ or $d$ is odd ($\pm 1$) and the other is even ($\pm 2$). Suppose that $b$ is in fact odd and $d$ is even. From (3), it would then follow that $c$ is even, but then the left side of (2) would be odd, which is impossible. A similar contradiction can be reached if $b$ is even and $d$ is odd.  □

**9.28**    Write an analysis of proof that corresponds to the condensed proof given in Exercise 9.23. Indicate which techniques are used and how they are applied. Fill in the details of any missing steps where appropriate.

*9.29    Write an analysis of proof that corresponds to the condensed proof given in Exercise 9.25. Indicate which techniques are used and how they are applied. Fill in the details of any missing steps where appropriate.

*9.30    Write an analysis of proof that corresponds to the condensed proof given below. Indicate which techniques are used and how they are applied. Fill in the details of any missing steps where appropriate.

> **Proposition.** If $a$ is a real number with $a < 0$, then, for every real number $y < 0$, the set $C = \{$real numbers $x : ax^2 \leq y\}$ is not bounded.
>
> **Proof.** Let $y < 0$ and suppose that $C$ is bounded. Then there is a real number $M > 0$ such that, for every element $x \in C$, $|x| < M$. However, $x = \max\left\{M, \sqrt{\frac{y}{a}}\right\}$ satisfies $|x| \geq M$ and $ax^2 \leq a\left(\sqrt{\frac{y}{a}}\right)^2 = y$ and so the proof is complete.  $\square$

# 10

## The Contrapositive Method

In the contradiction method described in the previous chapter, you work forward from the two statements $A$ and $NOT\ B$ to reach a contradiction. The challenge with this method is to determine what the contradiction is going to be. The technique presented in this chapter has the advantage of directing you toward one specific contradiction.

### 10.1  HOW AND WHEN TO USE THE CONTRAPOSITIVE METHOD

The **contrapositive method** is similar to contradiction in that you begin by assuming that $A$ and $NOT\ B$ are true. Unlike contradiction, however, you do not work forward from both $A$ and $NOT\ B$. Instead, you work forward only from $NOT\ B$. Your objective is to reach the contradiction that $A$ is false (hereafter written $NOT\ A$). Can you ask for a better contradiction than that? How can $A$ be true and false at the same time? To repeat, in the contrapositive method you assume that $A$ and $NOT\ B$ are true; you then work forward from $NOT\ B$ to reach the contradiction $NOT\ A$. Use the techniques in Chapter 8 to write the statement $NOT\ B$ to work forward from and the statement $NOT\ A$ to work backward from.

You can think of the contrapositive method as a passive form of contradiction in the sense that the assumption that $A$ is true passively provides the contradiction. In contrast, with the contradiction method, the assumption that $A$ is true is used actively to reach a contradiction. The next proposition demonstrates the contrapositive method.

**Definition 17** *A real-valued function f of one real variable is* **one-to-one** *if and only if for all real numbers x and y with $x \neq y$, $f(x) \neq f(y)$.*

**Proposition 15** *If m and b are real numbers with $m \neq 0$, then the function $f(x) = mx + b$ is one-to-one.*

**Analysis of Proof.**    The forward-backward method is used to begin the proof even though the hypothesis contains the keyword "not." A key question associated with $B$ is, "How can I show that a function is one-to-one?" Applying Definition 17 to the particular function $f(x) = mx + b$ means you must show that

**B1:** For all real numbers $x$ and $y$ with $x \neq y$, $mx + b \neq my + b$.

This new statement contains both the keywords "for all" and "not" (in "not equal"). When more than one group of keywords is present in a statement, apply appropriate techniques based on the occurrence of the keywords as they appear from left to right (just as you learned to do with nested quantifiers in Chapter 7). Because the first keywords from the left in the backward statement $B1$ are "for all," the choose method is used next. So, choose

**A1:** Real numbers $x$ and $y$ with $x \neq y$,

for which you must show that

**B2:** $mx + b \neq my + b$.

Recognizing the keyword "not" in $B2$, it is appropriate to proceed with either the contradiction or contrapositive method. In this case, the contrapositive method is used. Thus, you work forward from *NOT B2*:

**A2 (NOT B2):** $mx + b = my + b$.

and backward from *NOT A1* (the last statement in the forward process):

**B3 (NOT A1):** $x = y$.

The remainder of the proof is simple algebra applied to $A2$. Specifically, on subtracting $b$ from both sides of $A2$ you obtain:

**A3:** $mx = my$.

Finally, because $m \neq 0$ by the hypothesis $A$, you can divide both sides of $A3$ by $m$ and obtain $B3$, and so the proof is complete.

In the condensed proof that follows, note that no mention is made of the choose or contrapositive methods.

**Proof of Proposition 15.**    Let $x$ and $y$ be real numbers for which $mx + b = my + b$. It is shown that $x = y$. But this follows by subtracting $b$ from both sides of the equality and then dividing by $m$, noting that $m \neq 0$.  $\square$

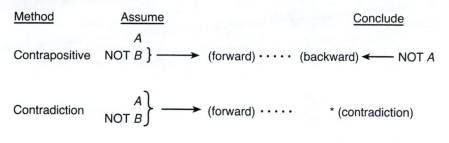

*Fig. 10.1* Comparing the contrapositive and contradiction methods.

## 10.2 COMPARING THE CONTRAPOSITIVE METHOD

As already mentioned, the contrapositive method is a type of proof by contradiction. Each of these methods has its advantages and disadvantages, as illustrated in Figure 10.1. The disadvantage of the contrapositive method is that you work forward from only one statement (namely, *NOT B*) instead of two. On the other hand, the advantage is that you know what you are looking for (namely, *NOT A*). Thus you can apply the backward process to *NOT A*. The option of working backward is not available in the contradiction method because you do not know what contradiction you are looking for.

It is also interesting to compare the contrapositive and forward-backward methods. With the forward-backward method, you work forward from *A* and backward from *B*; with the contrapositive method, you work forward from *NOT B* and backward from *NOT A* (see Figure 10.2).

From Figure 10.2, it is not hard to see why the contrapositive method might be better than the forward-backward method. Perhaps you can obtain more useful information by working forward from *NOT B* rather than from *A*. It might also be easier to work backward from *NOT A* rather than from *B*, as is done in the forward-backward method.

The forward-backward method arises from considering what happens to the truth of "*A* implies *B*" when *A* is true and when *A* is false (recall Table 1.1 on page 4). The contrapositive method arises from similar considerations regarding *B*. Specifically, if *B* is true, then, from Table 1.1, the statement "*A* implies *B*" is true. Hence, there is no need to consider the case when *B* is

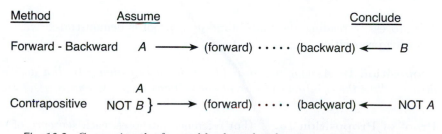

*Fig. 10.2* Comparing the forward-backward and contrapositive methods.

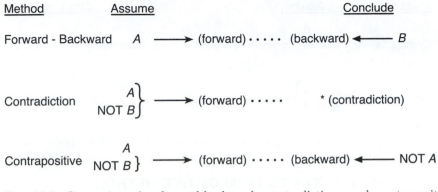

| Method | Assume | | Conclude |

*Fig. 10.3* Comparing the forward-backward, contradiction, and contrapositive methods.

true. So suppose $B$ is false. To ensure that "$A$ implies $B$" is true, according to Table 1.1, you must show that $A$ is false. Thus, with the contrapositive method, you assume that $B$ is false and try to conclude that $A$ is false.

The statement "$A$ implies $B$" is logically equivalent to "$NOT$ $B$ implies $NOT$ $A$" (see Table 3.2 on page 34). You can therefore think of the contrapositive method as the forward-backward method applied to the statement "$NOT$ $B$ implies $NOT$ $A$." Most condensed proofs that use the contrapositive method make no reference to the contradiction method.

There is one instance that indicates you should use, or at least consider seriously, the contradiction or contrapositive method. This occurs when $B$ contains the keyword "no" or "not," for then, $NOT$ $B$ has some useful information.

In the contradiction method, you work forward from the two statements $A$ and $NOT$ $B$ to obtain a contradiction. In the contrapositive method, you also reach a contradiction, but you do so by working forward from $NOT$ $B$ to reach the conclusion $NOT$ $A$ and so you should also work backward from $NOT$ $A$. A comparison of the forward-backward, contradiction, and contrapositive methods is given in Figure 10.3.

## 10.3   READING A PROOF

The process of reading and understanding a proof is demonstrated with the following proposition.

**Proposition 16** *Assume that $a$ and $b$ are integers with $a \neq 0$. If $a$ does not divide $b$, then the equation $ax^2 + bx + b - a = 0$ has no positive integer solution.*

**Proof of Proposition 16.**    (For reference purposes, each sentence of the proof is written on a separate line.)

**S1:** Suppose that $x > 0$ is an integer with $ax^2 + bx + b - a = 0$.
**S2:** Then $x = \frac{-b \pm (b-2a)}{2a}$.
**S3:** Because $x > 0$, it must be that $x = 1 - \frac{b}{a}$.
**S4:** But then $b = (1 - x)a$ and so $a|b$.

The proof is now complete. $\square$

**Analysis of Proof.**   An interpretation of statements $S1$ through $S4$ follows.

**Interpretation of S1:** *Suppose that $x > 0$ is an integer with*
$ax^2 + bx + b - a = 0$.
   The author is assuming *NOT B*; that is,

**A1 (NOT B):** $x > 0$ is an integer with $ax^2 + bx + b - a = 0$.

This means that the contradiction or contrapositive method is used. However, on reading $S4$, you see that the author reaches the statement *NOT A*:

**B1 (NOT A):** $a|b$,

thus indicating that this is a proof by the contrapositive method.

**Interpretation of S2:** *Then $x = \frac{-b \pm (b-2a)}{2a}$.*
   The author works forward from $A1$ using the quadratic formula to obtain the following:

**A2:** $x = \frac{-b \pm \sqrt{b^2 - 4a(b-a)}}{2a} = \frac{-b \pm (b-2a)}{2a}$.

**Interpretation of S3:** *Because $x > 0$, it must be that $x = 1 - \frac{b}{a}$.*
   The author works forward by algebra from $A2$ to see that the two roots are $x = -1$ and $x = 1 - \frac{b}{a}$. Because, from $A1$, $x > 0$, the author rules out the root $x = -1$ and correctly concludes that

**A3:** $x = 1 - \frac{b}{a}$.

**Interpretation of S4:** *But then $b = (1 - x)a$ and so $a|b$.*
   The author works backward from $B1$ by asking the key question, "How can I show that an integer (namely, $a$) divides another integer (namely, $b$)?" Using the definition to answer the question, the author needs to show the following:

**B2:** There is an integer $c$ such that $b = ca$.

Recognizing the quantifier "there is" in $B2$, the author uses the construction method to produce the value of $c$. Specifically, the author works forward from $A3$ by algebra to solve for $b$, obtaining $b = (1 - x)a$, from which the author notes—without mentioning—that the desired value of $c$ in $B2$ is $c = 1 - x$. Having reached *NOT A*, the contrapositive method is now complete, and so is the proof.

## Summary

The contrapositive method, being a type of proof by contradiction, is used when the last statement in the backward process contains the keyword "no" or "not." With the contrapositive method, you work toward the specific contradiction $NOT\ A$ by

1. Assuming that $A$ and $NOT\ B$ are true.

2. Working forward from $NOT\ B$ in an attempt to obtain $NOT\ A$.

3. Working backward from $NOT\ A$ in an attempt to reach $NOT\ B$.

You can also think of the contrapositive method as the forward-backward method applied to the statement "$NOT\ B$ implies $NOT\ A$" because this statement is logically equivalent to the statement "$A$ implies $B$."

Both the contrapositive and contradiction methods require that you be able to write the $NOT$ of a statement, which you have learned to do in Chapter 8. In the next chapter you will learn proof techniques for dealing with statements that contain special keywords associated with quantifiers.

## Exercises

**Note:** Solutions to those exercises marked with a $B$ are in the back of this book. Solutions to those exercises marked with a $W$ are located on the web at http://www.wiley.com/college/solow/.

**Note:** All proofs should contain an analysis of proof and a condensed version. Definitions for all mathematical terms are provided in the glossary at the end of the book.

$^B$**10.1**   If the contrapositive method is used to prove the following propositions, then what statement(s) will you work forward from and what statement(s) will you work backward from?

    a. If $n$ is an integer for which $n^2$ is even, then $n$ is even.

    b. Suppose that $S$ is a subset of the set $T$ of real numbers. If $S$ is not bounded, then $T$ is not bounded.

    c. If $p > 1$ is an integer such that, for every integer $n$ with $1 < n \leq \sqrt{p}$, $n$ does not divide $p$, then $p$ is prime.

**10.2**   If the contrapositive method is used to prove the following propositions, then what statement(s) will you work forward from and what statement(s) will you work backward from?

    a. Suppose that $n$ and $p$ are positive integers. If $n|p$, then $n \leq p$.

b. If $a$ is a positive real number, then for all real numbers $b$, $c$, and $M$ with $M > 0$, there is a real number $x$ such that $ax^2 + bx + c > M$.

c. Suppose that $f$ is a function of one real variable. If $x_1 < x_2 < x_3$ are three real numbers for which $f(x_1) > f(x_2) < f(x_3)$, then the function $f(x)$ is not linear.

**10.3**   Bob said, "If I study hard, then I will get at least a B in this course." After the course was over, Mary said, "You got an A in the course, so therefore you studied hard." Was Mary's statement correct? Why or why not? Explain.

$^B$**10.4**   In a proof by the contrapositive method that, "If $r$ is a real number with $r > 1$, then there is no real number $t$ with $0 < t < \pi/4$ such that $\sin(t) = r\cos(t)$," which of the following is a result of the forward process?

a. $r - 1 \geq 0$.

b. $\sin^2(t) = r^2(1 - \sin^2(t))$.

c. $1 - r < 0$.

d. $\tan(t) = 1/r$.

**10.5**   Suppose that $m$ and $n$ are integers with $m \neq 0$. In a proof by the contrapositive method that, "If $m$ does not divide $n$, then $mx^2 + nx + (n - m)$ has no positive integer root," which of the following is a result of the forward process?

a. $m$ divides $n$.

b. There is an integer $x > 0$ such that $mx^2 + nx + (n - m) \neq 0$.

c. There is an integer $x > 0$ such that $mx^2 + nx + (n - m) = 0$.

d. There is an integer $x \leq 0$ such that $mx^2 + nx + (n - m) = 0$.

**10.6**   When proving the proposition in the previous exercise using the contrapositive method, what is the subsequent key question and answer?

$^B$**10.7**   Suppose the contrapositive method is used to prove the proposition, "If the derivative of the function $f$ at the point $x$ is not equal to 0, then $x$ is not a local maximum of $f$." Which of the following is the correct key question? What is wrong with the other choices?

a. How can I show that the point $x$ is a local maximum of the function $f$?

b. How can I show that the derivative of a function $f$ at a point $x$ is 0?

c. How can I show that a point is a local maximum of a function?

d. How can I show that the derivative of a function at a point is 0?

**10.8**  Suppose that $f$ is a function of one real variable, $S$ is a set of real numbers, and that the contrapositive method is used to prove the proposition, "If no element $x \in S$ satisfies the property that $f(x) = 0$, then $f$ is not bounded above." Which of the following is a correct key question? What is wrong with the other choices?

a. How can I show that a function is not bounded above?

b. How can I show that there is an element $x \in S$ such that $f(x) = 0$?

c. How can I show that a function is bounded above?

d. How can I show that there is a point in a set where the value of a function is 0?

$^B$**10.9**  Is the contrapositive or contradiction method used in the following proof of the proposition that, "If $n$ and $p$ are positive integers and $n|p$, then $n \le p$"? If the contradiction method is used, identify the contradiction.

> **Proof.** Suppose that $n > p$. Then because $n|p$, there is an integer $c$ such that $p = cn$. Now $n > 0$ and $p > 0$, so $c > 0$. Therefore, $p = cn > cp \ge p$, and so the proof is complete.  □

**10.10**  Is the contrapositive or contradiction method used in the proof in Exercise 9.22? If the contradiction method is used, identify the contradiction.

**10.11**  Is the contrapositive or contradiction method used in the proof in Exercise 9.23? If the contradiction method is used, identify the contradiction.

$^B$**10.12**  Suppose $p > 1$ is an integer and that you want to prove by the contrapositive method that, "If there is no integer $m$ with $1 < m \le \sqrt{p}$ such that $m|p$, then $p$ is prime." Which of the construction, choose, and specialization techniques would you use subsequently in doing the proof? Explain.

**10.13**  Suppose that $a$, $b$, and $c$ are real numbers with $a > 0$ and that you want to prove by the contrapositive method that, "If there is no real number $x$ with $ax^2 + bx + c = 0$, then there is a real number $x$ such that $ax^2 + bx + c \ge 0$." Which of the construction, choose, and specialization techniques would you use subsequently in doing the proof? Explain.

$^*$**10.14**  Suppose you are using the contrapositive method to prove that, "If $a$ is a positive real number, then for all real numbers $b$, $c$, and $M$ with $M > 0$, there is a real number $x$ such that $ax^2 + bx + c > M$." Which of the construction, choose, and specialization techniques would you use subsequently in doing the proof? Explain.

$^B$**10.15**   Suppose that $p$ and $q$ are positive real numbers. Prove, by the contrapositive method, that, if $\sqrt{pq} \neq (p+q)/2$, then $p \neq q$.

**10.16**   Suppose that $a$ and $b$ are positive real numbers. Prove, by the contrapositive method, that, if $a \neq b$, then $(a+b)/2 > \sqrt{ab}$.

$^W$**10.17**   Prove, by the contrapositive method, that, if $c$ is an odd integer, then the equation $n^2 + n - c = 0$ has no integer solution for $n$.

*$^*$**10.18**   Suppose that $m$ and $n$ are integers with $m \neq 0$. Use the contrapositive method to prove that, if $m$ does not divide $n$, then $mx^2 + nx + n - m$ has no positive integer root.

**10.19**   Use the approach in the proof of Proposition 15 to prove that the function $f(x) = x^3$ is one-to-one.

$^W$**10.20**   Prove, by the contrapositive method, that, if no angle of a quadrilateral $RSTU$ is obtuse, then the quadrilateral $RSTU$ is a rectangle.

*$^*$**10.21**   Suppose that $v$ is an upper bound for a set $S$ of real numbers. Prove, by the contrapositive method, that, if there is no real number $\epsilon > 0$ such that, for every element $x \in S$, $x \leq v - \epsilon$, then there is no real number $u < v$ such that $u$ is an upper bound for $S$.

$^W$**10.22**   Prove, by the contrapositive method, that, if $x$ is a real number that satisfies the property that, for every real number $\epsilon > 0$, $x \geq -\epsilon$, then $x \geq 0$.

*$^*$**10.23**   Prove, by the contrapositive method, that, if $u$ is a least upper bound for a set $S$ of real numbers, then $\forall$ real number $\epsilon > 0$, $\exists$ an element $x \in S$ such that $x > u - \epsilon$.

$^B$**10.24**   Write an analysis of proof that corresponds to the condensed proof given below. Indicate which techniques are used and how they are applied. Fill in the details of any missing steps where appropriate.

> **Proposition.** Suppose that $p > 1$ is an integer. If there is no integer $m$ with $1 < m \leq \sqrt{p}$ such that $m|p$, then $p$ is prime.

> **Proof.**   Assume that $p$ is not prime; that is, that there is an integer $n$ with $1 < n < p$ such that $n|p$. In the event that $n \leq \sqrt{p}$, then $n$ is the desired integer $m$. Otherwise, $n > \sqrt{p}$. Because $n|p$, there is an integer $k$ such that $p = nk$. It then follows that $k$ is the desired value for $m$. This is because $k > 1$, for otherwise, $p \leq n$. Also, $k \leq \sqrt{p}$, for otherwise, $nk > \sqrt{p}\sqrt{p} = p$. □

*$^*$**10.25**   Write an analysis of proof that corresponds to the condensed proof given below. Indicate which techniques are used and how they are applied. Fill in the details of any missing steps where appropriate.

**Definition.** A set $S$ of real numbers is **bounded** if there is a real number $M > 0$ such that, for all elements $x \in S$, $|x| < M$.

**Proposition.** Suppose that $S$ and $T$ are sets of real numbers with $S \subseteq T$. If $S$ is not bounded, then $T$ is not bounded.

**Proof.** Suppose that $T$ is bounded. Hence, there is a real number $M' > 0$ such that, for all $x \in T$, $|x| < M'$. It is shown that $S$ is bounded. To that end, let $x \in S$. Because $S \subseteq T$, it follows that $x \in T$. But then $|x| < M'$ and so $S$ is bounded, thus completing the proof. $\square$

*10.26    Answer the given questions about the following proof that, "If $a > 0$ is an integer that is not a square, then $\sqrt{a}$ is irrational."

**Proof.** Suppose, to the contrary, that $\sqrt{a}$ is rational. Then there are integers $p$ and $q$ with $q \neq 0$ such that $\sqrt{a} = p/q$. This means that $p^2 = aq^2$ and so $q^2 | p^2$. But then it follows that $q|p$. Hence, there is an integer $b$ such that $p = bq$ and so $p^2 = b^2 q^2$. Thus $a = b^2$ and so the proof is complete. $\square$

a. By reading only the first sentence, which proof techniques might the author be using?

b. What technique is the author using in the second sentence?

c. Justify the statements $p^2 = aq^2$ and $q^2 | p^2$ in the third sentence.

d. In the fourth sentence, the author is using previous knowledge. State the proposition in if/then form that the author assumes has already been proved. Use symbols that do not overlap with the symbols in the proof.

e. Why is the author justified in claiming that the proof is complete in the last sentence?

# 11

## The Uniqueness Methods

You now have three major techniques for proving that "$A$ implies $B$": the forward-backward, the contrapositive, and the contradiction methods. You have also learned how and when to use the construction, choose, and specialization methods. Additional quantifier techniques are developed in this chapter. Specifically, you will know to use a **uniqueness method** when a statement contains the keyword "unique" (or "one and only one," or "exactly one") as well as the quantifier "there is" in the form:

> There is a *unique* object with a certain property such that something happens.

This statement can arise in either the forward or the backward process, so there are two proof techniques.

### 11.1 THE FORWARD UNIQUENESS METHOD

When you encounter a uniqueness statement in the forward process, you know that there is one and only one object, say $X$, with the certain property and for which the something happens. In addition to using $X$ with its certain property and something that happens, if you were to come across another object $Y$ with the same certain property and for which that something happens, you can conclude that $X$ and $Y$ are the same; that is, that $X = Y$, and that should help you to reach the conclusion that $B$ is true. The act of writing that $X = Y$ as a new statement in the forward process is called the **forward uniqueness method** and is illustrated with the following example.

**Proposition 17** *If a, b, and c are real numbers with the property that the equation $ax^2 + bx + c = 0$ has a unique real solution, then $b^2 - 4ac = 0$.*

**Analysis of Proof.**    Recognizing the keyword "unique" in the hypothesis, it is appropriate to use the forward uniqueness method. Working forward from the hypothesis that the equation $ax^2 + bx + c = 0$ has a real solution, you know by the quadratic formula that

**A1:** The following are solutions to the equation:

$$\bar{x} = \frac{-b + \sqrt{b^2 - 4ac}}{2a} \quad \text{and} \quad \bar{y} = \frac{-b - \sqrt{b^2 - 4ac}}{2a}.$$

In $A1$, both $\bar{x}$ and $\bar{y}$ are solutions to the equation $ax^2 + bx + c = 0$, so, by the hypothesis that there is a unique solution, the forward uniqueness method allows you to state that

**A2:** $\bar{x} = \bar{y}$.

The conclusion is obtained by working forward from $A1$ and $A2$, as follows:

$$\bar{x} = \frac{-b+\sqrt{b^2-4ac}}{2a} \quad = \quad \frac{-b-\sqrt{b^2-4ac}}{2a} = \bar{y} \quad \text{(from } A1 \text{ and } A2\text{)}$$

$$b^2 - 4ac \quad = \quad 0 \qquad\qquad\qquad \text{(algebra)}.$$

The proof is now complete.

**Proof of Proposition 17.**    Because the equation $ax^2 + bx + c = 0$ has a real solution, you know from the quadratic formula that the solutions are

$$\bar{x} = \frac{-b + \sqrt{b^2 - 4ac}}{2a} \quad \text{and} \quad \bar{y} = \frac{-b - \sqrt{b^2 - 4ac}}{2a}.$$

However, because the hypothesis states that the solution to the equation is unique, it follows (by the forward uniqueness method) that $\bar{x} = \bar{y}$, and so, by algebra, $b^2 - 4ac = 0$, thus completing the proof.  $\square$

## 11.2    THE BACKWARD UNIQUENESS METHOD

When the keyword "unique" appears in the backward process, you must show not only that there is an object with a certain property such that something happens, but also that there is only one such object. That is, with the **backward uniqueness method** you must prove two statements: (a) there is an object with the certain property such that the something happens and (b) there is only one such object. The first task is done either by the construction or the contradiction method. The second task is accomplished in one of the following two standard ways.

### The Direct Uniqueness Method

With the **direct uniqueness method**, you assume that there are two objects having the certain property and for which the something happens. If there really is only one such object then, using the two objects with their certain properties, the something that happens, and the information in $A$, you must conclude that the two objects are one and the same—that is, that they are equal. The forward-backward method is usually the best way to prove that the two objects are equal. This process is illustrated now.

**Proposition 18** *If $a$, $b$, $c$, $d$, $e$, and $f$ are real numbers with the property that $ad - bc \neq 0$, then there are unique real numbers $x$ and $y$ such that $ax + by = e$ and $cx + dy = f$.*

**Analysis of Proof.** Recognizing the keyword "unique" in the conclusion, you should use the backward uniqueness method. Accordingly, it is first necessary to construct real numbers $x$ and $y$ for which $ax + by = e$ and $cx + dy = f$. This is done in Proposition 4 on page 43 with the construction method. It remains to ensure that there is only one pair of numbers that satisfy the two equations. This is now established by the direct uniqueness method. Accordingly, you assume that $(x_1, y_1)$ and $(x_2, y_2)$ are two objects with the certain property and for which the something happens. In this case, that means that

**A1:** $ax_1 + by_1 = e$ and $cx_1 + dy_1 = f$ and
**A2:** $ax_2 + by_2 = e$ and $cx_2 + dy_2 = f$.

With these four equations and the hypothesis $A$, the forward-backward method is used to show that the two objects are the same; that is, that

**B1:** $(x_1, y_1) = (x_2, y_2)$.

A key question associated with $B1$ is, "How can I show that two ordered pairs of real numbers are equal?" Using Definition 4 on page 26 for equality of ordered pairs, one answer is to show that

**B2:** $x_1 = x_2$ and $y_1 = y_2$.

Both of these statements are obtained from the forward process by applying algebraic manipulations to the four equations in $A1$ and $A2$ and by using the hypothesis that $ad - bc \neq 0$, as indicated in the condensed proof that follows.

**Proof of Proposition 18.** The existence of the real numbers $x$ and $y$ satisfying the two equations is established in Proposition 4 on page 43. To see that there is only one such solution, assume that $(x_1, y_1)$ and $(x_2, y_2)$ are real numbers satisfying

(1) $ax_1 + by_1 = e$  (2) $cx_1 + dy_1 = f$

(3) $ax_2 + by_2 = e$  (4) $cx_2 + dy_2 = f$.

Subtracting (3) from (1) and (4) from (2) results in

$$(5)\ \ a(x_1 - x_2) + b(y_1 - y_2) = 0$$

$$(6)\ \ c(x_1 - x_2) + d(y_1 - y_2) = 0.$$

Multiplying (5) by $d$ and (6) by $b$ and then subtracting (6) from (5) yields

$$(7)\ \ (ad - bc)(x_1 - x_2) = 0.$$

Because, by hypothesis, $ad - bc \neq 0$, one has $x_1 - x_2 = 0$ and hence $x_1 = x_2$. A similar sequence of algebraic manipulations establishes that $y_1 = y_2$ and thus the uniqueness is proved. $\square$

## The Indirect Uniqueness Method

With the **indirect uniqueness method** for showing that there is only one object satisfying a certain property and for which something happens, you assume that there are *two different* objects having the certain property and for which the something happens. Now supposedly this cannot happen, so, by using the certain property, the something that happens, the information in $A$, and especially the fact that the objects are different, you must reach a contradiction. This process is demonstrated now.

**Proposition 19**  *If $r > 0$, then there is a unique real number $x$ such that $x^3 = r$.*

**Analysis of Proof.**    The appearance of the keyword "unique" in the conclusion suggests starting with the backward uniqueness method. Accordingly, the first step is to use the construction method to produce a real number $x$ such that $x^3 = r$. This part of the proof is omitted so as to focus on how the indirect uniqueness method is subsequently used to establish that there is only one such real number. To that end, suppose that

> **A1:** $x$ and $y$ are two different real numbers (that is, $x \neq y$)
> such that $x^3 = r$ and $y^3 = r$.

Using this information, and especially the fact that $x \neq y$, it is shown that $r = 0$, which contradicts the hypothesis that $r > 0$.

 To show that $r = 0$, work forward from A1. Specifically, because $x^3 = r$ and $y^3 = r$, it follows that

> **A2:** $x^3 = y^3$    or    $x^3 - y^3 = 0.$

On factoring, you have that

> **A3:** $(x - y)(x^2 + xy + y^2) = 0.$

Here is where you can use the fact that $x \neq y$ to divide both sides of $A3$ by $x - y \neq 0$, obtaining

**A4:** $x^2 + xy + y^2 = 0$.

Thinking of $A4$ as a quadratic equation of the form $ax^2 + bx + c = 0$, in which $a = 1$, $b = y$, and $c = y^2$, the quadratic formula states that

**A5:** $x = \frac{-y \pm \sqrt{y^2 - 4y^2}}{2} = \frac{-y \pm \sqrt{-3y^2}}{2}$.

Because $x$ is real and the foregoing formula requires taking the square root of $-3y^2$, it must be that

**A6:** $y = 0$,

and if $y = 0$, then a contradiction to $r > 0$ is reached from $A1$ because

**A7:** $r = y^3 = 0$.

**Proof of Proposition 19.** The proof that there is a real number $x$ such that $x^3 = r$ is omitted. To see that $x$ is the only such number, assume that $y \neq x$ also satisfies $y^3 = r$. Hence it follows that $0 = x^3 - y^3 = (x - y)(x^2 + xy + y^2)$. Because $x \neq y$, it must be that $x^2 + xy + y^2 = 0$. By the quadratic formula,

$$x = \frac{-y \pm \sqrt{y^2 - 4y^2}}{2} = \frac{-y \pm \sqrt{-3y^2}}{2}.$$

Now $x$ is real, so it must be that $y = 0$. But then $r = y^3 = 0$, thus contradicting the hypothesis that $r > 0$, and so the proof is complete. ☐

## Summary

In this chapter, you have learned the various uniqueness methods. Use a uniqueness method when you come across a statement in the form, "there is a unique object (or one and only one object, or exactly one object) with a certain property such that something happens."

When this statement occurs in the forward process, with the forward uniqueness method you

1. Look for two objects, say $X$ and $Y$, with that certain property and for which that something happens.

2. You can then write, as a new statement in the forward process, that $X$ and $Y$ are the same; that is, that $X = Y$. This statement should then help you establish that the conclusion $B$ is true.

Use the backward uniqueness method when, in the backward process, you need to show that "there is a unique object with a certain property such that something happens." Doing so requires two steps: first showing that there is one such object, say $X$, and then showing that there is only one such object. While the first task is accomplished with the construction or contradiction method, you can accomplish the second task in one of two ways. With the direct uniqueness method you

1. Assume that, in addition to the object $X$, $Y$ is also an object with the certain property and for which the something happens.

2. Use the certain property and something that happens for $X$ and $Y$ together with the hypothesis $A$ to show that $X$ and $Y$ are the same (that is, that $X = Y$).

With the indirect uniqueness method, you

1. Assume that $Y$ is a different object from $X$ with the certain property and for which the something happens.

2. Use the properties of $X$ and $Y$, the fact that they are different, and the hypothesis $A$ to reach a contradiction.

## Exercises

**Note:** Solutions to those exercises marked with a $B$ are in the back of this book. Solutions to those exercises marked with a $W$ are located on the web at http://www.wiley.com/college/solow/.

**Note:** All proofs should contain an analysis of proof and a condensed version. Definitions for all mathematical terms are provided in the glossary at the end of the book.

$^B$**11.1**     Suppose that each of the following statements arises in the forward process. When subsequently applying the forward uniqueness method with the given objects, indicate (i) what you would have to show about those objects and (ii) what you would then be able to conclude as a new statement in the forward process.

a.  Statement:     There is one and only one line through two given points $(x_1, y_1)$ and $(x_2, y_2)$ in the plane.

   Given objects:   Lines $y = mx + b$ and $y = cx + d$.

b.  Statement:     There is exactly one solution to the system of equations $ax + by = 0$ and $cx + dy = 0$, where $a$, $b$, $c$, and $d$ are given real numbers.

   Given objects:   Solutions $(x_1, y_1)$ and $(x_2, y_2)$.

c.  Statement:     There is a unique complex number $c + di$ for which $(a + bi)(c + di) = 1$, where $a$ and $b$ are given real numbers and $i = \sqrt{-1}$.

   Given objects:   Complex numbers $r + si$ and $t + ui$.

**11.2**    Suppose that each of the following statements arises in the forward process. When subsequently applying the forward uniqueness method with the given objects, indicate (i) what you would have to show about those objects and (ii) what you would then be able to conclude as a new statement in the forward process.

a.    Statement:    There is a unique maximizer of the function $ax^2 + bx + c$, where $a$, $b$, and $c$ are given real numbers with $a < 0$.
    Given objects:    Real numbers $x^*$ and $y^*$.

b.    Statement:    For a given function $f$ of one real variable, there is one and only one function $g$ such that, for every real number $x$, $f(g(x)) = g(f(x)) = x$.
    Given objects:    Functions $F$ and $G$.

c.    Statement:    For a given integer $n$, there is a unique integer $a > 0$ such that $a|n$ and for every integer $b > 0$ such that $b|n$, $b|a$.
    Given objects:    Integers $p$ and $q$.

$^B$**11.3**    Suppose that each of the statements in Exercise 11.1 appears in the backward process. How would you proceed to prove these statements using (i) the direct uniqueness method and (ii) the indirect uniqueness method?

**11.4**    Suppose that each of the statements in Exercise 11.2 appears in the backward process. How would you proceed to prove these statements using (i) the direct uniqueness method and (ii) the indirect uniqueness method?

*$^*$**11.5**    Answer the given questions about the following proof that, "If $u$ is an upper bound for a set $S$ of real numbers and $u \in S$, then $u$ is the unique element of $S$ with the property that, for every element $x \in S$, $x \le u$."

> **Proof.** Suppose that $v \in S$ also satisfies the property that, for every element $x \in S$, $x \le v$. Then, because $u \in S$, it must be that $u \le v$. Likewise, because for every element $x \in S$, $x \le u$ and $v \in S$, it must be that $v \le u$. It follows that $u = v$ and so the proof is complete.  $\square$

a. In this backward uniqueness method, why does the author not first establish that there is an object with the certain property such that something happens?

b. Does this proof use the direct or indirect uniqueness method? Explain.

c. What proof technique is used in the second sentence?

d. Why is the author justified in claiming that the proof is complete in the last sentence?

**11.6**    Write an analysis of proof that corresponds to the condensed proof given below. Indicate which techniques are used and how they are applied. Fill in the details of any missing steps where appropriate.

> **Proposition.** If $a$ and $c$ are real numbers with $a < 0$, then $x = 0$ is the unique maximizer of the function $f(x) = ax^2 + c$.
>
> **Proof.** It was already shown in Exercise 5.15 that $x = 0$ is a maximizer of $f$. Suppose now that $y$ is also a maximizer of $f$. Then, for every real number $z$, $ay^2 + c \geq az^2 + c$. In particular, $ay^2 + c \geq a(0)^2 + c = c$. But then it follows by algebra that $y^2 \leq 0$. Thus $y = 0$ and so the proof is complete. $\square$

**11.7**    Write an analysis of proof that corresponds to the condensed proof given below. Indicate which techniques are used and how they are applied. Fill in the details of any missing steps where appropriate.

> **Proposition.** If $a$, $b$, $c$, and $d$ are real numbers for which the equations $ax + by = 0$ and $cx + dy = 0$ have a unique solution, then $ad - bc \neq 0$.
>
> **Proof.** Assume that $ad - bc = 0$. Note that, because the two equations have a unique solution, it cannot be that both $c$ and $d$ are 0. It then follows that $x = d$ and $y = -c$ is a solution to the two equations, as is $x = 0$ and $y = 0$. By the uniqueness, it must be that $d = 0$ and $c = 0$. This contradiction completes the proof. $\square$

$^B$**11.8**    Prove that, if $x$ is a real number $> 2$, then there is a unique real number $y < 0$ such that $x = 2y/(1 + y)$.

**11.9**    Prove that, if $a$ and $b$ are integers with $a \neq 0$ such that $a|b$, then there is a unique integer $k$ such that $b = ka$. (See Definition 1 on page 26.)

$^W$**11.10**    Prove, by the indirect uniqueness method, that, if $m$ and $b$ are real numbers with $m \neq 0$, then there is a unique number $x$ such that $mx + b = 0$.

**11.11**    Prove, by the indirect uniqueness method, that there is a unique integer $n$ for which $2n^2 - 3n - 2 = 0$.

$^B$**11.12**    Prove that, if $a$ and $b$ are real numbers, at least one of which is not 0, and $i = \sqrt{-1}$, then there is a unique complex number, say $c + di$, such that $(a + bi)(c + di) = 1$.

$^*$**11.13**    Prove that, if $f$ is a function of one real variable such that for every real number $y$, there is a unique real number $x$ such that $f(x) = y$, then the function $f$ is one-to-one.

# 12

---

# *Induction*

In Chapter 5, you learned to use the choose method when the quantifier "for all" appears in the statement $B$. There is one special form of $B$ containing the quantifier "for all" for which a separate technique known as **induction** is likely to be more successful.

## 12.1  HOW TO USE INDUCTION

You should consider induction seriously (even before the choose method) when $B$ has the form:

For every integer $n \geq 1$, "something happens,"

where the something that happens is some statement, $P(n)$, that depends on the integer $n$. The following is an example:

$$\text{For all integers } n \geq 1, \underbrace{\sum_{k=1}^{n} k = \frac{n(n+1)}{2}}_{P(n)}, \quad \text{where } \sum_{k=1}^{n} k = 1 + \cdots + n.$$

When considering induction, the keywords to look for are "integer" and "$\geq 1$."

One way to attempt proving such statements is to make an infinite list of statements, one for each of the integers starting from $n = 1$, and then prove each statement separately. While the first few statements on the list are usually easy to verify, the issue is how to check statement number $n$ and beyond. For the foregoing example, the list is:

$$P(1): \qquad \sum_{k=1}^{1} k = \frac{1(1+1)}{2} \qquad \text{or} \qquad 1 = 1$$

$$P(2): \qquad \sum_{k=1}^{2} k = \frac{2(2+1)}{2} \qquad \text{or} \qquad 1 + 2 = 3$$

$$P(3): \qquad \sum_{k=1}^{3} k = \frac{3(3+1)}{2} \qquad \text{or} \qquad 1 + 2 + 3 = 6$$

$$\vdots$$

$$P(n): \qquad \sum_{k=1}^{n} k = \frac{n(n+1)}{2}$$

$$P(n+1): \qquad \sum_{k=1}^{n+1} k = \frac{(n+1)[(n+1)+1]}{2} \quad = \quad \frac{(n+1)(n+2)}{2}$$

$$\vdots$$

Induction is a clever method for proving that each of these statements in the infinite list is true. Think of induction as a proof machine that starts with $P(1)$ and progresses down the list, proving each statement as it proceeds. Here is how the machine works.

The machine is started by your verifying that $P(1)$ is true, which is easy to do for the foregoing example. Then $P(1)$ is fed into the machine. The machine uses the fact that $P(1)$ is true and automatically proves that $P(2)$ is true. You then put $P(2)$ into the machine. The machine uses the fact that $P(2)$ is true to show that $P(3)$ is true, and so on (see Figure 12.1).

Observe that, by the time the machine is going to prove that $P(n+1)$ is true, it will already have shown that $P(n)$ is true (from the previous step). Thus, in designing the machine, you can assume that $P(n)$ is true; your job is to make sure that $P(n+1)$ is also true. In summary, the following steps constitute a proof by induction.

### The Steps of Induction

**Step 1.** Verify that $P(1)$ is true.

**Step 2.** Use the assumption that $P(n)$ is true to prove that $P(n+1)$ is true.

To perform Step 1, replace $n$ everywhere in $P(n)$ by 1. To verify that the resulting statement is true usually requires only some minor rewriting.

Step 2 is more challenging. You must reach the conclusion that $P(n+1)$ is true by using the assumption that $P(n)$ is true. There is a standard way of doing this. Begin by writing the statement $P(n+1)$, which you want to conclude is true. Because you are assuming that $P(n)$ is true, you should

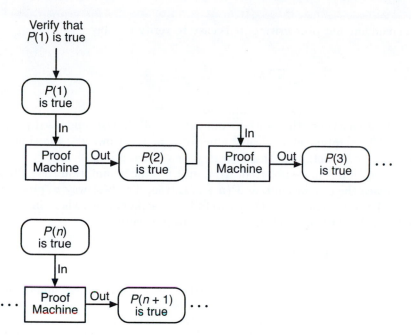

*Fig. 12.1* The proof machine for induction.

somehow try to rewrite the statement $P(n + 1)$ in terms of $P(n)$, for then you can make use of the assumption that $P(n)$ is true. Using the assumption that $P(n)$ is true is referred to as **using the induction hypothesis.** On establishing that $P(n+1)$ is true, the proof is complete. The steps of induction are illustrated with the following proposition.

**Proposition 20** *For every integer $n \geq 1$, $\sum_{k=1}^{n} k = \frac{n(n+1)}{2}$.*

**Analysis of Proof.** When using the method of induction, it is helpful to write the statement $P(n)$. In this case:

$$P(n): \sum_{k=1}^{n} k = \frac{n(n+1)}{2}.$$

The first step in a proof by induction is to verify $P(1)$. Replacing $n$ everywhere by 1 in $P(n)$, you obtain

$$P(1): \sum_{k=1}^{1} k = \frac{1(1+1)}{2}.$$

With a small amount of rewriting, it is easy to verify this because

$$\sum_{k=1}^{1} k = 1 = \frac{1(1+1)}{2}.$$

This step is often so easy that authors omit the details in the condensed proof by saying, "The statement is clearly true for $n = 1$." However, you should include appropriate details of this step in your written proofs.

The second step is more involved. You must use the assumption that $P(n)$ is true to reach the conclusion that $P(n+1)$ is true. The best way to proceed is to write the statement $P(n+1)$ by carefully replacing $n$ everywhere in $P(n)$ with $n+1$ and rewriting a bit, if necessary. In this case:

$$P(n+1): \sum_{k=1}^{n+1} k = \frac{(n+1)[(n+1)+1]}{2} = \frac{(n+1)(n+2)}{2}.$$

To reach the conclusion that $P(n+1)$ is true, begin with the left side of the equality in $P(n+1)$ and try to make that side look like the right side. In so doing, you should use the information in $P(n)$ by relating the left side of the equality in $P(n+1)$ to the left side of the equality in $P(n)$. Then you will be able to use the right side of the equality in $P(n)$. In this example,

$$P(n+1): \sum_{k=1}^{n+1} k = \left( \sum_{k=1}^{n} k \right) + (n+1).$$

Now you can use the assumption that $P(n)$ is true by replacing $\sum_{k=1}^{n} k$ in $P(n+1)$ with $n(n+1)/2$ from $P(n)$, obtaining

$$P(n+1): \sum_{k=1}^{n+1} k = \left( \sum_{k=1}^{n} k \right) + (n+1) = \frac{n(n+1)}{2} + (n+1).$$

All that remains is a bit of algebra to rewrite $\frac{n(n+1)}{2} + (n+1)$ as $\frac{(n+1)(n+2)}{2}$, thus obtaining the right side of the equality in $P(n+1)$. The algebraic steps are:

$$\frac{n(n+1)}{2} + (n+1) = (n+1)\left( \frac{n}{2} + 1 \right) = \frac{(n+1)(n+2)}{2}.$$

Your ability to relate $P(n+1)$ to $P(n)$ so as to use the induction hypothesis that $P(n)$ is true determines the success of a proof by induction. If you are unable to relate $P(n+1)$ to $P(n)$, then you might wish to consider a different proof technique. On the other hand, if you can relate $P(n+1)$ to $P(n)$, you will find that induction is easier to use than almost any other technique. To illustrate this fact, you are asked in the exercises to prove Proposition 20 without using induction. Compare your proof with the condensed proof that follows.

**Proof of Proposition 20.**   The statement is clearly true for $n = 1$. Assume the statement is true for $n$; that is, that $\sum_{k=1}^{n} k = n(n+1)/2$. Then

$$\sum_{k=1}^{n+1} k \;=\; \left( \sum_{k=1}^{n} k \right) + (n+1)$$

$$=\; \frac{n(n+1)}{2} + (n+1)$$

$$=\; (n+1)\left( \frac{n}{2} + 1 \right)$$

$$=\; \frac{(n+1)(n+2)}{2}.$$

Thus the statement is true for $n+1$, and so the proof is complete.  □

Note that induction does not help you to discover the correct form of the statement $P(n)$. Rather, induction only verifies that a given statement $P(n)$ is true for all integers $n$ greater than or equal to some initial one.

## 12.2   SOME VARIATIONS ON INDUCTION

From the foregoing discussion, you know that, in the second step in a proof by induction, you use the assumption that $P(n)$ is true to show that $P(n+1)$ is true.   From a notational point of view, some authors prefer to use the assumption that $P(n-1)$ is true to show that $P(n)$ is true.   These two approaches are identical—either can be used, depending on your notational preference. What is important is that you establish that, if a general statement on the infinite list is true, then the next statement is also true.

When using induction, the first value for $n$ need not be 1.   For instance, you can use induction to prove that "for all integers $n \geq 5$, $2^n > n^2$." The only modification is that, to start the proof, you must verify $P(n)$ for the first given value of $n$. In this case, that first value is $n = 5$, so you have to check that $2^5 > 5^2$ (which is true because $2^5 = 32$ while $5^2 = 25$). The second step of the induction proof remains the same—you still have to show that, if $P(n)$ is true (that is, $2^n > n^2$), then $P(n+1)$ is also true (that is, $2^{n+1} > (n+1)^2$). In so doing, you can also use the fact that $n \geq 5$, if necessary.

Another modification to the basic induction method arises when you are having difficulty relating $P(n+1)$ to $P(n)$. Suppose, however, that you can relate $P(n+1)$ to $P(j)$, where $j < n$. In this case, you would like to use the fact that $P(j)$ is true, but can you assume that $P(j)$ is in fact true? The answer is yes. To see why, recall the analogy of the proof machine (look again at Figure 12.1) and observe that, by the time the machine has to show that $P(n+1)$ is true, the machine has already proved that all of the statements $P(1), \ldots, P(j), \ldots, P(n)$ are true. Thus, when trying to show that $P(n+1)$ is true, you can assume that $P(n)$ *and all preceding statements are true.* Such a proof is referred to as **generalized induction** and is illustrated now.

## 12.3   READING A PROOF

The process of reading and understanding a proof is demonstrated with the following proposition.

**Proposition 21** *Any integer $n \geq 2$ can be expressed as a finite product of primes (see Definition 2 on page 26).*

**Proof of Proposition 21.**    (For reference purposes, each sentence of the proof is written on a separate line.)

**S1:**  The statement is clearly true for $n = 2$.

**S2:**  Now assume the statement is true for all integers between 2 and $n$; that is, that any integer $j$ with $2 \leq j \leq n$ can be expressed as a finite product of primes.

**S3:**  If $n + 1$ is prime, the statement is true for $n + 1$.

**S4:**  Otherwise, there are integers $p$ and $q$ with $1 < p, q < n+1$ such that $n + 1 = pq$.

**S5:**  But by the induction hypothesis, $p$ and $q$ can be expressed as a finite product of primes, and therefore so can $n + 1$.

The proof is now complete.   □

**Analysis of Proof.**   An interpretation of statements $S1$ through $S5$ follows.

**Interpretation of S1:** *The statement is clearly true for $n = 2$.*
    The author is performing the first step of induction by mentioning that the statement is true for the first value of $n$; namely, $n = 2$. The statement is true because 2 is itself prime.

**Interpretation of S2:** *Now assume the statement is true for all integers between 2 and $n$; that is, that any integer $j$ with $2 \leq j \leq n$ can be expressed as a finite product of primes.*
    The author is performing the second step of generalized induction by assuming that the statement is true for all integers between 2 and $n$. It remains to show that the statement is true for $n + 1$.

**Interpretation of S3:** *If $n + 1$ is prime, the statement is true for $n + 1$.*
    The author notes that the statement is true for $n + 1$ when $n + 1$ is prime, which is clearly correct. Presumably, the author will also show that the statement is also true when $n + 1$ is not prime.

**Interpretation of S4:** *Otherwise, there are integers $p$ and $q$ with $1 < p, q < n + 1$ such that $n + 1 = pq$.*
    The author is showing that the statement is true when $n + 1$ is not prime. Specifically, the author is using the fact that, when $n + 1$ is not prime, $n + 1$

can be expressed as the product of two integers, say $p$ and $q$, strictly between 1 and $n + 1$.

**Interpretation of S5:** *But by the induction hypothesis, $p$ and $q$ can be expressed as a finite product of primes, and therefore so can $n + 1$.*

The author is applying the induction hypothesis to $p$ and $q$, which is valid because $p$ and $q$ are integers for which $2 \leq p, q \leq n$ (see $S4$). Doing so yields that $p$ and $q$ are each a finite product of primes. The author then notes that, as a result, $n + 1 = pq$ is also a finite product of the primes that constitute $p$ and the product of the primes that constitute $q$. The generalized induction proof is now complete because the author has correctly established that the statement is true for $n + 1$.

## Summary

In this chapter, you have learned to use mathematical induction. Use induction when the statement you are trying to prove has the form, "For every integer $n \geq n_0$, $P(n)$," where $P(n)$ is some statement that depends on $n$. To apply the method of induction,

1. Verify that the statement $P(n)$ is true for $n_0$. (To do this, replace $n$ everywhere in $P(n)$ by $n_0$, rewrite the resulting statement, and try to establish that $P(n_0)$ is true.)

2. Assume that $P(n)$ is true.

3. Write the statement $P(n + 1)$ by replacing $n$ everywhere in $P(n)$ with $n + 1$. (Some rewriting may be necessary to simplify the expression for $P(n + 1)$.)

4. Reach the conclusion that $P(n + 1)$ is true. To do so, relate $P(n + 1)$ to $P(n)$ and then use the fact that $P(n)$ is true. The key to using induction rests on your ability to relate $P(n + 1)$ to $P(n)$.

## Exercises

**Note:** Solutions to those exercises marked with a $B$ are in the back of this book. Solutions to those exercises marked with a $W$ are located on the web at http://www.wiley.com/college/solow/.

**Note:** Unless otherwise stated, do all proofs by induction and write only a condensed version. Definitions for all mathematical terms are provided in the glossary at the end of the book.

$^B$**12.1**    For which of the following statements is induction applicable? When induction is not applicable, explain why not.

a. For every positive integer $n$, 8 divides $5^n + 2(3^{n-1}) + 1$.

b. There is an integer $n \geq 0$ such that $2n > n^2$.

c. For every integer $n \geq 1$, $1(1!) + \cdots + n(n!) = (n+1)! - 1$. (Recall that $n! = n(n-1)\cdots 1$.)

d. For every integer $n \geq 4$, $n! > n^2$.

e. For every real number $n \geq 1$, $n^2 \geq n$.

**12.2**    Upon learning about the method of induction, a student said, "I do not understand something. After showing that the statement is true for $n = 1$, you want me to assume that $P(n)$ is true and to show that $P(n+1)$ is true. How can I assume that $P(n)$ is true—after all, aren't we trying to *show* that $P(n)$ is true?" Answer this question.

$^W$**12.3**    With regard to induction,

a. Why and when would you want to use induction instead of the choose method?

b. Why is it not possible to use induction on statements of the form, "For every real number with a certain property, something happens?"

**12.4**    Suppose that $A(n)$ and $B(n)$ are statements that depend on a positive integer $n$. Explain how you would use induction to prove that, "For every integer $n \geq 1$, if $A(n)$ is true, then $B(n)$ is true." For Step 2 of induction, indicate what statement(s) you would assume is true and what statement(s) you would have to show is true.

$^B$**12.5**    Describe a modified induction procedure for proving that:

a. For every integer $n \leq n_0$, $P(n)$ is true.

b. For every integer $n$, $P(n)$ is true.

**12.6**    Describe a modified induction procedure for proving that, for every positive odd integer $n$, $P(n)$ is true.

$^B$**12.7**    Prove that, for every integer $n \geq 1$, $1(1!) + \cdots + n(n!) = (n+1)! - 1$.

**12.8**    Prove that, for every integer $k \geq 1$, $1 + 2 + 2^2 + \cdots + 2^{k-1} = 2^k - 1$.

$^W$**12.9**    Prove that, for every integer $n \geq 5$, $2^n > n^2$.

**12.10**    Prove that, for every integer $n \geq 1$,

$$\frac{1}{n!} \leq \frac{1}{2^{n-1}}.$$

**12.11**    Prove that $\$X$ invested at $i\%$ interest compounded once annually for $n \geq 0$ years will result in $\$ (1+r)^n X$, where $r = \frac{i}{100}$.

$^B$**12.12**    Prove that a set of $n \geq 1$ elements has $2^n$ subsets (including the empty set).

**12.13**    Prove that, if $x > 1$ is a given real number, then, for every integer $n \geq 2$, $(1+x)^n > 1+nx$.

**12.14**    Prove that, for every integer $n \geq 2$, if $C_1, \ldots, C_n$ are convex sets, then $\cap_{i=1}^n C_i$ is a convex set.

$^B$**12.15**    Prove, *without using induction*, that, for any integer $n \geq 1$, $\sum_{k=1}^n k = n(n+1)/2$.

**12.16**    A machine is filled with an odd number of chocolate candies and an odd number of caramel candies. For 25 cents, the machine dispenses two candies. Prove that, before being empty, the machine will dispense at least one pair that consists of one chocolate candy and one caramel candy.

$^W$**12.17**    Prove that, if $i = \sqrt{-1}$, then for every integer $n \geq 1$, the complex number $[\cos(x) + i\sin(x)]^n = \cos(nx) + i\sin(nx)$. Do this by showing that (1) the statement is true for $n = 1$ and (2) if the statement is true for $n - 1$, then the statement is true for $n$.

**12.18**    Prove that, in a line of at least two people, if the first person is a woman and the last person is a man, then somewhere in the line there is a man standing immediately behind a woman.

$^B$**12.19**    In the following condensed proof, explain how the author relates $P(n+1)$ to $P(n)$ and where the induction hypothesis is used.

**Proposition.** $\forall$ integer $n \geq 2$, $\prod_{k=2}^n (1 - \frac{1}{k^2}) = \frac{n+1}{2n}$. (Here, $\prod$ stands for the product of all the numbers to the right of the symbol, similar to the way $\sigma$ stands for the sum.)

**Proof.** The statement is true for $n = 2$ because $1 - \frac{1}{4} = \frac{3}{4} = \frac{2+1}{2(2)}$. Assume the statement is true for $n$. Then

$$\prod_{k=2}^{n+1} (1 - \tfrac{1}{k^2}) = \left( \prod_{k=2}^{n} (1 - \tfrac{1}{k^2}) \right) \left( 1 - \tfrac{1}{(n+1)^2} \right)$$

$$= \frac{n+1}{2n} - \frac{1}{2n(n+1)}$$

$$= \frac{n+2}{2(n+1)}.$$

The proof is now complete.  $\square$

$^{W}$**12.20**    In the following condensed proof, explain how the author relates $P(n+1)$ to $P(n)$ and where the induction hypothesis is used.

**Proposition.** For every integer $n \geq 1$, the derivative of $x^n$ is $nx^{n-1}$.

**Proof.** The statement is true for $n = 1$ because, $(x)' = 1 = 1x^0$. Assume now that $(x^n)' = nx^{n-1}$. Then, for $x^{n+1}$,

$$
\begin{aligned}
(x^{n+1})' &= [(x)(x^n)]' \\
&= (x)'x^n + x\,(x^n)' \\
&= x^n + x\,(nx^{n-1}) \\
&= (n+1)x^n.
\end{aligned}
$$

The proof is now complete.  □

**12.21**    In the following condensed proof, explain how the author relates $P(n+1)$ to $P(n)$ and where the induction hypothesis is used.

**Proposition.** For every integer $n \geq 2$, if $x_1, \ldots, x_n$ are real numbers strictly between 0 and 1, then $(1-x_1)\cdots(1-x_n) > 1 - x_1 - \cdots - x_n$.

**Proof.** When $n = 2$, the statement is true because $(1 - x_1)(1 - x_2) = 1 - x_1 - x_2 + x_1 x_2 > 1 - x_1 - x_2$. Assume the statement is true for $n$. Then for $n + 1$,

$$
\begin{aligned}
&(1 - x_1)(1 - x_2)\cdots(1 - x_{n+1}) \\
=\ & [(1 - x_1)(1 - x_2)\cdots(1 - x_n)](1 - x_{n+1}) \\
>\ & (1 - x_1 - \cdots - x_n)(1 - x_{n+1}) \\
=\ & 1 - x_1 - \cdots - x_n - (1 - x_1 - x_2 - \cdots - x_n)(x_{n+1}) \\
=\ & 1 - x_1 - \cdots - x_n - x_{n+1} + (x_1 + x_2 + \cdots + x_n)(x_{n+1}) \\
>\ & 1 - x_1 - \cdots - x_n - x_{n+1}.
\end{aligned}
$$

The proof is now complete.  □

$^*$**12.22**    Answer the given questions about the following proof that, "$\forall$ integer $n \geq 2$, if $x_1, \ldots, x_n$ are real numbers, then $\sqrt{x_1^2 + \cdots + x_n^2} \leq |x_1| + \cdots + |x_n|$."

**Proof.** You can easily verify that the statement is true when $n = 2$. Now assume that the statement is true for $n$. Then for $n + 1$, letting $z^2 = x_1^2 + \cdots + x_n^2$, it follows that

$$
\begin{aligned}
\sqrt{x_1^2 + \cdots + x_n^2 + x_{n+1}^2} &= \sqrt{z^2 + x_{n+1}^2} \\
&\leq |z| + |x_{n+1}| \qquad\qquad (1) \\
&= \sqrt{x_1^2 + \cdots + x_n^2} + |x_{n+1}| \\
&\leq |x_1| + \cdots + |x_n| + |x_{n+1}|. \quad (2)
\end{aligned}
$$

The proof is now complete.  □

a. Justify the first sentence by proving that the statement is true for $n = 2$.

b. Justify the inequality in (1).

c. Justify the inequality in (2).

*12.23    Answer the given questions about the following proof that, "If $f$ is a function of one variable for which there is a real number $\alpha$ with $0 < \alpha < 1$ such that, for all real numbers $x$ and $y$, $|f(x) - f(y)| \leq \alpha|x - y|$ and $x_*$ is a real number for which $f(x_*) = x_*$, then for any real number $x_0$ and integer $n \geq 1$, $|f^n(x_0) - x_*| \leq \alpha^n|x_0 - x_*|$, where $f^n(x) = f(f^{n-1}(x))$ and $f^1(x) = f(x)$."

**Proof.** Let $x_0$ be a real number. Then for $n = 1$, you have $|f^1(x_0) - x_*| = |f(x_0) - f(x_*)| \leq \alpha|x_0 - x_*|$. Now assume that $|f^n(x_0) - x_*| \leq \alpha^n|x_0 - x_*|$, then for $n + 1$,

$$
\begin{aligned}
|f^{n+1}(x_0) - x_*| &= |f(f^n(x_0)) - f(x_*)| \\
&\leq \alpha|f^n(x_0) - x_*| \quad (1) \\
&\leq \alpha(\alpha^n|x_0 - x_*|) \quad (2) \\
&= \alpha^{n+1}|x_0 - x_*|.
\end{aligned}
$$

The proof is now complete. $\square$

a. What proof technique is the author using in the first sentence and why?

b. Justify the inequality in the second sentence. What technique is the author using?

c. Justify the inequality in (1).

d. Justify the inequality in (2).

B12.24    What, if anything, is wrong with the following proof that all horses have the same color?

**Proof.** Let $n$ be the number of horses. When $n = 1$, the statement is clearly true; that is, one horse has the same color, whatever color it is. Assume that any group of $n$ horses has the same color. Now consider a group of $n + 1$ horses. Taking any $n$ of them, the induction hypothesis states that they all have the same color, say, brown. The only issue is the color of the remaining "uncolored" horse. Consider, therefore, any other group of $n$ of the $n+1$ horses that contains the uncolored horse. Again, by the induction hypothesis, all of the horses in the new group have the same color. Then, because all of the colored horses in this group are brown, the uncolored horse must also be brown. $\square$

$^W$**12.25**     What, if anything, is wrong with the following condensed proof?

**Proposition.** If $r$ is a real number with $|r| \leq 1$, then for all integers $n \geq 1$, $1 + r + r^2 + \cdots + r^{n-1} = (1 - r^n)/(1 - r)$.

**Proof.** The statement is clearly true for $n = 1$. Assume it is true for $n$. Then, for $n + 1$, one has,

$$
\begin{aligned}
1 + \cdots + r^n &= (1 - r^n)/(1 - r) + r^n \\
&= (1 - r^n + r^n - r^{n+1})/(1 - r) \\
&= (1 - r^{n+1})/(1 - r).
\end{aligned}
$$

The proof is now complete.  □

*12.26     What, if anything, is wrong with the following condensed proof?

**Proposition.** For every integer $n \geq 2$, if $S_1, S_2, \ldots, S_n$ are convex sets of real numbers, then $S_1 \cup S_2 \cup \cdots \cup S_n$ is a convex set.

**Proof.**  To see that the statement is true for $n = 2$, let $x, y \in S_1 \cup S_2$ and let $t$ be a real number with $0 \leq t \leq 1$. Because $x, y \in S_1 \cup S_2$ and because $S_1$ is convex, $tx + (1 - t)y \in S_1$. Likewise, because $S_2$ is convex, $tx + (1-t)y \in S_2$. Thus, $tx + (1-t)y \in S_1 \cup S_2$ and so the union of two convex sets is a convex set.  Assume now that the statement is true for $n$ and let $S_1, S_2, \ldots, S_{n+1}$ be convex sets of real numbers. You then have that

$$ S_1 \cup \cdots \cup S_{n+1} = (S_1 \cup \cdots \cup S_n) \cup S_{n+1}. $$

By the induction hypothesis, $S = S_1 \cup \cdots \cup S_n$ is convex. Finally, it has already been shown that the union of two convex sets is convex and therefore $S \cup S_{n+1}$ is convex, completing the proof.  □

*12.27     For a positive integer $n$, define the following function:

$$
f(n) = \begin{cases} n/2, & \text{if } n \text{ is even} \\ \\ 3n + 1, & \text{if } n \text{ is odd} \end{cases}
$$

Attempt to prove by induction that, for any integer $n \geq 1$, there is an integer $k > 0$ such that $f^k(n) = 1$, where $f^k(n)$ is the result of successively applying the foregoing function $k$ times, starting with the integer $n$, that is, $f^k(n) = f^{k-1}(f(n))$, and $f^1(n) = f(n)$. Explain why you are having difficulty in doing so.

# 13

## The Either/Or Methods

The **either/or methods** presented in this chapter arise when you come across the keywords "either/or" in the form "either $C$ or $D$ is true," (where $C$ and $D$ are statements). These keywords can appear in the forward and backward processes and so two different techniques are available.

### 13.1  PROOF BY CASES

To illustrate one of these methods, suppose that the keywords either/or arise in the hypothesis of a proposition whose form is "$C$ OR $D$ implies $B$." According to the forward-backward method, you can assume that $C$ OR $D$ is true; you must conclude that $B$ is true. The only question is whether you should assume that $C$ is true or whether you should assume that $D$ is true. Because you do not know which of these is correct, you should proceed with a **proof by cases**; that is, you should do two proofs. In the first case you assume that $C$ is true and prove that $B$ is true; in the second case you assume that $D$ is true and prove that $B$ is true. This technique is illustrated now. Observe how the words "either/or" are introduced intentionally in the middle of the proof so as to use a proof by cases.

**Proposition 22** *If $a$ is a negative real number, then $\bar{x} = -b/(2a)$ is a maximizer of the function $ax^2 + bx + c$.*

**Analysis of Proof.** No keywords appear in the hypothesis or conclusion, so the forward-backward method is used to begin the proof. Working backward, a key question is, "How can I show that a number [namely, $\bar{x} = -b/(2a)$] is a maximizer of a function?" Applying the definition given in Exercise 5.1(a) on page 61, it is necessary to show that

**B1:** For every real number $x$, $a\bar{x}^2 + b\bar{x} + c \geq ax^2 + bx + c$.

Recognizing the keywords "for all" in the backward process, the choose method is used to choose

**A1:** A real number $x$,

for which it must be shown that

**B2:** $a\bar{x}^2 + b\bar{x} + c \geq ax^2 + bx + c$.

Subtracting $ax^2 + bx + c$ from both sides of $B2$ and factoring out $\bar{x} - x$, it must be shown that

**B3:** $(\bar{x} - x)[a(\bar{x} + x) + b] \geq 0$.

If $\bar{x} - x = 0$, then $B3$ is true. Thus you can assume that

**A2:** $\bar{x} - x \neq 0$.

It is important to note here that you can rewrite $A2$ as follows so as to contain the keywords "either/or" explicitly:

**A3:** Either $\bar{x} - x > 0$ or $\bar{x} - x < 0$.

At this point, recognizing the keywords "either/or" in the forward process, it is time to use a proof by cases. Accordingly, you should do two proofs—first assume that $\bar{x} - x > 0$ and prove that $B3$ is true; then assume that $\bar{x} - x < 0$ and again prove that $B3$ is true. These two proofs are done now.

**Case 1:** Assume that

**A4:** $\bar{x} - x > 0$.

In this case you can divide both sides of $B3$ by the positive number $\bar{x} - x$; thus, it must be shown that

**B4:** $a(\bar{x} + x) + b \geq 0$.

Working forward from the fact that $\bar{x} = -b/(2a)$ and $a < 0$ (see the hypothesis), it follows from $A4$ that

**A5:** $2a\bar{x} + b > 0$,

and so

**A6:** $a(\bar{x} + x) + b = ax + b/2 = (2ax + b)/2 > 0$.

Thus $B3$ is true and this completes the first case.

**Case 2:** Now assume that

**A4:** $\bar{x} - x < 0$.

In this case you can divide both sides of B3 by the negative number $\bar{x} - x$; thus, it is necessary to show that

**B4:** $a(\bar{x} + x) + b \leq 0$.

Working forward from the fact that $\bar{x} = -b/(2a)$ and $a < 0$ (see the hypothesis), it follows from A4 that

**A5:** $2ax + b < 0$,

and so

**A6:** $a(\bar{x} + x) + b = ax + b/2 = (2ax + b)/2 < 0$.

Thus B4 is true and this completes the second case and the entire proof.

Observe that the proof of the foregoing second case is almost identical to that of the first case except for a reversal of sign in several places. Most mathematicians would not, in general, write the details for both cases. Rather, when they recognize the similarity of the two cases, they would say, "Assume, without loss of generality, that Case 1 occurs ...." They would then proceed to present the details for that case, omitting the second case altogether. In other words, the words "assume, **without loss of generality**, ..." mean that the author will present only one of the cases in detail; you will have to provide the details of the other case for yourself, as is illustrated in the following condensed proof of Proposition 22.

**Proof of Proposition 22.**    Let $x$ be a real number. (The word "let" indicates that the choose method is used.) It is shown that $a\bar{x}^2 + b\bar{x} + c \geq ax^2 + bx + c$, or equivalently, that $(\bar{x} - x)[a(\bar{x} + x) + b] \geq 0$. This is true if $\bar{x} - x = 0$, so assume that $\bar{x} - x \neq 0$. Then either $\bar{x} - x > 0$ or $\bar{x} - x < 0$. Assume, without loss of generality, that $\bar{x} - x > 0$. Because $a < 0$ and $\bar{x} = -b/(2a)$, it follows that $[a(\bar{x} + x) + b] > 0$, and the proof is complete.    $\square$

## 13.2    PROOF BY ELIMINATION

Suppose now that the keywords "either/or" arise in the conclusion of a proposition whose form is "$A$ implies $C$ *OR* $D$." With the forward-backward method, you assume $A$ is true and need to conclude that either $C$ is true or else $D$ is true. The only question is whether you should try to show that $C$ is true or whether you should try to show that $D$ is true. In some of these proofs, as you work forward, you might also encounter the keywords "either/or" in the form "either $A1$ or $A2$." Proceeding with a proof by cases, it might happen that, in the first case, when you assume $A1$ is true, you can work forward to establish that $C$ is true. Then, in the second case, when you assume that $A2$ is true, you can work forward to establish that $D$ is true, thus completing the proof. This approach is illustrated now.

**Proposition 23** *If* $x^2 - 5x + 6 \geq 0$, *then either* $\underbrace{x \leq 2}_{C}$ *or* $\underbrace{x \geq 3}_{D}$.

**Analysis of Proof.** Recognizing the keywords "either/or" in the conclusion, you can now turn to the forward process and see if the keywords "either/or" arise. Working forward from $A$ by factoring, it follows that

   **A1:** $(x-2)(x-3) \geq 0$.

The only way the product of the two real numbers in $A1$ can be $\geq 0$ is if

   **A2:** Either $x - 2 \leq 0$ and $x - 3 \leq 0$
   or $x - 2 \geq 0$ and $x - 3 \geq 0$.

Recognizing the keywords "either/or" in the forward statement $A2$, a proof by cases is appropriate. Accordingly, assume first that

**Case 1:** $x - 2 \leq 0$ and $x - 3 \leq 0$.
   In this case, it follows from the first inequality that $x \leq 2$, which establishes that statement $C$ in the conclusion is true. Now assume that

**Case 2:** $x - 2 \geq 0$ and $x - 3 \geq 0$.
   In this case, it follows from the second inequality that $x \geq 3$, which establishes that statement $D$ in the conclusion is true.
   The proof is now complete.

**Proof of Proposition 23.** From the hypothesis that $x^2 - 5x + 6 \geq 0$, either $x - 2 \leq 0$ and $x - 3 \leq 0$ or else $x - 2 \geq 0$ and $x - 3 \geq 0$. In the former case, $x \leq 2$, while in the latter case, $x \geq 3$, thus establishing the conclusion and completing the proof. $\square$

   An alternative approach to proving a proposition that contains the keywords "either/or" in the conclusion in the form "either $C$ OR $D$" is to use a method called a **proof by elimination**. Specifically, to eliminate the uncertainty of trying to determine whether to prove that $C$ is true or that $D$ is true, suppose you were to make the *additional assumption* that $C$ is not true. Clearly it had better turn out that, in this case, $D$ is true. Thus, with a proof by elimination, you assume that $A$ is true and $C$ is false; you must then conclude that $D$ is true, as is illustrated now with Proposition 23.

**Analysis of Proof.** Recognizing the keywords "either/or" in the conclusion, you can proceed with a proof by elimination. Thus, assume that

**A:**     $x^2 - 5x + 6 \leq 0$     and
**A1 (NOT C):**   $x > 2.$

It is your job to conclude that

**B1 (D):**   $x \geq 3.$

Working forward from $A$ by factoring, it follows that

**A2:** $(x - 2)(x - 3) \geq 0.$

Dividing both sides of $A2$ by $x - 2 > 0$ (see $A1$), you have

**A3:** $x - 3 \geq 0.$

Adding 3 to both sides of $A3$ yields $B1$, thus completing the proof.

**Proof of Proposition 23.**   Assume that $x^2 - 5x + 6 \geq 0$ and $x > 2$. It follows that $(x - 2)(x - 3) \geq 0$. Because $x > 2$, $x - 2 > 0$, and so it must be that $x \geq 3$, as desired.   □

Note that you can do a proof by elimination of "$A$ implies $C$ OR $D$" equally well by assuming that $A$ is true and $D$ is false and then concluding that $C$ is true. Try this approach with Proposition 23.

## 13.3   READING A PROOF

The process of reading and understanding a proof is demonstrated with the following proposition.

**Proposition 24** *If $p$ and $b$ are positive integers for which $p$ is prime and $p$ does not divide $b$, then the only positive integer that divides both $p$ and $b$ is 1.*

**Proof of Proposition 24.**   (For reference purposes, each sentence of the proof is written on a separate line.)

**S1:**   Clearly, 1 divides both $p$ and $b$.
**S2:**   To see that 1 is the only such integer, let $d > 0$ be an integer that divides both $b$ and $p$.
**S3:**   Thus, there is an integer $k$ such that $b = kd$.
**S4:**   Because $p$ is prime and $d$ divides $p$, it must be that $d = 1$ or $d = p$.
**S5:**   Now $d \neq p$, otherwise, $b = kp$ and $p$ would divide $b$.
**S6:**   It therefore follows that $d = 1$.

The proof is now complete.   □

**Analysis of Proof.**   An interpretation of statements $S1$ through $S6$ follows.

**Interpretation of S1:** *Clearly, 1 divides both p and b.*

The author recognizes the keywords "one and only one" in the conclusion and therefore uses a uniqueness method. Accordingly, the author first states that 1 divides both $p$ and $b$ (which is true because 1 divides every integer).

**Interpretation of S2:** *To see that 1 is the only such integer, let $d > 0$ be an integer that divides both b and p.*

This statement indicates that the author is using the direct uniqueness method and therefore assumes that

**A1:** $d > 0$ divides both $b$ and $p$.

To complete the direct uniqueness method, the author must show that the two objects, $d$ and 1, are the same; that is, that

**B1:** $d = 1$.

Indeed the author reaches this conclusion in $S6$.

**Interpretation of S3:** *Thus, there is an integer $k$ such that $b = kd$.*

The author is working forward by definition from the fact that $d$ divides $b$ (see $A1$) to claim that

**A2:** There is an integer $k$ such that $b = kd$.

**Interpretation of S4:** *Because p is prime and d divides p, it must be that $d = 1$ or $d = p$.*

The author is working forward by definition from the hypothesis that $p$ is prime, so,

**A3:** The only positive integers that divide $p$ are 1 and $p$.

The author rewords $A3$ to read

**A4:** For every positive integer $a$ that divides $p$, $a = 1$ or $a = p$.

In this form, the author recognizes the quantifier "for all" in the forward process and specializes $A4$ to $a = d > 0$, which does divide $p$ (see $A1$). The result of this specialization, as the author claims in $S4$, is

**A5:** $d = 1$ or $d = p$.

**Interpretation of S5:** *Now $d \neq p$, otherwise, $b = kp$ and $p$ would divide b.*

Recognizing the keywords "either/or" in $A5$, the author proceeds with a proof by cases and assumes first that $d = p$. However, the author shows by contradiction that this is impossible. Specifically, the author works forward by substituting $d = p$ into $A2$, which results in $b = kp$. By definition, this means that $p$ divides $b$, which contradicts the hypothesis that $p$ does not divide $d$.

**Interpretation of S6:** *It therefore follows that $d = 1$.*

Having ruled out the first case of $d = p$ in $A5$, the author now proceeds to the second case, which is that $d = 1$, as stated in $S6$. However, the fact that $d = 1$ completes the direct uniqueness method (see $B1$) and hence the proof.

## Summary

Use an either/or method when you encounter these keywords in the forward or backward process. A proof by cases is used in the forward process to show that "$C$ OR $D$ implies $B$." To do so you must do two proofs; that is,

Case 1: Prove that $C$ implies $B$.

Case 2: Prove that $D$ implies $B$.

A proof by elimination is used in the backward process to show that "$A$ implies $C$ OR $D$," as follows:

1. Assume that $A$ and $NOT$ $C$ are true.

2. Work forward from $A$ and $NOT$ $C$ to establish that $D$ is true.

3. Work backward from $D$.

(You could equally well assume that $A$ and $NOT$ $D$ are true, and work forward to prove that $C$ is true. You can also work backward from $C$ in this case.)

Observe that, with a proof by elimination, only one proof is needed to show that "$A$ implies $C$ OR $D$"—you can prove either "$A$ AND $(NOT$ $C)$ implies $D$," or "$A$ AND $(NOT$ $D)$ implies $C$"; either one of these two proofs by itself suffices. In contrast, with a proof by cases, two separate proofs are needed to show that "$C$ OR $D$ implies $B$"—you must prove both that "$C$ implies $B$" and that "$D$ implies $B$."

## Exercises

**Note:** Solutions to those exercises marked with a $B$ are in the back of this book. Solutions to those exercises marked with a $W$ are located on the web at http://www.wiley.com/college/solow/.

**Note:** All proofs should contain an analysis of proof and a condensed version. Definitions for all mathematical terms are provided in the glossary at the end of the book.

**13.1**    Rewrite the following statements so that the words "either/or" appear explicitly (where $a$ and $b$ are integers, $x$ and $y$ are real numbers, and $S$ and $T$ are sets).

   a.   $a|b$ and $b|a$ implies $a = \pm b$.      b.  $|x| > 3$.

   c.   $xy = 0$.                   d.  $z \in S \cup T$.

$^B$**13.2**   A student said, "I started to use a proof by cases to show that "$C$ OR $D$ implies $B$," but when I assumed that $C$ was true, instead of establishing that $B$ was true, I reached a contradiction. What does this mean?"

   a. Answer the student's question.

   b. Identify where this situation arises in the condensed proof of Proposition 6 on page 57.

$^B$**13.3**   Answer the following questions pertaining to the either/or methods.

   a. Describe how a proof by elimination is applied to prove a statement of the form, "If $A$, then $C$ OR $D$ OR $E$."

   b. Describe how a proof by cases is applied to prove a statement of the form, "If $C$ OR $D$ OR $E$, then $B$."

**13.4**   If the contradiction method is used on each of the following problems, what technique will you use to work forward from $NOT\ B$? Explain.

   a. $A$ implies $(C\ AND\ D)$.

   b. $A$ implies $[(NOT\ C)\ AND\ (NOT\ D)]$.

$^B$**13.5**   Explain where, why, and how an either/or method is used in the condensed proof of the proposition in Exercise 5.18 on page 65. Is this a proof by cases or by elimination?

$^W$**13.6**   Explain where, why, and how an either/or method is used in the condensed proof presented in Exercise 9.27 on page 111. Is this a proof by cases or by elimination?

**13.7**   Explain where, why, and how an either/or method is used in the proof of Proposition 16 on page 116. Is this a proof by cases or by elimination?

**13.8**   Explain where, why, and how an either/or method is used in the proof of Proposition 21 on page 136. Is this a proof by cases or by elimination?

**13.9**   Explain where, why, and how an either/or method is used in the following proof. Is this a proof by cases or by elimination?

      **Proposition.** If $i = \sqrt{-1}$ and $a + bi$ and $c + di$ are complex numbers for which $(a + bi)(c + di) = 1$, then $a \neq 0$ or $b \neq 0$.

      **Proof.** Because $(a + bi)(c + di) = 1$, $ac + adi + bci - bd = 1$. That is, $ac - bd = 1$ and $ad + bc = 0$. If $a = 0$, then $-bd = 1$. It then follows that $b \neq 0$ and so the proof is complete. $\square$

$^{B}$**13.10**     Consider the proposition, "If $x$ is a real number that satisfies the property that $x^3 + 3x^2 - 9x - 27 \geq 0$, then $|x| \geq 3$."

    a. Reword the proposition so that it is of the form "$A$ implies $C$ *OR* $D$."

    b. Prove the proposition by assuming that $A$ and *NOT* $C$ are true.

$^{W}$**13.11**     Prove the proposition in the previous exercise by assuming that $A$ and *NOT* $D$ are true.

$^{W}$**13.12**     Prove that, if $a$ and $b$ are integers for which $a|b$ and $b|a$, then $a = \pm b$.

*$^{*}$**13.13**     Write an analysis of proof that corresponds to the condensed proof given below. Indicate which techniques are used and how they are applied. Fill in the details of any missing steps where appropriate.

> **Proposition.** If $n$ is a positive integer, then either $n$ is prime, or $n$ is a square, or $n$ divides $(n-1)!$.
>
> **Proof.** If $n = 1$, then $n = 1^2$ is a square and the proposition is true. Similarly, if $n = 2$, then $n$ is prime and again the proposition is true. So suppose that $n > 2$ is neither prime nor a square. Because $n > 2$ is not prime, there are integers $a$ and $b$ with $1 < a < n$ and $1 < b < n$ such that $n = ab$. Also, because $n$ is not square, $a \neq b$. This means that $a$ and $b$ are integers with $2 \leq a \neq b \leq n-1$. That is, $a$ and $b$ are two different terms of $(n-1)(n-2)\cdots 1$. Thus, $ab = n$ divides $(n-1)!$. $\square$

$^{B}$**13.14**     Prove that, if $a$, $b$, and $c$ are integers for which $a|b$ or $a|c$, then $a|(bc)$.

**13.15**     Prove that, if $m$ and $n$ are integers, then either 4 divides $mn$ or else 4 does not divide $n$.

*$^{*}$**13.16**     Prove that, if $S$ and $T$ are subsets of a universal set $U$, then $(S \cap T)^c = S^c \cup T^c$, where $X^c = \{x \in U : x \notin X\}$. Indicate clearly where and why an either/or method arises.

*$^{*}$**13.17**     Answer the given questions about the following proof that, "If $n$ and $m$ are positive integers for which $n \leq 2m$, then either $n$ is prime or $m \geq \sqrt{n}$."

> **Proof.** Assume that $n$ is not prime. Then there are integers $a$ and $b$ with $1 < a, b < n$ such that $n = ab$. Now it is easy to show that $a$ and $b \leq n/2$. Thus, $n = ab \leq n^2/4 \leq m^2$. The proof is completed by taking the positive square root of both sides of the foregoing inequality. $\square$

    a. What proof technique is the author using in the first sentence and why?

b. Justify the third sentence by proving that, if $c$ and $n$ are positive integers with $c < n$ and $c|n$, then $c \le n/2$. (You need not provide an analysis of proof.)

c. Justify the inequality $n^2/4 \le m^2$ in the fourth sentence.

d. Justify the statement in the last sentence and explain why the author is correct in claiming that the proof is complete.

*13.18    Answer the given questions about the following proof that, "If $a$, $b$, and $p$ are integers for which $p$ is prime and $p|(ab)$, then $p|a$ or $p|b$."

> **Proof.** Assume that $p$ does not divide $a$. Then, because $p$ is prime, by a previously proved result, it follows that there are integers $m$ and $n$ such that $mp + na = 1$. Multiplying through by $b$ yields $mpb + nab = b$. Furthermore, because $p|(ab)$, there is an integer $c$ such that $ab = cp$. You now have $b = mpb + ncp = (mb + nc)p$. But this means that $p|b$ and so the proof is complete.  □

a. What proof technique is the author using in the first sentence and why? What statement should the author show to complete the proof and where does the author do this?

b. In the second sentence, the author is using previous knowledge. State this previous proposition in an "if...then..." form using symbols that are different from the ones in the proposition here.

c. Justify the equality $b = mpb + ncp$ in the fifth sentence.

d. Justify the statement in the last sentence and explain why the author is correct in claiming that the proof is complete.

# 14

## The Max/Min Methods

The final techniques you will learn are the **max/min methods** that arise in problems dealing with the smallest and largest element of a set of real numbers. Observe that some sets of real numbers do not have a smallest and largest element, such as {real numbers $s : 0 < s < 1$}. However, for the moment, suppose that $S$ is a nonempty set of real numbers having both a smallest member, say $x$, and a largest member, say $y$. In this case one writes $x = \min\{s : s \in S\}$ and $y = \max\{s : s \in S\}$ which, by definition, means that $x$ and $y$ are elements of $S$ such that, for every element $s \in S$, $s \geq x$ and $s \leq y$.

## 14.1 HOW TO USE THE MAX/MIN METHODS

For some sets, such as {real numbers $s : 0 \leq s \leq 1$}, it is easy to identify the smallest element ($x = 0$) and the largest element ($y = 1$). In other cases, you might have to prove, for example, that a particular element $x \in S$ is or is not the smallest element of $S$. According to the definition,

$x = \min\{s : s \in S\}$ if and only if for every element $s \in S$, $s \geq x$.

Likewise, negating the foregoing definition using the techniques in Chapter 8,

$x \neq \min\{s : s \in S\}$ if and only if there is an element $s \in S$ such that $s < x$.

Thus, you can see that working with the minimum (and maximum) of a set involves working with the quantifiers *there is* and *for all*. Indeed, the

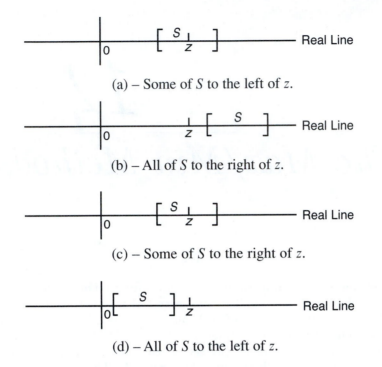

Fig. 14.1   Possible positions of a set $S$ relative to a real number $z$.

idea behind the max/min methods is to convert the given statement into an equivalent one containing a quantifier—then you can use the appropriate choose, construction, or specialization method.

To see how this conversion is done, suppose that, for a given real number $z$, you are interested in the position of the set $S$ relative to the number $z$. For instance, you might want to prove one of the following statements:

| Problem Statement | Math Statement | Associated Fig. |
|---|---|---|
| 1. Some of $S$ is to the left of $z$. | $\min\{s : s \in S\} \leq z$ | Figure 14.1(a) |
| 2. All of $S$ is to the right of $z$. | $\min\{s : s \in S\} \geq z$ | Figure 14.1(b) |
| 3. Some of $S$ is to the right of $z$. | $\max\{s : s \in S\} \geq z$ | Figure 14.1(c) |
| 4. All of $S$ is to the left of $z$. | $\max\{s : s \in S\} \leq z$ | Figure 14.1(d) |

The technique associated with converting the first two to statements with quantifiers is discussed here and the remaining two are left as exercises.

Consider, therefore, the problem of trying to show that the smallest member of $S$ is $\leq z$. An equivalent problem containing a quantifier is obtained by considering the foregoing statement 1. Because *some* of $S$ should be to the left of $z$, you need to show that there is an elements $s \in S$ such that $s \leq z$, which can be done with the construction method, as is illustrated now.

**Proposition 25** *If $a$, $b$, and $c$ are real numbers with $a > 0$ and $b \neq 0$, then* $\min\{ax^2 + bx + c : x \text{ is a real number}\} < c$.

**Analysis of Proof.**  From the form of $B$, the max/min method is used. According to the foregoing discussion, you can convert the statement $B$ to,

**B1:** There is a real number $x$ such that $ax^2 + bx + c < c$,

or equivalently, by subtracting $c$ from both sides of the inequality, you must show that

**B2:** There is a real number $x$ such that $ax^2 + bx = x(ax + b) < 0$.

Once in this form, it is clear that you should use the construction method to produce the real number $x$.

Turning to the forward process, you know that $b \neq 0$ and so,

**A1:** Either $b < 0$ or $b > 0$.

Recognizing the keywords "either/or" in the forward process, a proof by cases is appropriate. Accordingly,

**Case 1:** Assume that $b < 0$.
   In this case, because $a > 0$ from the hypothesis,

**A2:** $-b/a > 0$.

So, constructing $x$ as any real number with $-b/a > x > 0$, $B2$ is true because

**A3:** $x(ax + b) < 0$ (as $x > 0$ and $ax + b < 0$).

**Case 2:** Assume that $b > 0$.
   In this case, because $a > 0$ from the hypothesis,

**A4:** $-b/a < 0$.

So, constructing $x$ as any real number with $-b/a < x < 0$, $B2$ is true because

**A5:** $x(ax + b) < 0$ (as $x < 0$ and $ax + b > 0$).

This completes the proof by cases and the entire proof as well.

**Proof of Proposition 25.** It will be shown that there is a real number $x$ such that $ax^2 + bx + c < c$, for then, $\min\{ay^2 + by + c : y \text{ is a real number}\} < ax^2 + bx + c < c$ (this is the max/min method). To that end, assume, without loss of generality, that $b < 0$ (here is where a proof by cases is used). Then because $a > 0$, any real number $x$ for which $-b/a > x > 0$ satisfies $ax^2 + bx + c < c$ and so the proof is complete. $\square$

Turning now to the problem of showing that the smallest member of $S$ is $\geq z$, the approach is slightly different. To proceed, consider the foregoing statement 2. Because *all* of $S$ should be to the right of $z$, an equivalent problem is to show that, for every element $s \in S$, $s \geq z$. The choose method is used to do so, as illustrated in the next section.

## 14.2    READING A PROOF

As you know, some sets of numbers do not have a largest and/or smallest element. One condition when a subset of the **natural numbers**—denoted by $N = \{\text{integers } n : n > 0\}$—does have a smallest element is given in the following axiom (recall from Chapter 3 that an axiom is a statement that is accepted as being true without a supporting proof):

**The Least Integer Principle**—Every nonempty set of positive integers has a least element; that is, if $S$ is a nonempty subset of $N$, then there is an element $x \in S$ such that, for every element $s \in S$, $s \geq x$.

The use of the Least Integer Principle is illustrated in the following proof. Can you identify where the max/min method arises?

**Proposition 26** *If $x < y$ are positive real numbers, then there is a rational number $r$ such that $x < r < y$.*

**Proof of Proposition 26.** (For reference purposes, each sentence of the proof is written on a separate line.)

**S1:**    Let $n$ be a positive integer such that $n(y - x) > 1$.

**S2:**    Now consider the set $T = \{\text{integers } k > 0 : k > nx\}$.

**S3:**    It is clear that $T \neq \emptyset$ and so, by the Least Integer Principle, $T$ has a least element, say $m$.

**S4:**    As $m \in T$, $m > nx$ and it is now shown that $m < ny$.

**S5:**    Suppose, to the contrary, that $m \geq ny > 1 + nx > 1$.

**S6:**    But then $m - 1 \in T$ because $m - 1 \geq ny - 1 > nx > 0$.

**S7:**    The fact that $m - 1 < m$ contradicts $m$ being the least element of $T$.

**S8:**    It now follows that $nx < m < ny$ and so $r = m/n$ satisfies $x < r < y$ and, as such, $r$ is the desired rational number.

The proof is now complete. $\square$

**Analysis of Proof.**    An interpretation of statements $S1$ through $S8$ follows.

**Interpretation of S1:** *Let $n$ be a positive integer such that $n(y - x) > 1$.*

The author is creating an integer $n$ with the property that $n(y - x) > 1$, which is possible because $y - x > 0$ and so any integer $> \frac{1}{y-x}$ will do. What is not clear is why the author is doing this. To learn the answer, ask yourself what proof technique you would use to begin this proof. Recognizing the keywords "there is" in the conclusion, the author should use the construction method to construct a rational number $r$ with $x < r < y$. If this is so, then the integer $n$ in statement $S1$ should be used to construct the rational number $r$, which, according to the definition, requires constructing an integer numerator and nonzero integer denominator. Indeed, as indicated in $S8$, $n$ is the denominator of $r$. It remains to construct the integer numerator $m$ and show that $n$ and $m$ satisfy the needed properties that $n \neq 0$ (where does the author do this?) and $x < r = m/n < y$, which is what statements $S2$ through $S7$ are about.

**Interpretation of S2:** *Now consider the set $T = \{integers\ k > 0 : k > nx\}$.*

The author has defined a set $T$ of positive integers that, hopefully, will be used to construct the integer numerator of the rational number $r$.

**Interpretation of S3:** *It is clear that $T \neq \emptyset$ and so, by the Least Integer Principle, $T$ has a least element, say $m$.*

The author is now working forward from the previous knowledge of the Least Integer Principle. In so doing, the author recognizes the quantifier "for all" in the forward process and, as such, specializes the for-all statement in the Least Integer Principle to the specific set $T$. To do so, however, the author must verify that $T$ satisfies the certain property of being nonempty in that for-all statement. Indeed, the author mentions that $T$ is nonempty, which is true because any integer $> nx$ is in $T$. The result of specialization is the least integer $m$ of the set $T$. As seen in statement $S8$, $m$ is the numerator of the rational number $r$; that is, in $S8$, the author constructs $r = m/n$. According to the construction method, the author must show that $x < r < y$, which is what the remaining statements $S4$ through $S7$ are about.

**Interpretation of S4:** *As $m \in T$, $m > nx$ and it is now shown that $m < ny$.*

The fact that $m \in T$ is true because $m$ is the least integer of $T$, which, by definition, means that $m \in T$. By the defining property of $T$ (see $S2$), this in turn means that $m > nx$, as the author states in $S4$. It is not clear why the author then says that "...it will be shown that $m < ny$." The answer is in $S8$ because, if indeed $nx < m < ny$, then dividing through by $n > 0$ yields the desired conclusion that $x < m/n = r < y$, as the author notes in $S8$.

**Interpretation of S5:** *Suppose, to the contrary, that $m \geq ny > 1 + nx > 1$.*

The author is using the contradiction method to show that $m < ny$ and therefore correctly assumes that $m \geq ny$. The author then notes that $ny > 1 + nx$, which follows by simple algebra from statement $S1$. The final inequality that $1 + nx > 1$ is true because $nx > 0$. The foregoing inequalities should eventually be used to reach a contradiction. Can you identify the contradiction?

**Interpretation of S6:** *But then $m - 1 \in T$ because $m - 1 \geq ny - 1 > nx > 0$.*

The author is showing that $m - 1 \in T$, again hopefully to reach a contradiction. To show that $m - 1 \in T$, the author verifies that $m - 1$ satisfies the defining property of $T$, namely, that $m - 1 > 0$ and $m - 1 > nx$. To do this, the author notes that $m - 1 \geq ny - 1$, which is true because $m \geq ny$ (see $S5$). Finally, $ny - 1 > nx$ by applying algebra to $S1$ and the author notes that $nx > 0$.

**Interpretation of S7:** *The fact that $m - 1 < m$ contradicts $m$ being the least element of $T$.*

Here, at last, the author reaches a contradiction and this is accomplished with the max/min method. Specifically, by the max/min method, because $m = \min\{t : t \in T\}$, it follows that, for every element $t \in T$, $t \geq m$. The author has shown that this statement is not true—that is, that there is an element $t \in T$ such that $t < m$. Indeed, the author has used the construction method to produce the integer $t = m - 1 \in T$ (see $S6$), which is clearly less than $m$. This contradiction establishes that $m < ny$.

**Interpretation of S8:** *It now follows that $nx < m < ny$ and so $r = m/n$ satisfies $x < r < y$ and, as such, $r$ is the desired rational number.*

The author now completes the construction of the rational number $r$, namely, $r = m/n$. However, the author must show that this value of $r$ is correct—that is, that $n \neq 0$ and that $x < r < y$. The author actually mentions in $S1$ that $n$ is positive, which is true because $n$ is chosen this way, and so $n \neq 0$. To see that $x < r < y$, the author has established that $nx < m < ny$ from $S4$ and the subsequent proof by contradiction in $S5$, $S6$, and $S7$. As stated in $S8$, dividing through by $n > 0$, it follows that $x < m/n = r < y$, and so indeed the proof is complete.

## Summary

Use a max/min method when you need to show that the largest or smallest element of a set is $\leq$ or $\geq$ some fixed number. To do so, convert the statement into an equivalent statement containing a quantifier and then apply the choose, construction, or specialization method, whichever is appropriate, based on whether the quantifier appears in the forward or backward process.

## Exercises

**Note:** Solutions to those exercises marked with a $B$ are in the back of this book. Solutions to those exercises marked with a $W$ are located on the web at http://www.wiley.com/college/solow/.

**Note:** All proofs should contain an analysis of proof and a condensed version. Definitions for all mathematical terms are provided in the glossary at the end of the book.

$^B$**14.1**    Convert the following max/min problems to an equivalent statement containing a quantifier, in which $S$ is a set of real numbers and $z$ is a given real number.

    a. $\max\{s : s \in S\} \leq z$.

    b. $\max\{s : s \in S\} \geq z$.

**14.2**    Convert the following max/min problems to an equivalent statement containing a quantifier, in which $a$, $b$, $c$, and $u$ are given real numbers, and $x$ is a variable.

    a. $\min\{cx : ax \leq b \text{ and } x \geq 0\} \leq u$.

    b. $\max\{cx : ax \leq b \text{ and } x \geq 0\} \geq u$.

    c. $\min\{ax : b \leq x \leq c\} \geq u$.

    d. $\max\{ax : b \leq x \leq c\} \leq u$.

**14.3**    Convert each of the following statements into an equivalent statement having a quantifier.

    a. The maximum of a function $f$ over all real numbers $x$ with $0 \leq x \leq 1$ is less than or equal to the real number $y$.

    b. The minimum of a function $f$ over all real numbers $x$ with $0 \leq x \leq 1$ is less than or equal to the real number $y$.

$^B$**14.4**    Prove that $\min\{x(x-2) : x \text{ is a real number}\} \geq -1$.

**14.5**    Prove that $\max\{\text{real numbers } x : x \leq 2^{-x}\} \geq 0.5$.

**14.6**    Prove that, if $S$ and $T$ are sets of real numbers such that $S$ has a smallest element, $S \subset T$, and $t^*$ is a real number such that, for each element $t \in T$, $t \geq t^*$, then $\min\{s : s \in S\} \geq t^*$.

$^W$**14.7**    Suppose that $a$, $b$, and $c$ are given real numbers and that $x$ and $u$ are variables. Prove that $\min\{cx : ax \geq b, x \geq 0\}$ is at least as large as $\max\{ub : ua \leq c, u \geq 0\}$.

**\*14.8**    Suppose that $S$ and $T$ are subsets of real numbers for which $S$ has a smallest element and $T$ has a largest element. Prove that, if $S \cap T \neq \emptyset$, then $\max\{t : t \in T\} \geq \min\{s : s \in S\}$.

**\*14.9**    Prove that $\min\{s : s \in S\} = -\max\{-s : s \in S\}$, where $S$ is a set of real numbers. (Hint: To show that the two numbers $\min\{s : s \in S\}$ and $-\max\{-s : s \in S\}$ are equal, show that the first number is $\leq$ the second number and vice versa. Then use the max/min methods to do so.)

**14.10**    Answer the given questions about the following proof that, if $a$, $b$, and $c$ are real numbers with $a < 0$, then $\max\{ax^2 + bx + c : x \text{ is a real number}\} \leq (4ac - b^2)/(4a)$.

> **Proof.** Let $x$ be a real number. It then follows that
>
> $$ax^2 + bx + c = a\left(x + \frac{b}{2a}\right)^2 + \frac{4ac - b^2}{4a} \leq \frac{4ac - b^2}{4a}.$$
>
> The proof is now complete.  □

a. What proof techniques are used in the first sentence and why did the author use those techniques?

b. Justify the inequality in the second sentence.

**\*14.11**    Answer the given questions about the following proof that induction works. That is, suppose that $P(n)$ is some statement that depends on the integer $n$ and consider the following proof that, if $P(1)$ is true and for all $n \geq 1$, $P(n)$ implies $P(n+1)$, then for all integers $n \geq 1$, $P(n)$ is true.

> **Proof.** Suppose to the contrary that $n \geq 1$ is an integer for which $P(n)$ is false. Let $T = \{\text{integers } m \geq 1 : P(m) \text{ is false}\}$. Note that $T \neq \emptyset$ and so, by the Least Integer Principle, $T$ has a smallest element, say $k$. Now $k \geq 2$ and $P(k)$ is false. But then, from the hypothesis, it follows that $P(k-1)$ is false. This means that $k - 1 \in T$, which contradicts the fact that $k$ is the smallest element of $T$ and this completes the proof.  □

a. The first sentence indicates that this is a proof by contradiction. What is the contradiction?

b. Justify the statement that $T \neq \emptyset$ in the third sentence.

c. Why is the author justified in claiming that $k \geq 2$ in the fourth sentence?

d. Why does the author need $k \geq 2$ instead of $k \geq 1$? Explain.

e. Justify the fifth sentence. What proof technique has the author used?

# 15

---

# *Summary*

The list of proof techniques is now complete. The techniques presented here are not the only ones, but they do constitute the basic set. You will come across others as you are exposed to more mathematics—perhaps you will develop some of your own. In any event, there are many fine points and tricks that you will pick up with experience. A final summary of how and when to use each of the various techniques for proving the proposition "*A* implies *B*" is in order.

## 15.1  THE FORWARD-BACKWARD METHOD

With the forward-backward method, you assume that $A$ is true and your job is to prove that $B$ is true. Through the forward process, you derive from $A$ a finite sequence of statements, $A1, A2, \ldots, An$, that are necessarily true as a result of assuming that $A$ is true. This sequence is guided by the backward process whereby, through asking and answering the key question, you derive from $B$ a new statement, $B1$, with the property that, if $B1$ is true, then so is $B$. This backward process is then applied to $B1$, resulting in a new statement, $B2$, and so on. The objective is to link the forward sequence to the backward sequence by generating a statement in the forward sequence that is the same as the last statement obtained in the backward sequence. Then, like a column of dominoes, you can do the proof by going forward along the sequence from $A$ all the way to $B$.

## 15.2   THE CONSTRUCTION METHOD

When obtaining the sequence of statements, watch for quantifiers to appear, for then the construction, choose, and/or specialization methods are likely to be useful in doing the proof. For instance, when the quantifier "there is" arises in the backward process in the standard form:

> There is an "object" with a "certain property" such that "something happens,"

consider using the construction method to produce the desired object. With the construction method, you work forward from the assumption that $A$ is true to construct (produce, or devise an algorithm to produce, and so on) the object. However, the actual proof consists of showing that the object you construct satisfies the certain property and also that the something happens.

## 15.3   THE CHOOSE METHOD

When the quantifier "for all" arises in the backward process in the standard form:

> For all "objects" with a "certain property," "something happens,"

consider using the choose method. Here, your objective is to design a model proof for establishing that the something happens for a general object with the certain property. If successful, then you could, in theory, repeat this proof for each object with the certain property. You therefore choose an object that has the certain property. You must conclude that, for the chosen object, the something happens. Once you choose the object, work forward from the fact that the chosen object has the certain property (together with the information in $A$) and backward from the something that happens.

## 15.4   THE SPECIALIZATION METHOD

When the quantifier "for all" arises in the forward process in the standard form:

> For all "objects" with a "certain property," "something happens,"

you will probably want to use the specialization method. To do so, watch for one of these objects to arise, often in the backward process. By using specialization, you can then conclude, as a new statement in the forward process, that the something does happen for that particular object. That fact should then be helpful in reaching the conclusion that $B$ is true. When using specialization, be sure to verify that the particular object does satisfy the certain property, for only then does the something happen.

If statements contain more than one quantifier—that is, they are nested—process them in the order in which they appear from left to right. As you read the first quantifier in the statement, identify its objects, certain property, and something that happens. Then apply an appropriate technique based on whether the statement is in the forward or backward process, and whether the quantifier is "for all" or "there is." This process is repeated until all quantifiers are dealt with.

## 15.5   THE CONTRADICTION METHOD

When the statement $B$ contains the keyword "no" or "not," or when the forward-backward method fails, you should consider the contradiction method. With this approach, you assume that $A$ is true and $B$ is false. This gives you two facts from which you must derive a contradiction to something that you know to be true. Where the contradiction arises is not always obvious but is obtained by working forward from the statements $A$ and $NOT$ $B$.

## 15.6   THE CONTRAPOSITIVE METHOD

In the event that the contradiction method fails, there is still hope with the contrapositive method. To use the contrapositive approach, write the statements $NOT$ $B$ and $NOT$ $A$ using the techniques of Chapter 8. Then, by beginning with the assumption that $NOT$ $B$ is true, your job is to conclude that $NOT$ $A$ is true. This is best accomplished with the forward-backward method, working forward from $NOT$ $B$ and backward from $NOT$ $A$. Remember to watch for quantifiers to appear in the forward and backward processes, for if they do, then the corresponding construction, choose, and/or specialization methods may be useful.

## 15.7   THE UNIQUENESS METHODS

Use a uniqueness method when you come across a statement in the form, "there is a unique object (or one and only one object, or exactly one object) with a certain property such that something happens." When this statement occurs in the forward process, with the forward uniqueness method you

1. Assume that there is an object $X$ with the certain property and for which the something happens.

2. Look for another object $Y$ with that certain property and for which that something happens. You can then write, as a new forward statement, that $X$ and $Y$ are the same; that is, that $X = Y$. This statement should then help you establish that the conclusion $B$ is true.

Use the backward uniqueness method when you need to show that there is a unique object with a certain property such that something happens. Doing so requires two steps: first showing that there is one such object, say $X$, and then showing that there is *only* one such object. While the first task is accomplished with the construction or contradiction method, you can accomplish the second task in one of two ways. With the direct uniqueness method you (1) assume that, in addition to the object $X$, $Y$ is also an object with the certain property and for which the something happens and (2) use the properties of $X$ and $Y$ together with the hypothesis $A$ to show that $X$ and $Y$ are the same (that is, that $X = Y$). With the indirect uniqueness method you (1) assume that $Y$ is a different object from $X$ with the certain property and for which the something happens and (2) use the properties of $X$ and $Y$, the fact that they are different, and the hypothesis $A$ to reach a contradiction.

## 15.8   THE INDUCTION METHOD

Consider using the induction method (even before the choose method) when the statement $B$ has the form:

> For every integer $n \geq$ some initial integer, a statement $P(n)$ is true.

The first step of induction is to verify that $P(n)$ is true for the first possible value of $n$. The second step requires you to show that, if $P(n)$ is true, then $P(n+1)$ is true. The success of a proof by induction rests on your ability to relate $P(n+1)$ to $P(n)$ so that you can use the assumption that $P(n)$ is true. In other words, to perform the second step of induction, write the statement $P(n)$, replace $n$ everywhere by $n+1$ to obtain $P(n+1)$, and then see if you can express $P(n+1)$ in terms of $P(n)$. Only then will you be able to use the assumption that $P(n)$ is true to reach the conclusion that $P(n+1)$ is true.

## 15.9   EITHER/OR METHODS

Use a proof by cases when the keywords "either/or" arise in the forward process in the form, "If $C$ OR $D$, then $B$." Two proofs are required. In the first case you assume that $C$ is true and then prove that $B$ is true; in the second case you assume that $D$ is true and then prove that $B$ is true.

Use a proof by elimination when the keywords "either/or" arise in the backward process in the form, "If $A$, then $C$ OR $D$." To do so, assume that $A$ is true and $C$ is not true (that is, $A$ and $NOT$ $C$); you should then show that $D$ is true. This is best accomplished with the forward-backward method. Alternatively, you can assume that $A$ and $NOT$ $D$ are true; in this case you have to show that $C$ is true.

## 15.10   THE MAX/MIN METHODS

When a statement indicates that the smallest (largest) element of a set of real numbers is less (greater) than or equal to a particular real number, say $z$, then you should use a max/min method. Doing so involves rewriting the statement in an equivalent form using the quantifier "for all" or "there is," whichever is appropriate. Once in this form, you can then apply the choose, construction, or specialization method.

### How to Read and Do Proofs

A final summary of how to read and do proofs is in order.

### How to Read a Proof

A written proof is nothing more than a sequence of applications of the individual techniques you have now learned. However, due to the way in which they are written, there are three reasons why proofs are challenging to read:

1. The author does not always refer to the techniques by name.

2. Several steps are combined in a single sentence with little or no justification.

3. The steps of a proof are not necessarily presented in the order in which they were performed when the proof was done.

To read a proof, you have to reconstruct the author's thought processes. Doing so requires that you identify which techniques are used and how they apply to the particular problem. Begin by trying to determine what technique is used to start the proof. Then try to follow the methodology associated with that technique. Watch for quantifiers to appear, for then the author is likely to use the corresponding choose, induction, construction, and/or specialization methods. The inability to follow a particular step of a written proof is often due to the lack of sufficient detail. To fill in the gaps, learn to ask yourself how you would proceed to do the proof. Then try to see if the written proof matches your thought process.

The following is a final example of how to read a proof, where it is shown that, when you divide an integer $b$ by an integer $a \geq 1$, you get a unique whole number and a unique remainder. The subsequent analysis-of-proof provides an explanation of which techniques are used and how they are applied to the specific problem.

**Proposition 27** *If $a$ and $b$ are integers with $a \geq 1$, then there are unique integers $q$ and $r$ such that $b = aq + r$, where $0 \leq r < a$.*

**Proof of Proposition 27.**    (For reference purposes, each sentence of the proof is written on a separate line.)

**S1:**    The values of $q$ and $r$ are obtained from the least element of the following set of integers: $M = \{\text{integers } w \geq 0 :$ there is an integer $k$ such that $w = b - ak\}$.

**S2:**    Now if $b \geq 0$, then for $k = 0$, you have $w = b \in M$.

**S3:**    While if $b < 0$, then for $k = b$, you have $w = b - ak = b - ab = b(1 - a) \geq 0$ as $b < 0$ and $1 - a \leq 0$, so $w \in M$.

**S4:**    Now if $0 \in M$, then 0 is the least element of $M$; otherwise, $M$ has a least element by the Least-Integer Principle and so, in either case, let $r$ be the least element of $M$.

**S5:**    Because $r \in M$, (1) $r \geq 0$ and (2) there is an integer $q$ such that $r = b - aq$, or equivalently, $b = aq + r$.

**S6:**    To see that $r < a$, assume that $r \geq a$.

**S7:**    Then $w = b - a(q + 1)$ is a smaller element of $M$ than $r$.

**S8:**    You can see that $w \in M$ because, for $k = q + 1$, you have $w = b - ak = b - a(q + 1) = b - aq - a = r - a \geq 0$.

**S9:**    Also, $w < r$ because $w = b - a(q + 1) = b - aq - a < b - aq = r$.

**S10:**    It remains to show that these values of $q$ and $r$ are unique, so, suppose that $m$ and $n$ are also integers for which $b = am + n$ and $0 \leq n < a$.

**S11:**    But then, $a(q - m) = n - r$ and $-a < n - r < a$.

**S12:**    From this you have $-a < a(q - m) < a$ and so $q - m = 0$.

**S13:**    Finally, substituting $m = q$ in $a(q - m) = n - r$ yields $n - r = 0$, that is, $n = r$.

The proof is now complete.    $\square$

**Analysis of Proof.**    An interpretation of statements $S1$ through $S13$ follows.

**Interpretation of S1:**    *The values of $q$ and $r$ are obtained from the least element of the following set of integers: $M = \{$integers $w \geq 0$ : there is an integer $k$ such that $w = b - ak \}$.*

The author recognizes the keyword "unique" in the conclusion and is therefore using a uniqueness method (is it the direct or indirect method?). As such, the first step is to construct the integers $q$ and $r$, which the author states are going to come from the least element of the set $M$. Thus, the author is using previous knowledge of the Least Integer Principle (see page 156).

**Interpretation of S2:**    *Now if $b \geq 0$, then for $k = 0$, you have $w = b \in M$.*

To use the Least Integer Principle it must be shown that $M \neq \emptyset$, which is what the author is doing in $S1$, when $b \geq 0$ and in $S2$, when $b < 0$. Specifically, because either $b \geq 0$ or $b < 0$, the author uses a proof by cases to show that $M \neq \emptyset$. In the first case, when $b \geq 0$, the author constructs $k = 0$ in $S1$ to show that $w = b \geq 0$ satisfies the defining property of $M$.

**Interpretation of S3:** *While if $b < 0$, then for $k = b$, you have $w = b - ak = b - ab = b(1 - a) \geq 0$ as $b < 0$ and $1 - a \leq 0$, so $w \in M$.*

The author is completing the proof by cases. Specifically, for the case when $b < 0$, the author shows that $w = b - ab$ satisfies the defining property of $M$. To do so, the author constructs $k = b$ and shows that $w = b - ak = b - ab = b(1 - a) \geq 0$, the last inequality being true because $b < 0$ and $a \geq 1$ from the hypothesis. Thus, from $S1$ and $S2$, the author has shown that $M \neq \emptyset$.

**Interpretation of S4:** *Now if $0 \in M$, then $0$ is the least element of $M$; otherwise, $M$ has a least element by the Least-Integer Principle and so, in either case, let $r$ be the least element of $M$.*

Note that to use the Least Integer Principle requires that the set consist of strictly positive integers. The author mentally observes that the set $M$ consists of nonnegative integers and therefore works forward to note that either $0 \in M$ or $0 \notin M$. Proceeding with a proof by cases, in the first case, when $0 \in M$, the author states that $0$ is the least integer of $M$, which is true because all elements of $M$ are $\geq 0$. In the second case, when $0 \notin M$, the author uses the Least Integer Principle to claim that $M$ has a least element, which is true because, in this case, $M$ consists of strictly positive integers and it has already been shown in $S2$ and $S3$ that $M \neq \emptyset$. Thus, in either case, $M$ has a smallest element, $r$. Here is where the author has finally constructed the integer $r$. It remains to construct the integer $q$, which the author does next.

**Interpretation of S5:** *Because $r \in M$, (1) $r \geq 0$ and (2) there is an integer $q$ such that $r = b - aq$, or equivalently, $b = aq + r$.*

The author uses the defining property of $M$ and the fact that $r \in M$ to claim that there is an integer $q$ such that $r = b - aq$, or equivalently, $b = aq + r$. The author has now constructed the integers $r$ in $S4$ and $q$ in $S5$. According to the construction method, the author must show that these values of $r$ and $q$ are correct—that is, that $0 \leq r < a$. The first inequality is true by the defining property of $M$, as stated in $S5$. The fact that $r < a$ is shown next.

**Interpretation of S6:** *To see that $r < a$, assume that $r \geq a$.*

The author is using the contradiction method to show that $r < a$ and, accordingly, assumes that this is not true—that is, that $r \geq a$.

**Interpretation of S7:** *Then $w = b - a(q + 1)$ is a smaller element of $M$ than $r$.*

The author is making a claim that, if true, is in fact a contradiction. Specifically, from $S4$, the author has shown that $r$ is the smallest element of $M$. However, if the author's claim in $S7$ is true, then $w$ would be a smaller element of $M$ than $r$, which cannot happen. The author must therefore show that $w = b - a(q + 1) \in M$ (which is done in $S8$) and that $w < r$ (which is done in $S9$).

**Interpretation of S8:** *You can see that $w \in M$ because, for $k = q + 1$, you have $w = b - ak = b - a(q + 1) = b - aq - a = r - a \geq 0$.*

The author is showing that $w \in M$ by using the defining property of $M$. Thus, the author constructs the integer $k = q + 1$ and shows, by algebra and the fact that $r = b - aq$ from $S5$, that $w = b - ak \geq 0$.

**Interpretation of S9:** *Also, $w < r$ because $w = b - a(q + 1) = b - aq - a < b - aq = r$.*

The author is showing that $w < r$ by algebra and the fact that $r = b - aq$ from $S5$. The author has now shown that $w \in M$ in $S8$ and $w < r$ in $S9$. This establishes the contradiction that $r$ is not the smallest element of $M$, as stated in $S4$. This contradiction means that $r < a$ and so the author has now completed the construction method by establishing that the constructed values of $q$ and $r$ satisfy the desired properties that $b = aq + r$ (see $S5$) and $0 \leq r < a$ (see $S5$ and $S6$).

**Interpretation of S10:** *It remains to show that these values of $q$ and $r$ are unique, so, suppose that $m$ and $n$ are also integers for which $b = am + n$ and $0 \leq n < a$.*

The author is now completing the proof by using the direct uniqueness method to show that $q$ and $r$ are unique. Thus, the author assumes that $m$ and $n$ are also integers that satisfy the same properties as $q$ and $r$, that is, that $b = am + n$ and $0 \leq n < a$. According to the direct uniqueness method, the author must show that $m$ and $n$ are the same as $q$ and $r$, that is, that $m = q$ and $n = r$. This is accomplished in $S11$, $S12$, and $S13$.

**Interpretation of S11:** *But then, $a(q - m) = n - r$ and $-a < n - r < a$.*

The author applies algebra to the properties of these integers. Specifically, subtracting $b = am + n$ from $b = aq + r$ and bringing the term $r - n$ to the other side leads to $a(q - m) = n - r$. To see that $-a < n - r$, the author uses the fact that $0 \leq n$ and $r < a$. Specifically, $0 \leq n$ and $-a < -r$, so adding these two inequalities results in $-a < n - r$. Similarly, because $0 \leq r$ and $n < a$, you have $-r \leq 0$ and $n < a$, so adding these two inequalities results in $n - r < a$.

**Interpretation of S12:** *From this you have $-a < a(q - m) < a$ and so $q - m = 0$.*

In $S11$, the author replaces the expression $n - r$ in the inequalities with $a(q - m)$ to obtain $-a < a(q - m) < a$. The author then mentally (1) divides the foregoing inequality through by $a > 0$ to obtain $-1 < q - m < 1$ and (2) notes that the only integer strictly between $-1$ and $1$ is $0$ and hence concludes that $q - m = 0$.

**Interpretation of S13:** *Finally, substituting $m = q$ in $a(q - m) = n - r$ yields $n - r = 0$, that is, $n = r$.*

The author uses the fact that $q - m = 0$ from $S12$ to state that $m = q$ and then substitutes that into $a(q - m) = n - r$ (see $S11$) to obtain $n - r = 0$, which means that $n = r$. This together with the fact that $m = q$ obtained in $S12$ completes the direct uniqueness method and hence the proof.

## How to Read a Proof

When trying to prove that "$A$ implies $B$," let the form of $A$ and $B$ guide you as much as possible. For example, you should scan the statement $B$ for certain keywords, which often indicate how to proceed. If you come across the quantifier "there is," then consider the construction method, whereas the quantifier "for all" suggests using the choose or induction method. When the statement $B$ contains the word "no" or "not," you will probably want to use the contrapositive or contradiction method. Other keywords to look for are "unique," "either/or," and "maximum" and "minimum," for then the corresponding uniqueness, either/or, and max/min methods are appropriate. If no keywords appear in the hypothesis or conclusion, then you should proceed with the forward-backward method. Here is a final example of doing a proof in which the following definitions are used.

**Definition 18** *A function $f$ of one real variable is a* **convex function** *if and only if, for all real numbers $x$, $y$, and $t$ with $0 \leq t \leq 1$, it follows that* $f(tx + (1 - t)y) \leq tf(x) + (1 - t)f(y)$.

**Definition 19** *A function $f$ of one variable is* **increasing** *if and only if, for all real numbers $x$ and $y$ with $x \leq y$, $f(x) \leq f(y)$.*

**Proposition 28** *If $g$ is a convex function and $f$ is an increasing convex function, then the function $f \circ g$ is a convex function.* [Recall that $f \circ g$ is the function defined for all real numbers $x$ by $(f \circ g)(x) = f(g(x))$.]

**Analysis of Proof.**    The forward-backward method is used to begin the proof because the hypothesis $A$ and the conclusion $B$ do not contain keywords (such as "for all" or "there is"). Working backward, you are led to the key question, "How can I show that a function (namely, $f \circ g$) is convex?" Using Definition 18, you must show that

> **B1:** For all real numbers $x$, $y$, and $t$ with $0 \leq t \leq 1$, it follows that $(f \circ g)(tx + (1 - t)y) \leq t(f \circ g)(x) + (1 - t)(f \circ g)(y)$.

Because the backward statement $B1$ contains the quantifier "for all," you should now proceed with the choose method. Accordingly, you should choose

> **A1:** Real numbers $x$, $y$, and $t$ with $0 \leq t \leq 1$,

for which you must show that

> **B2:** $(f \circ g)(tx + (1 - t)y) \leq t(f \circ g)(x) + (1 - t)(f \circ g)(y)$.

This is accomplished by working forward. Specifically, from the hypothesis that $g$ is convex, by definition, this means that

**A2:** For all real numbers $u$, $v$, and $s$ with $0 \le s \le 1$, it follows that $g(su + (1 - s)v) \le sg(u) + (1 - s)g(v)$.

Recognizing the keywords "for all" in the forward statement $A2$, you should now use specialization. Specifically, specializing $A2$ with $u = x$, $v = y$, and $s = t$ and noting from $A1$ that $0 \le t \le 1$, the result is that

**A3:** $g(tx + (1 - t)y) \le tg(x) + (1 - t)g(y)$.

Working forward from the hypothesis that $f$ is an increasing function, by Definition 19, this means that

**A4:** For all real numbers $u$ and $v$ with $u \le v$, $f(u) \le f(v)$.

Recognizing the keywords "for all" in the forward statement $A4$, you should now use specialization. Specifically, specializing $A4$ with $u = g(tx + (1 - t)y)$ (the left side of the inequality in $A3$) and $v = tg(x) + (1 - t)g(y)$ (the right side of the inequality in $A3$), and noting from $A3$ that, for these values of $u$ and $v$, $u \le v$, the result is that

**A5:** $f(g(tx + (1 - t)y)) \le f(tg(x) + (1 - t)g(y))$.

Now work forward from the hypothesis that $f$ is a convex function. Accordingly, by definition, this means that

**A6:** For all real numbers $u$, $v$, and $s$ with $0 \le s \le 1$, it follows that $f(su + (1 - s)v) \le sf(u) + (1 - s)f(v)$.

Recognizing the keywords "for all" in the forward statement $A6$, you should now use specialization. Specifically, specializing $A6$ with $u = g(x)$, $v = g(y)$, and $s = t$ and noting from $A1$ that $0 \le t \le 1$, the result is that

**A7:** $f(tg(x) + (1 - t)g(y)) \le tf(g(x)) + (1 - t)f(g(y))$.

Combining the inequalities in $A5$ and $A7$, you have

**A8:** $f(g(tx + (1 - t)y)) \le tf(g(x)) + (1 - t)f(g(y))$,

or equivalently, by the definition of $f \circ g$,

**A9:** $(f \circ g)(tx + (1 - t)y) \le t(f \circ g)(x) + (1 - t)(f \circ g)(y)$.

The proof is now complete because $A9$ is the same as $B2$.

**Proof of Proposition 28.**    To see that $f \circ g$ is convex, let $x$, $y$, and $t$ be real numbers, with $0 \le t \le 1$, for which it must be shown that

$$(f \circ g)(tx + (1 - t)y) \le t(f \circ g)(x) + (1 - t)(f \circ g)(y). \tag{15.1}$$

Now because $g$ is convex, by definition,

$$g(tx + (1 - t)y) \le tg(x) + (1 - t)g(y). \tag{15.2}$$

Applying $f$ to both sides of (15.2) and using the hypothesis that $f$ is an increasing function yields

$$f(g(tx + (1 - t)y)) \le f(tg(x) + (1 - t)g(y)). \tag{15.3}$$

Also, because $f$ is convex, by definition, the right side of (15.3) satisfies

$$f(tg(x) + (1 - t)g(y)) \le tf(g(x)) + (1 - t)f(g(y)). \tag{15.4}$$

Combining the inequalities in (15.3) and (15.4), you have

$$f(g(tx + (1 - t)y)) \le tf(g(x)) + (1 - t)f(g(y)). \tag{15.5}$$

The proof is now complete because (15.5) is the same as (15.1).  □

You now have all of the techniques you need to do a proof, but keep in mind that this is a creative endeavor. For example, you know when to use the construction method, but actually constructing the object can require a great deal of creativity. Nevertheless, learning to do proofs is like learning a language—the more you practice, the easier it gets. If all of these techniques fail, you may wish to stick to Greek—after all, it's all Greek to me.

### Exercises

**Note:** Solutions to those exercises marked with a $B$ are in the back of this book. Solutions to those exercises marked with a $W$ are located on the web at http://www.wiley.com/college/solow/.

**Note:** All proofs should contain an analysis of proof and a condensed version. Definitions for all mathematical terms are provided in the glossary at the end of the book.

$^B$**15.1**    For each of the following statements, indicate which technique you would use to begin the proof and explain why.

    a. If $p$ and $q$ are odd integers, then the equation $x^2 + 2px + 2q = 0$ has no rational solution for $x$.

    b. For every integer $n \ge 4$, $n! > n^2$.

    c. If $f$ and $g$ are convex functions, then $f + g$ is a convex function.

    d. If $a$, $b$, and $c$ are real variables, then the maximum value of $ab + bc + ac$ subject to the condition that $a^2 + b^2 + c^2 = 1$ is less than or equal to 1.

    e. In a plane, there is one and only one line perpendicular to a given line $L$ through a point $P$ on the line.

**15.2**    For each of the following statements, indicate which technique you would use to begin the proof and explain why.

a. If $p > 1$ is an integer that is not prime, then there is an integer $m$ with $1 < m \le \sqrt{p}$ such that $m|p$.

b. If $f$ and $g$ are continuous functions at the point $x$, then so is the function $f + g$.

c. If $a$ and $b$ are integers for which $a|b$ and $b|a$, then $a = \pm b$.

d. If $f$ and $g$ are functions such that (1) for all real numbers $x$, $f(x) \le g(x)$ and (2) there is no real number $M$ such that, for all $x$, $f(x) \le M$, then there is no real number $M > 0$ such that, for all real numbers $x$, $g(x) \le M$.

e. If $f$ and $g$ are continuous functions at the point $x$, then, for every real number $\epsilon > 0$, there is a real number $\delta > 0$ such that, for all real numbers $y$ with $|x - y| < \delta$, $|f(x) + g(x) - (f(y) + g(y))| < \epsilon$.

$^B$**15.3**    For each of the problems in Exercise 15.1, state how the technique you chose to begin the proof would be applied to that problem. Indicate what you would assume, what you would conclude, and how you go about doing it.

**15.4**    For each of the problems in Exercise 15.2, state how the technique you chose to begin the proof would be applied to that problem. Indicate what you would assume, what you would conclude, and how you go about doing it.

$^W$**15.5**    Describe how you would use each of the following techniques to prove that, "For every integer $n \ge 4$, $n! > n^2$." State what you would assume and what you would conclude.

a.    Induction method.                    b.    Choose method.

c.    Forward-Backward method.            d.    Contradiction method.

**15.6**    Suppose that $S$ and $T$ are sets of real numbers with $S \subseteq T$. Describe how you would use each of the following techniques to prove that, "If $\epsilon > 0$ is a real number such that, for every element $x \in T$, $x \le \epsilon$, then for every element $x \in S$, $x \le \epsilon$." State what you would assume and what you would conclude.

a.    Choose method.                       b.    Specialization method.

c.    Contradiction method.               d.    Contrapositive method.

**15.7**    Reword the following proposition in such a way that it would be appropriate, based on keywords, to use the given technique to begin the proof: "If $X = \{(1 + \frac{1}{n})^n : n > 0 \text{ is an integer}\}$, then, for every element $x \in X$, $x \le 3$."

a.  Max/Min method.                    b.  Contradiction method.

c.  Induction method.

**15.8**  Suppose the forward-backward method is used to start each of the following proofs. List all of the techniques that are likely to be used subsequently in the proof.

a. $(C \; AND \; D)$ implies $(E \; OR \; F)$.

b. $(C \; OR \; D)$ implies $(E \; AND \; F)$.

c. If $X$ is an object such that, for all objects $Y$ with a certain property, something happens, then there is an object $Z$ with a certain property such that something else happens.

d. If for all objects $X$ with a certain property, something happens, then there is an object $Y$ with a certain property such that, for all objects $Z$ with a certain property, something else happens.

**15.9**  Repeat the previous exercise assuming that the contrapositive method is used to start each proof.

**15.10**  Identify all of the techniques that are used in the condensed proof in Exercise 13.13 on page 151.

**15.11**  Identify all of the techniques that are used in the condensed proof in Exercise 10.24 on page 121.

*__15.12__  Identify all of the techniques that are used in the condensed proof in Exercise 6.25 on page 78.

**15.13**  Prove that, if $ABC$ is a right triangle with sides of integer length $a$ and $b$ and hypotenuse of integer length $c$, then $\frac{1}{2}ax^2 + cx + b$ has a rational root.

**15.14**  Prove that, if $S$ is a set of real numbers and $x = \max\{s : s \in S\}$, then $x$ is the only element of $S$ such that, for every element $s \in S$, $s \leq x$.

*__15.15__  Prove that, if $p$ is a polynomial and $q$ is a polynomial, then $p + q$ is a polynomial [where for every real number $x$, $(p + q)(x) = p(x) + q(x)$]. (Recall that a polynomial is a function $f$ of one real variable for which there is an integer $n \geq 0$ and real numbers $a_0, a_1, \ldots, a_n$ such that, for every real number $x$, $f(x) = a_0 + a_1x^1 + \cdots + a_nx^n$.)

**15.16**  Let $f$ be a function of one real variable. Prove that, if $x_1 < x_2 < x_3$ are real numbers for which $f(x_1) < f(x_2)$ and $f(x_2) > f(x_3)$, then $f$ is not linear. (Recall that a function $f$ is linear if and only if there are real numbers $m$ and $b$ such that, for every real number $x$, $f(x) = mx + b$.)

**15.17**    Answer the given questions pertaining to the following condensed proof that, for a function $f$ of one variable, if $x_1 < x_2 < x_3$ are real numbers for which $f(x_1) < f(x_2)$ and $f(x_2) > f(x_3)$, then $f$ is not a convex function.

> **Proof.** Suppose, to the contrary, that $f$ is a convex function. Then consider the real number $t = \frac{x_3 - x_2}{x_3 - x_1}$. It is not hard to show that $0 < t < 1$ and that $x_2 = tx_1 + (1 - t)x_3$. Therefore, by the fact that $f$ is a convex function, it follows that
>
> $$\begin{aligned} f(x_2) &= \quad f(tx_1 + (1 - t)x_3) \quad \leq \ tf(x_1) + (1 - t)f(x_3) \\ &< \ tf(x_2) + (1 - t)f(x_2) \ = \ f(x_2). \end{aligned}$$
>
> The foregoing contradiction completes the proof.  $\square$

a. The first sentence indicates that the author is using the contradiction method. What is the contradiction?

b. Why is the author creating a value for $t$? What proof technique is this value of $t$ being used for?

c. Justify the third sentence by showing that $0 < t < 1$ and also that $x_2 = tx_1 + (1 - t)x_3$.

d. Justify the strict inequality that $tf(x_1) + (1 - t)f(x_3) < tf(x_2) + (1 - t)f(x_2)$.

# Appendix A
# Examples of Proofs from Discrete Mathematics

*Discrete mathematics* is the study of techniques for solving problems involving a finite number of irreducible parts. The objective of this appendix is not to teach discrete mathematics, but rather to provide examples of how the various techniques you have learned are used in doing proofs in this subject. You will also see how to read and understand such written proofs as they might appear in a textbook or other mathematical literature. It is assumed that the reader is familiar with the basic notions of sets and functions; however, the material in this appendix is completely self-contained.

## A.1  EXAMPLES FROM SET THEORY

A **set** is a collection of related items, each of which is an **element of the set**; for example, the following set consists of the elements 70, 75, and 85:

$$S = \{70, 75, 85\}.$$

Notationally, an element $x$ in $S$ is written $x \in S$, and $x \notin S$ means that $x$ is not an element of $S$. The collection of all items that could possibly be included in the set is called the **universal set**. For example, when working with sets of exam scores, the universal set is $U = \{0, 1, \ldots, 100\}$. Also, the **empty set**, denoted by $\emptyset$, is the set with no elements.

It is possible to compare two sets whose elements come from the same universal set in the following ways:

**Definition 20** *A set $A$ is a* **subset** *of a set $B$, written $A \subseteq B$ or $A \subset B$, if and only if, for all elements $x \in A$, $x \in B$.*

**Definition 21** *A set $A$ is* **equal to** *a set $B$, written $A = B$, if and only if $A \subseteq B$ and $B \subseteq A$.*

It is also possible to create new sets from existing sets, as described in the following definitions, in which $U$ is the universal set:

**Definition 22** *The* **complement of a set** *$A$ is the set $A^c = \{x \in U : x \notin A\}$.*

**Definition 23** *The* **union of two sets** *$A$ and $B$ is the set $A \cup B = \{x \in U : x \in A \text{ or } x \in B\}$.*

**Definition 24** *The* **intersection of two sets** *$A$ and $B$ is the set $A \cap B = \{x \in U : x \in A \text{ and } x \in B\}$.*

## How to Do a Proof

You will now see how the foregoing concepts are used in doing a proof. In the example that follows, pay particular attention to how the form of the statement under consideration suggests the technique to use.

**Proposition 29** *If $A$, $B$, and $C$ are sets, then $A \cup (B \cap C) = (A \cup B) \cap (A \cup C)$.*

**Analysis of Proof.** If you do not see any keywords—such as *there is, for all, no, not,* and so on—in the hypothesis or conclusion of the proposition, it is reasonable to begin with the forward-backward method (see Chapter 2). A key question associated with the conclusion is, "How can I show that a set [namely, $A \cup (B \cap C)$] is equal to another set [namely, $(A \cup B) \cap (A \cup C)$]?" Using Definition 21 for equality of two sets, one answer is to show that

**B1:** $A \cup (B \cap C) \subseteq (A \cup B) \cap (A \cup C)$ and
$(A \cup B) \cap (A \cup C) \subseteq A \cup (B \cap C)$.

Only the first statement in $B1$ is proved here—the second statement is left for you to prove in the exercises. A key question associated with the first statement in $B1$ is, "How can I show that a set [namely, $A \cup (B \cap C)$] is a subset of another set [namely, $(A \cup B) \cap (A \cup C)$]?" Using Definition 20 for a subset to answer this question, you must now show that

**B2:** For every element $x \in A \cup (B \cap C)$, $x \in (A \cup B) \cap (A \cup C)$.

You should now recognize the keywords "for every" in $B2$ and thus turn to the choose method (see Chapter 5). Accordingly, you would write, "Let $x \in A \cup (B \cap C)$. It will be shown that $x \in (A \cup B) \cap (A \cup C)$." Thus, you should choose

**A1:** An element $x \in A \cup (B \cap C)$,

for which it must be shown that

**B3:** $x \in (A \cup B) \cap (A \cup C)$.

The idea now is to complete the choose method by working forward from $A1$ and backward from $B3$. To that end, one key question associated with $B3$ is, "How can I show that an element (namely, $x$) is in the intersection of two sets (namely, $A \cup B$ and $A \cup C$)?" Using Definition 24 to answer this question, you must show that

**B4:** $x \in A \cup B$ and $x \in A \cup C$.

The two statements in $B4$ are established by working forward from $A1$, where you know that $x$ belongs to the union of the two sets $A$ and $B \cap C$. Applying Definition 23 for the union of two sets, you can say that

**A2:** Either $x \in A$ or $x \in B \cap C$.

Recognizing the keywords "either/or" in the forward statement $A2$, you should now use a proof by cases (see Section 12.1). Accordingly, you must first assume that $x \in A$ and show that $B4$ is true, and then assume that $x \in B \cap C$ and show that $B4$ is true, as is done now in Cases 1 and 2.

**Case 1:** Assume that $x \in A$ (for which it will be shown that $B4$ is true).
    Because $x \in A$, $x \in A$ or $x \in B$, so $x \in A \cup B$, which is the first statement in $B4$. Likewise, because $x \in A$, $x \in A$ or $x \in C$, so $x \in A \cup C$, which is the second statement in $B4$. Thus, when $x \in A$, $B4$ is true. This leaves …

**Case 2:** Assume that $x \in B \cap C$ (for which it will be shown that $B4$ is true).
    Working forward from $x \in B \cap C$, by Definition 24 for the intersection of two sets, $x \in B$ and $x \in C$. Now because $x \in B$, $x \in A$ or $x \in B$, so $x \in A \cup B$ by Definition 23. Likewise, because $x \in C$, $x \in A$ or $x \in C$, so $x \in A \cup C$ by Definition 23. Thus, when $x \in B \cap C$, $B4$ is true.

The proof by cases is now complete, as is the proof that $A \cup (B \cap C) \subseteq (A \cup B) \cap (A \cup C)$. Do not forget that, from $B1$, you still have to show that $(A \cup B) \cap (A \cup C) \subseteq A \cup (B \cap C)$, as you are asked to do in the exercises.

**Proof of Proposition 29.**    The proposition is established by showing that $A \cup (B \cap C) \subseteq (A \cup B) \cap (A \cup C)$ and $(A \cup B) \cap (A \cup C) \subseteq A \cup (B \cap C)$. To that end, let $x \in A \cup (B \cap C)$. (The word "let" here indicates that the choose method is used.) It will be shown that $x \in (A \cup B) \cap (A \cup C)$. Because $x \in A \cup (B \cap C)$, by definition, $x \in A$ or $x \in B \cap C$. Assume first that $x \in A$. (This is Case 1 of the proof by cases.) It then follows that $x \in A$ or $x \in B$; that is, that $x \in A \cup B$. Likewise, because $x \in A$, $x \in A$ or $x \in C$; that is, $x \in A \cup C$. It has thus been shown that, if $x \in A$, $x \in A \cup B$ and $x \in A \cup C$; that is, that $x \in (A \cup B) \cap (A \cup C)$. Turning to the case when $x \in B \cap C$ (this is Case 2), it follows that $x \in B$ and $x \in C$. This means that $x \in A$ or $x \in B$, so $x \in A \cup B$, and also that $x \in A$ or $x \in C$, so $x \in A \cup C$. Thus, in this case, it has also been shown that $x \in (A \cup B) \cap (A \cup C)$. Having completed the proof that $A \cup (B \cap C) \subseteq (A \cup B) \cap (A \cup C)$, it remains to show that $(A \cup B) \cap (A \cup C) \subseteq A \cup (B \cap C)$, which is left to the exercises. $\square$

A summary of suggestions on how to do a proof is given at the end of this appendix. Now you will see how to read a proof.

## How to Read a Proof

Reading proofs can be challenging because the author does not always refer to the techniques by name, several steps may be combined in a single sentence with little or no justification, and the steps of a proof are not always presented in the order in which they were performed when the proof was done. To read a proof, you have to reconstruct the author's thought processes. Doing so requires that you identify which techniques are used and how they apply to the particular problem. The next example demonstrates how to do so.

**Proposition 30** *If $A$ and $B$ are two sets from the universal set $U$, then* $(A \cup B)^c = A^c \cap B^c$.

**Proof of Proposition 30.**    (For reference purposes, each sentence of the proof is written on a separate line.)

**S1:**    It is first shown that $(A \cup B)^c \subseteq A^c \cap B^c$.

**S2:**    To that end, let $x \in (A \cup B)^c$.

**S3:**    Because $x \in (A \cup B)^c$, $x \notin A \cup B$.

**S4:**    It now follows that $x \notin A$ and $x \notin B$; that is, $x \in A^c \cap B^c$.

**S5:**    The remaining proof that $A^c \cap B^c \subseteq (A \cup B)^c$ is left to the exercises.

The proof is now complete. $\square$

**Analysis of Proof.**   An interpretation of statements $S1$ through $S5$ follows.

**Interpretation of S1:** *It is first shown that* $(A \cup B)^c \subseteq A^c \cap B^c$.

The reason you may find this sentence challenging to understand is that the author has not mentioned what technique is used to begin the proof. To determine this, ask yourself what technique *you* would use to get started. If you do not see any keywords—such as *there is, for all, no, not* and so on—in the hypothesis or conclusion of the proposition, it is reasonable to begin with the forward-backward method (which is what this author has done, without telling you). Working backward, a key question associated with the conclusion is, "How can I show that a set (namely, $(A \cup B)^c$) is equal to another  set (namely, $A^c \cap B^c$)?" The author then uses Definition 21 for equality of two sets to answer this question, so it must be shown that

**B1:** $(A \cup B)^c \subseteq A^c \cap B^c$ and $A^c \cap B^c \subseteq (A \cup B)^c$.

Recognizing that there are two separate statements in $B1$ to prove, you can now understand why, in $S1$, the author says, "It is first shown that $(A \cup B)^c \subseteq A^c \cap B^c$." Can you see where in this proof the author addresses the other statement from $B1$; namely, $A^c \cap B^c \subseteq (A \cup B)^c$?

**Interpretation of S2:** *To that end, let* $x \in (A \cup B)^c$.

Once again, the author has failed to mention what proof techniques are used and has also skipped several steps, which you must recreate. Proceeding backward from the first statement in $B1$, a key question is, "How can I show that a set (namely, $(A \cup B)^c$) is a subset of another set (namely, $A^c \cap B^c$)?" The author answers this using Definition 20, so it must now be shown that

**B2:** For every element $x \in (A \cup B)^c$, $x \in A^c \cap B^c$.

Recognizing the keywords "for every" in $B2$, you should consider using the choose method (see Chapter 5), as the author does. Accordingly, after identifying in $B2$ the objects (elements $x$), the certain property ($x \in (A \cup B)^c$), and the something that happens ($x \in A^c \cap B^c$), you should choose

**A1:** An element $x \in (A \cup B)^c$,

for which it must be shown that

**B3:** $x \in A^c \cap B^c$.

(Note that the symbol $x$ has been used both for the general object in $B2$ as well as the specific chosen object in $A1$, even though these two objects are different.) Rereading $S2$, you can now see that the words "let $x \in (A \cup B)^c$" indicate that the choose method is used. If this interpretation is correct, then the next sentences of the proof should be directed toward establishing $B3$ to complete the choose method.

**Interpretation of S3:** *Because $x \in (A \cup B)^c$, $x \notin A \cup B$.*

Following the choose method, the author is now working forward from $A1$ by Definition 22 for the complement of a set to claim that

**A2:** $x \notin A \cup B$.

The author should still be trying to show that $B3$ is true.

**Interpretation of S4:** *It now follows that $x \notin A$ and $x \notin B$; that is, $x \in A^c \cap B^c$.*

The author continues to work forward from $A2$ by rewriting. Specifically, using Definition 23 for the union of two sets, another way to write $A2$ is $\mathrm{NOT}(x \in A \ \mathrm{OR} \ x \in B)$. Using the rules for writing the NOT of a statement containing the keyword OR (see Chapter 8), it follows from $A2$ that

**A3:** $x \notin A$ AND $x \notin B$.

Now the author works forward from $A3$ by using Definition 22 for the complement of a set to state that

**A4:** $x \in A^c$ AND $x \in B^c$.

Finally, the author works forward from $A4$ by using Definition 24 for the intersection of two sets to claim correctly in $S4$ that

**A5:** $x \in A^c \cap B^c$.

You can see that $A5$ is the same as $B3$, so the choose method is now complete, and the author has correctly shown that $(A \cup B)^c \subseteq A^c \cap B^c$.

**Interpretation of S5:** *The remaining proof that $A^c \cap B^c \subseteq (A \cup B)^c$ is left to the exercises.*

Recall from $B1$ that the author must show that both $(A \cup B)^c \subseteq A^c \cap B^c$ and $A^c \cap B^c \subseteq (A \cup B)^c$. In $S4$, it has been shown that $(A \cup B)^c \subseteq A^c \cap B^c$. It thus remains to show that $A^c \cap B^c \subseteq (A \cup B)^c$, which you are asked to do in the exercises at the end of this appendix.

A summary of how to read proofs is given at the end of this appendix.


## A.2    EXAMPLES FROM FUNCTIONS

You have already seen many examples of *functions*, such as the function $f(x) = 2x + 3$. A function transforms an input value, $x$, to an output value, $f(x)$, as stated formally in the following definition.

**Definition 25** *Given two sets, the* **domain** *$A$ and the* **codomain** *$B$, a* **function** *$f : A \to B$ (read "f is a function from A to B") is a collection of ordered pairs $(x, f(x))$, where, for each element $x \in A$, there is a unique element $f(x) \in B$.*

As seen, a function $f : A \rightarrow B$ associates to each input value $x \in A$ a unique output value $f(x) \in B$. An important problem in many applications is to find an input value $x \in A$ for which $f(x)$ is a specific desired output value, $y$:

**Problem of Solving f(x) = y:** Given a function $f : A \rightarrow B$ and a desired output value $y \in B$, find an input value $x \in A$ such that $f(x) = y$.

It is important to realize that, for a given $y$, there may or may not be any value of $x$ for which $f(x) = y$. For example, consider the function $f : R \rightarrow R$ defined by $f(x) = x^2$, where $R$ is the set of all real numbers. For $y = -1$, there is no real number $x$ such that $f(x) = x^2 = -1$. In view of this example, the objective now is to develop conditions on the function $f$ and the sets $A$ and $B$ so that it is possible to prove that, for every value $y \in B$, there is one and only one input value $x \in A$ such that $f(x) = y$. To that end, for any given $y \in B$, you first need the ability to find at least one $x \in A$ for which $f(x) = y$, which motivates the following definition.

**Definition 26** *A function $f : A \rightarrow B$ is* **surjective** (*or* **onto**) *if and only if, for every element $y \in B$, there is an element $x \in A$ such that $f(x) = y$.*

Definition 26 ensures the existence of an input value $x \in A$ such that $f(x) = y$ but does not ensure that there is only one such input. The following definition is used to establish the uniqueness of such an input value.

**Definition 27** *A function $f : A \rightarrow B$ is* **injective** (*or* **one to one**) *if and only if, for all elements $u, v \in A$, with $u \neq v$, it follows that $f(u) \neq f(v)$.*

**How to Do a Proof**

The following proposition is used to illustrate how to do a proof involving functions. Pay particular attention to how the form of the statement under consideration suggests the technique to use.

**Proposition 31** *The function $f(x) = x^3$ is injective.*

**Analysis of Proof.** If you do not see any keywords—such as *there is, for all, no, not* and so on—in the proposition, it is reasonable to begin with the forward-backward method. A key question associated with the conclusion is, "How can I show that a function (namely, $f(x) = x^3$) is injective?" Using Definition 27, one answer is to show that

**B1:** For all real numbers $u$ and $v$, with $u \neq v$, $u^3 \neq v^3$.

Recognizing the keywords "for all" in the backward statement $B1$, you should now use the choose method to choose

**A1:** Real numbers $u$ and $v$ with $u \neq v$,

for which it must be shown that

**B2:** $u^3 \neq v^3$.

Recognizing the keyword "not" in $B2$, you should now consider using the contradiction method (see Chapter 9) or the contrapositive method (see Chapter 10) to show that $A1$ implies $B2$. Here, the contrapositive method is used. Accordingly, you should assume that

**A2 (NOT B2):** $u^3 = v^3$,

and you must show that

**B3 (NOT A1):** $u = v$.

To that end, work forward from $A2$ by subtracting $v^3$ from both sides and then factoring out $u - v$ to obtain

**A3:** $(u - v)(u^2 + uv + v^2) = 0$.

Now the only way the product of the two numbers $u - v$ and $u^2 + uv + v^2$ in $A3$ can be 0 is if one of them is 0; that is,

**A4:** Either $u - v = 0$ or $u^2 + uv + v^2 = 0$.

Recognizing the keywords "either/or" in the forward statement $A4$, you should now proceed with a proof by cases (see Section 12.1). That is, you should first assume that $u - v = 0$ and show that $B3$ is true, and then assume that $u^2 + uv + v^2 = 0$ and also show that $B3$ is true, as is done now.

**Case 1:** Assume that $u - v = 0$ (for which it will be shown that $B3$ is true).
    In this case, adding $v$ to both sides immediately results in $B3$.

**Case 2:** Assume that $u^2 + uv + v^2 = 0$ (for which it will be shown that $B3$ is true).
    In this case, to show that $u = v$ (see $B3$), think of $u^2 + uv + v^2 = 0$ as a quadratic equation of the form $au^2 + bu + c = 0$, in which $a = 1$, $b = v$ and $c = v^2$. According to the quadratic formula, you have

$$u = \frac{-b \pm \sqrt{b^2 - 4ac}}{2a} = \frac{-v \pm \sqrt{v^2 - 4v^2}}{2} = \frac{-v \pm \sqrt{-3v^2}}{2}. \tag{A.1}$$

The only way the expression for $u$ in (A.1) can result in a real number is if $v = 0$, in which case, from (A.1), $u = 0$, and so $u = v = 0$ and $B3$ is true.
    In either case $u = v$ and $B3$ is true, thus completing the proof.

**Proof of Proposition 31.** To show that $f(x) = x^3$ is an injective function, by definition, it must be shown that, for all real numbers $u$ and $v$ with $u \neq v$, $u^3 \neq v^3$. To that end, let $u$ and $v$ be real numbers with $u \neq v$. (The word "let" here indicates that the choose method is used.) It will be shown that $u^3 \neq v^3$. Equivalently, it is shown that, if $u^3 = v^3$, then $u = v$. (This is the contrapositive method.) Now if $u^3 = v^3$, then $(u - v)(u^2 + uv + v^2) = 0$, so either $u - v = 0$ or $u^2 + uv + v^2 = 0$. In the former case $u = v$, and in the latter case, from the quadratic formula,

$$u = \frac{-v \pm \sqrt{-3v^2}}{2}.$$

The only way the foregoing expression for $u$ can result in a real number is if $v = 0$, in which case, $u = v = 0$. Thus, in either case $u = v$, and so the proof is complete. $\square$

A summary of how to do proofs is given at the end of this appendix.

### How to Read a Proof

It is now proved that, if $f$ is both surjective and injective, then it is possible to solve the problem $f(x) = y$ uniquely, as stated in the following proposition. When reading the proof, be prepared to reconstruct the author's thought processes. Doing so requires that you identify which techniques are used and how they apply to the particular problem.

**Proposition 32** *If $A$ and $B$ are sets and $f : A \to B$ is a surjective and injective function, then for any element $y \in B$, there is a unique element $x \in A$ such that $f(x) = y$.*

**Proof of Proposition 32.**   (For reference purposes, each sentence of the proof is written on a separate line.)

**S1:**   Let $y \in B$, for which it is first shown that there is an element $x \in A$ such that $f(x) = y$.

**S2:**   Because $f$ is surjective, by definition, for every element $t \in B$, there is an element $s \in A$ such that $f(s) = t$.

**S3:**   In particular, for $y \in B$, there is an element $x \in A$ such that $f(x) = y$.

**S4:**   To address the uniqueness of $x$, suppose to the contrary that $w \neq x$ is also an element of $A$ that satisfies $f(w) = y$.

**S5:**   But this contradicts the fact that $f$ is injective because you would have $f(x) = f(w) = y$.

The proof is now complete. $\square$

**Analysis of Proof.**   An interpretation of statements $S1$ through $S5$ follows.

**Interpretation of S1:** *Let $y \in B$, for which it is first shown that there is an element $x \in A$ such that $f(x) = y$.*

Can you determine which technique the author is using to begin the proof? The answer is the choose method because the author has recognized (as you hopefully did also) that the first keywords in the conclusion of the proposition are "for any." Accordingly, the words "Let $y \in B$" in $S1$ indicate that the author chooses

**A1:** An element $y \in B$,

for which (as also stated in $S1$) it must be shown that

**B1:** There is a unique element $x \in A$ such that $f(x) = y$.

The author should now work forward from $A1$ (and the hypothesis $A$) to show that $B1$ is true. What technique(s) would you use to do so? Has the author used those same techniques?

**Interpretation of S2:** *Because $f$ is surjective, by definition, for every element $t \in B$, there is an element $s \in A$ such that $f(s) = t$.*

Looking at $B1$, you should recognize, as the author has, the keywords "there is" and "unique" and consider using the backward uniqueness method and either the direct or the indirect uniqueness method (see Section 11.1). (Reading the rest of the proof, can you determine which of these techniques the author has chosen?) With either uniqueness method, the first step is to construct the desired object—in this case, an element $x$ for which $f(x) = y$. Thus, in $S2$ the author turns to the forward process to do so. Specifically, the author works forward from the hypothesis that the function $f$ is surjective using Definition 26, with the symbols $t$ in place of $y$ and $s$ in place of $x$ to avoid overlapping notation, and states in $S2$ that

**A2:** For every element $t \in B$, there is an element $s \in A$ such that $f(s) = t$.

**Interpretation of S3:** *In particular, for $y \in B$, there is an element $x \in A$ such that $f(x) = y$.*

The author now works forward from $A2$. Can you determine which technique is used? The answer is specialization (see Chapter 6). This is because the first keywords from the left in the forward statement $A2$ are "for every." According to that method, the author must identify one specific object with which to apply specialization. The words, "In particular, for $y \in B$ ..." in $S3$ indicate that the specific object is $y$, which, as noted by the author, satisfies the certain property in $A2$ of being in the set $B$. Finally, the result of specialization is the something that happens in $A2$, obtained by replacing $t$ in $A2$ with the specific object $y$, resulting in

**A3:** There is an element $s \in A$ such that $f(s) = y$.

This is exactly what the author has written in the last part of $S3$, except that the author has used the symbol $x$ in place of $s$ in $A3$, thus resulting in

**A4:** There is an element $x \in A$ such that $f(x) = y$.

The author has now constructed the object $x$ for $B1$—namely, the $x$ in $A4$. To complete the construction method, the author must show that the constructed object satisfies the certain property and the something that happens in $B1$—namely, $x \in A$, $f(x) = y$, and that $x$ is unique. The author assumes that you can see from $A4$ that $x \in A$ and $f(x) = y$, so it remains only to show the uniqueness of $x$.

**Interpretation of S4:** *To address the uniqueness of $x$, suppose to the contrary that $w \neq x$ is also an element of $A$ that satisfies $f(w) = y$.*

The words "suppose to the contrary" in $S4$ indicate that the author is using the indirect uniqueness method. Accordingly, the author should, and does, assume that there is another object (namely, $w$) different from the already-constructed object $x$ that satisfies the certain property and the something that happens in $B1$. Thus, as the author states in $S4$, suppose

**A5:** $w \neq x$ is also an element of $A$ that satisfies $f(w) = y$.

According to the indirect uniqueness method, the author must now work forward from $A4$ and $A5$ (and especially the fact that $w \neq x$) to reach a contradiction. Can you determine what that contradiction is by reading $S5$?

**Interpretation of S5:** *But this contradicts the fact that $f$ is injective because you would have $f(x) = f(w) = y$.*

The author is now indicating that $A4$ and $A5$ contradict the fact that $f$ is injective. Can you determine precisely how the contradiction arises? The answer is that the author has worked forward from the hypothesis that $f$ is injective, using Definition 27 to note that

**A6:** For all $u, v \in A$ with $u \neq v$, it follows that $f(u) \neq f(v)$.

Recognizing the keywords "for all" in the forward process, the author specializes $A6$ with $u = w$ (from $A5$) and $v = x$ (from $A4$), both of which are in the set $A$ (see $A4$ and $A5$) and also satisfy $w \neq x$ (see $A5$). By specialization, you can conclude, from the something that happens in $A6$, that

**A7:** $f(w) \neq f(x)$.

But you also know from $A4$ and $A5$ that $f(x) = y$ and $f(w) = y$, yielding

**A8:** $f(x) = f(w)$.

But $A8$ contradicts $A7$, and so, as the author says, the proof is now complete.

## Summary

Creating proofs is not a precise science. Some general suggestions are provided here. When trying to prove that "$A$ implies $B$," you should consciously choose a technique based on keywords that appear in $A$ and $B$. For example, if the quantifiers "there is" and "for all" appear, then consider using the corresponding construction, choose, induction, and specialization methods. If no keywords are apparent, then it is probably best to proceed with the forward-backward method. Remember that, as you proceed through a proof, different techniques are needed as the form of the statement under consideration changes. If you are unsuccessful at completing a proof, there are several avenues to pursue before giving up. You might try asking yourself why $B$ cannot be false, thus leading you to the contradiction (or contrapositive) method. If you are really stuck, it can sometimes be advantageous to leave the problem for a while because, when you return, you might see a new approach. Undoubtedly, you will learn many tricks of your own as you solve more and more problems.

Reading proofs can be challenging because the author does not always refer to the techniques by name, several steps may be combined in a single sentence with little or no justification, and the steps of a proof are not necessarily presented in the order in which they were performed when the proof was done. To read a proof, you have to reconstruct the author's thought processes. Doing so requires that you identify which techniques are used and how they apply to the particular problem. Begin by trying to determine which technique is used to start the proof. Then try to follow the methodology associated with that technique. Watch for quantifiers to appear, for then it is likely that the corresponding construction, choose, induction, and specialization methods are used. The inability to follow a particular step of a written proof is often because of the lack of sufficient detail. To fill in the gaps, learn to ask yourself how you would proceed to do the proof. Then try to see if the written proof matches your thought process.

## Exercises

**Note:** Solutions to those exercises marked with a $B$ are in the back of this book. Solutions to those exercises marked with a $W$ are located on the web at http://www.wiley.com/college/solow/.

**Note:** All proofs should contain an analysis of proof and a condensed version. Definitions for all mathematical terms are provided in the glossary at the end of the book.

$^B$**A.1**    Suppose that $A$ and $B$ are sets. What is a common key question associated with trying to prove each of the following statements: $A \cup \emptyset = A$, $A \cap \emptyset = \emptyset$, $(A \cup B)^c = A^c \cap B^c$, and $(A \cap B)^c = A^c \cup B^c$? Provide an answer to this key question that does not use symbols.

$^{W}$**A.2**   In the proof of Proposition 26, the key question associated with the statement $(A \cup B) \cap (A \cup C) \subseteq A \cup (B \cap C)$ is, "How can I show that a set is a subset of another set?" State a different key question.

**A.3**   Complete Proposition 27 by proving that $A^c \cap B^c \subseteq (A \cup B)^c$.

$^{B}$**A.4**   Write an analysis of proof that corresponds to the following condensed proof. Indicate which techniques are used and how they are applied. Fill in the details of any missing steps where appropriate.

> **Proposition.** If $A$ and $B$ are sets whose elements come from the universal set $U$, then $(A \cap B)^c \subseteq A^c \cup B^c$.

> **Proof.**   Let $x \in (A \cap B)^c$. This means that $x \notin A \cap B$; that is, $x \notin A$ or $x \notin B$. But then $x \in A^c$ or $x \in B^c$, and so $x \in A^c \cup B^c$, thus completing the proof. $\square$

**A.5**   Write an analysis of proof that corresponds to the following condensed proof. Indicate which techniques are used and how they are applied. Fill in the details of any missing steps where appropriate.

> **Proposition.** If $A$ and $B$ are sets whose elements come from the universal set $U$, then $A^c \cup B^c \subseteq (A \cap B)^c$.

> **Proof.**   Let $x \in A^c \cup B^c$, so $x \in A^c$ or $x \in B^c$; that is, $x \notin A$ or $x \notin B$. But then $x \notin A \cap B$; that is, $x \in (A \cap B)^c$, and so the proof is complete. $\square$

**A.6**   Complete Proposition 26 by proving that $(A \cup B) \cap (A \cup C) \subseteq A \cup (B \cap C)$.

**A.7**   Prove that, if $A$, $B$, and $C$ are sets, then $A \cap (B \cup C) = (A \cap B) \cup (A \cap C)$.

$^{W}$**A.8**   The *Cartesian product* of two sets $S$ and $T$ is the set of all ordered pairs $(s, t)$, where $s \in S$ and $t \in T$; that is, $S \times T = \{(s,t) : s \in S \text{ and } t \in T\}$. Prove, by induction, that, if $S$ and $T$ each have $n \geq 1$ elements, then $S \times T$ has $n^2$ elements.

**A.9**   The *power set* of a set $T$, denoted by $2^T$, is the set of all subsets of $T$; that is, $2^T = \{S : S \subseteq T\}$. (Note that $\emptyset$ is an element of $2^T$.) Prove, by induction, that, if $T$ has $n \geq 1$ elements, then $2^T$ has $2^n$ elements.

$^{B}$**A.10**   Suppose that $A$ and $B$ are sets and that $f : A \to B$. Write what it means for $f$ not to be surjective. (See Definition 26 in Section A.2.)

**A.11**   Suppose that $A$ and $B$ are sets and that $f : A \to B$. Write what it means for $f$ not to be injective. (See Definition 27 in Section A.2.)

$^{B}$**A.12**   Is the function $f : R \to \{x \in R : x \geq 0\}$ defined by $f(x) = |x|$ surjective or not? In either case, prove your claim.

**A.13**    Is the function $f : R \to \{x \in R : x \geq 0\}$ defined by $f(x) = |x|$ injective or not? In either case, prove your claim.

$^W$**A.14**    Suppose that $a$ and $b$ are real numbers and that $f : R \to R$ is defined by $f(x) = ax + b$. Find a condition on $a$ so that $f$ is injective. State and prove a proposition to support your claim.

**A.15**    Suppose that $a$ and $b$ are real numbers and that $f : R \to R$ is defined by $f(x) = ax + b$. Find a condition on $a$ so that $f$ is surjective. State and prove a proposition to support your claim.

$^B$**A.16**    Write an analysis of proof that corresponds to the following condensed proof. Indicate which techniques are used and how they are applied. Fill in the details of any missing steps where appropriate.

> **Definition.** A real number $x$ is a **fixed point** of a function $f : R \to R$ if and only if $f(x) = x$.
>
> **Proposition.** If $f : R \to R$ is a function with the property that there is a real number $\alpha$ with $0 \leq \alpha < 1$ such that, for all real numbers $x, y \in R$, $|f(y) - f(x)| \leq \alpha|y - x|$, and $x_* \in R$ is a fixed point of $f$, then $x_*$ is the unique fixed point of $f$.
>
> **Proof.** To see that $x_*$ is the unique fixed point of $f$, suppose that $y_*$ is also a fixed point of $f$ and that $y_* \neq x_*$. Using $\alpha$ from the hypothesis, it then follows that
>
> $$|x_* - y_*| = |f(x_*) - f(y_*)| \leq \alpha|x_* - y_*|.$$
>
> However, dividing both sides of the foregoing inequality by $|x_* - y_*| \neq 0$ leads to the contradiction that $\alpha \geq 1$.  $\square$

**A.17**    Write an analysis of proof that corresponds to the following condensed proof. Indicate which techniques are used and how they are applied. Fill in the details of any missing steps where appropriate.

> **Proposition.** Suppose that $f : R \to R$ is a function with the same property as in Exercise A.16. If $x_* \in R$ is a fixed point of $f$, $x_0 \in R$, and for every integer $n = 1, 2, \ldots$, $x_n = f(x_{n-1})$, then for every integer $n = 0, 1, \ldots$, $|x_n - x_*| \leq \alpha^n|x_0 - x_*|$.
>
> **Proof.** The conclusion is true for $n = 0$ because $|x_0 - x_*| = \alpha^0|x_0 - x_*|$. Assume now that $|x_n - x_*| \leq \alpha^n|x_0 - x_*|$. Then
>
> $$\begin{aligned} |x_{n+1} - x_*| &= |f(x_n) - f(x_*)| \\ &\leq \alpha|x_n - x_*| \\ &\leq \alpha(\alpha^n|x_0 - x_*|) \\ &= \alpha^{n+1}|x_0 - x_*|. \end{aligned}$$
>
> The proof is now complete.  $\square$

# Appendix B
# Examples of Proofs from Linear Algebra

*Linear algebra* is the study of techniques for solving problems involving ordered lists of numbers (or other objects). The objective of this appendix is not to teach linear algebra, but rather to provide examples of how the various techniques you have learned are used in doing proofs in this subject. You will also see how to read and understand such written proofs as they might appear in a textbook or other mathematical literature. It is assumed that the reader is familiar with the basic notions of vectors and matrices; however, the material in this appendix is completely self-contained.

## B.1 EXAMPLES FROM VECTORS

An $n$-**vector** $\mathbf{x} = (x_1, \ldots, x_n)$, also called a **vector**, is an ordered list of $n$ real numbers. The positive integer $n$ is the **dimension of the vector**, and, for each $i = 1, \ldots, n$, the number $x_i$ is called **component** $i$ of $\mathbf{x}$. For example,

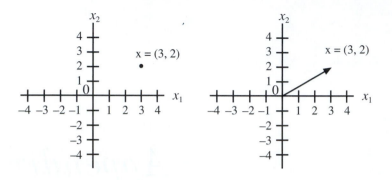

*Fig. B.1*   Visualizing a vector as a point and as an arrow.

the following 2-vector consists of the two components $x_1 = 3$ and $x_2 = 2$:

$$\mathbf{x} = (x_1, x_2) = (3, 2).$$

You can picture a vector in 2 dimensions as a point in the plane or as an arrow in the plane whose tail is located at the origin of the coordinate system and whose head is at the components of the vector (see Figure B.1).

Notationally, the set of all $n$-vectors is written $R^n$. One particular vector that arises frequently is the **zero vector**, written $\mathbf{0} = (0, \ldots, 0)$, whose dimension is understood from the context in which the vector is used. It is also possible to compare two vectors of the same dimension for equality, as follows.

**Definition 28** *Two $n$-vectors* $\mathbf{x} = (x_1, \ldots, x_n)$ *and* $\mathbf{y} = (y_1, \ldots, y_n)$ *are* **equal**, *written* $\mathbf{x} = \mathbf{y}$, *if and only if, for each component* $i = 1, \ldots, n$, $x_i = y_i$.

You can create new vectors from existing vectors. For example, given the $n$-vectors $\mathbf{x} = (x_1, \ldots, x_n)$ and $\mathbf{y} = (y_1, \ldots, y_n)$ and the real number $t$ (also called a **scalar**), you can create the following $n$-vectors:

$$\mathbf{x} \pm \mathbf{y} = (x_1 \pm y_1, \ldots, x_n \pm y_n)$$
$$t\mathbf{x} = (tx_1, \ldots, tx_n).$$

One other operation that arises in many applications is the **dot product of two $n$-vectors $\mathbf{x}$ and $\mathbf{y}$**, written $\mathbf{x} \bullet \mathbf{y}$, which is the following scalar:

$$\mathbf{x} \bullet \mathbf{y} = x_1 y_1 + \cdots + x_n y_n.$$

The foregoing operations satisfy the properties in Table B.1, many of which you are asked to prove in the exercises at the end of this appendix.

Another property of $n$-vectors that is useful in solving certain problems is now defined. Specifically, the two vectors in Figure B.2 *open up properly*,

(a) $\mathbf{x} + \mathbf{y} = \mathbf{y} + \mathbf{x}$             (b) $(\mathbf{x} + \mathbf{y}) + \mathbf{z} = \mathbf{x} + (\mathbf{y} + \mathbf{z})$

(c) $\mathbf{x} + \mathbf{0} = \mathbf{0} + \mathbf{x} = \mathbf{x}$             (d) $\mathbf{x} + (-\mathbf{x}) = \mathbf{0}$

(e) $t(u\mathbf{x}) = (tu)\mathbf{x}$             (f) $t(\mathbf{x} + \mathbf{y}) = (t\mathbf{x}) + (t\mathbf{y})$

(g) $(t + u)\mathbf{x} = (t\mathbf{x}) + (u\mathbf{x})$             (h) $1\mathbf{x} = \mathbf{x}$

(i) $0\mathbf{x} = \mathbf{0}$             (j) $\mathbf{x} \bullet \mathbf{y} = \mathbf{y} \bullet \mathbf{x}$

(k) $\mathbf{x} \bullet (\mathbf{y} + \mathbf{z}) = (\mathbf{x} \bullet \mathbf{y}) + (\mathbf{x} \bullet \mathbf{z})$     (l) $t(\mathbf{x} \bullet \mathbf{y}) = (t\mathbf{x}) \bullet \mathbf{y} = \mathbf{x} \bullet (t\mathbf{y})$

*Table B.1*   Properties of operations for $n$-vectors $\mathbf{x}$, $\mathbf{y}$, and $\mathbf{z}$ and scalars $t$ and $u$.

whereas those in Figure B.3 do not. A formal definition of when a collection of $n$-vectors opens up properly follows.

**Definition 29** *The $n$-vectors* $\mathbf{x}^1, \ldots, \mathbf{x}^k$ *are* **linearly independent** *if and only if, for all real numbers* $t_1, \ldots, t_k$ *for which* $t_1 \mathbf{x}^1 + \cdots + t_k \mathbf{x}^k = \mathbf{0}$, *it follows that* $t_1 = \cdots = t_k = 0$.

## How to Do a Proof

You will now see how the foregoing concepts are used in doing a proof. In the example that follows, pay particular attention to how the form of the statement under consideration suggests the technique to use.

**Proposition 33** *If* $\{\mathbf{x}^1, \ldots, \mathbf{x}^k\}$ *is a set of $k$ linearly independent $n$-vectors, then any nonempty subset contains linearly independent vectors.*

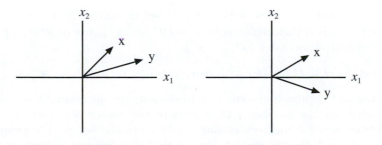

*Fig. B.2*   Examples of vectors that open up properly.

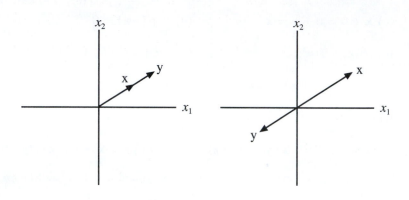

*Fig. B.3*   Examples of vectors that do not open up properly.

**Analysis of Proof.**   Recognizing the keywords "(for) any" in the conclusion, it is reasonable to begin with the choose method (see Chapter 5). Accordingly, after identifying in the for-all statement the objects (nonempty subsets of the vectors), the certain property (none), and the something that happens (the vectors are linearly independent), you should choose

**A1:**  A nonempty subset $S$ of the vectors $\{\mathbf{x}^1, \ldots, \mathbf{x}^k\}$,

for which you must show that

**B1:** The vectors in $S$ are linearly independent.

Before working forward from $A1$ and backward from $B1$, you need a notational method to represent the vectors in the chosen set $S$. There are various ways to do so, for example:

1. Let $j$ with $1 \leq j \leq k$ be the number of vectors in the chosen sub-set, which is then denoted by $S = \{\mathbf{x}^{i_1}, \ldots, \mathbf{x}^{i_j}\}$, where $i_1, \ldots, i_j$ are $j$ different elements of $\{1, \ldots, k\}$.

2. Let $j$ with $1 \leq j \leq k$ be the number of vectors in the chosen sub-set, which is then denoted by $S = \{\mathbf{y}^1, \ldots, \mathbf{y}^j\}$, where $\mathbf{y}^1, \ldots, \mathbf{y}^j$ are $j$ different elements of $\{\mathbf{x}^1, \ldots, \mathbf{x}^k\}$.

3. Let $\emptyset \neq I \subseteq \{1, \ldots, k\}$ with corresponding vectors $S = \{\mathbf{x}^i\}_{i \in I}$.

Here, the first approach is used, and, to simplify the notation further, observe that, whatever the number $j$ of vectors in the chosen set, there are many possible choices for the vectors that could be in the set $S$. For example, if $k = 3$ and $j = 2$, then $S$ could be any of the following sets of vectors:

$$S = \{\mathbf{x}^1, \mathbf{x}^2\} \quad \text{or} \quad S = \{\mathbf{x}^1, \mathbf{x}^3\} \quad \text{or} \quad S = \{\mathbf{x}^2, \mathbf{x}^3\}.$$

Recognizing the keywords "either/or" in the forward process, you can now proceed by cases, depending on which vectors are in $S$. For notational convenience, only the case in which $S = \{\mathbf{x}^1, \ldots, \mathbf{x}^j\}$ is used for doing the proof, with the understanding that the proof for the other cases is similar in spirit. In summary, using this specific notation, you have chosen

**A2:** An integer $j$ with $1 \leq j \leq k$ and $S = \{\mathbf{x}^1, \ldots, \mathbf{x}^j\}$,

for which it must now be shown that $B1$ is true.

   A key question associated with $B1$ is, "How can I show that some vectors (namely, $\mathbf{x}^1, \ldots, \mathbf{x}^j$) are linearly independent?" Using Definition 29 to answer this question, you must now show that

**B2:** For all scalars $t_1, \ldots, t_j$ such that $t_1 \mathbf{x}^1 + \cdots + t_j \mathbf{x}^j = \mathbf{0}$, it follows that $t_1 = \cdots = t_j = 0$.

Recognizing the keywords "for all" in the backward statement $B2$, you should now use the choose method to choose

**A3:** Real numbers $s_1, \ldots, s_j$ such that $s_1 \mathbf{x}^1 + \cdots + s_j \mathbf{x}^j = \mathbf{0}$,

for which you must show that

**B3:** $s_1 = \cdots = s_j = 0$.

   To do so, work forward from the hypothesis that the vectors $\mathbf{x}^1, \ldots, \mathbf{x}^k$ are linearly independent, which by Definition 29 means that

**A4:** For all scalars $t_1, \ldots, t_k$ for which $t_1 \mathbf{x}^1 + \cdots + t_k \mathbf{x}^k = \mathbf{0}$, it follows that $t_1 = \cdots = t_k = 0$.

Recognizing the keywords "for all" in the forward statement $A4$, you should now use specialization (see Chapter 6). To do so, you can use the specific values of $s_1, \ldots, s_j$ from $A3$ for $t_1, \ldots, t_j$ in $A4$, but what specific values should you use for $t_{j+1}, \ldots, t_k$? The answer is 0. Recall that, before you can specialize $A4$ to $t_1 = s_1, \ldots, t_j = s_j, t_{j+1} = 0, \ldots, t_k = 0$, you must verify that these values satisfy the certain property in $A4$—namely, that $t_1 \mathbf{x}^1 + \cdots + t_k \mathbf{x}^k = \mathbf{0}$. However, this is true because

$$
\begin{aligned}
t_1 \mathbf{x}^1 + \cdots + t_k \mathbf{x}^k &= t_1 \mathbf{x}^1 + \cdots + t_j \mathbf{x}^j + t_{j+1} \mathbf{x}^{j+1} + \cdots + t_k \mathbf{x}^k \\
&= s_1 \mathbf{x}^1 + \cdots + s_j \mathbf{x}^j + 0\mathbf{x}^{j+1} + \cdots + 0\mathbf{x}^k \\
&= s_1 \mathbf{x}^1 + \cdots + s_j \mathbf{x}^j \quad \text{(from (i) and (c) in Table B.1)} \\
&= \mathbf{0} \quad \text{(from } A3\text{).}
\end{aligned}
$$

Having verified the certain property in $A4$, the result of specialization is the something that happens in $A4$ for the values $t_1 = s_1, \ldots, t_j = s_j$, $t_{j+1} = 0, \ldots, t_k = 0$, resulting in

**A5:** $s_1 = \cdots = s_j = 0$.

Because $A5$ is the same as $B3$, the proof is now complete.

**Proof of Proposition 33.**    Let $S$ be any subset of the vectors $\mathbf{x}^1, \ldots, \mathbf{x}^k$ and, without loss of generality, assume that $S = \{\mathbf{x}^1, \ldots, \mathbf{x}^j\}$. (The words "without loss of generality" indicate that a proof by cases is used and the remaining cases are left for the reader to verify.) To see that the vectors in $S$ are linearly independent, let $s_1, \ldots, s_j$ be scalars with $s_1 \mathbf{x}^1 + \cdots + s_j \mathbf{x}^j = \mathbf{0}$. (The word "let" here indicates that the choose method is used.) It must be shown that $s_1 = \cdots = s_j = 0$. However, this follows from the hypothesis that the vectors $\mathbf{x}^1, \ldots, \mathbf{x}^k$ are linearly independent, which by definition means that, for all real numbers $t_1, \ldots, t_k$ for which $t_1 \mathbf{x}^1 + \cdots + t_k \mathbf{x}^k = \mathbf{0}$, it follows that $t_1 = \cdots = t_k = 0$. In particular, for $t_1 = s_1, \ldots, t_j = s_j$, $t_{j+1} = 0, \ldots, t_k = 0$, you have that $t_1 \mathbf{x}^1 + \cdots + t_k \mathbf{x}^k = \mathbf{0}$, and so each $t_i = 0$; that is, $s_1 = \cdots = s_j = 0$, thus completing the proof. (The words "In particular" here indicate that specialization is used.) $\square$

A summary of suggestions on how to do a proof is given at the end of this appendix. Now you will see how to read a proof.

### How to Read a Proof

Reading proofs can be challenging because the author does not always refer to the techniques by name, several steps may be combined in a single sentence with little or no justification, and the steps of a proof are not necessarily presented in the order in which they were performed when the proof was done. To read a proof, you have to reconstruct the author's thought processes. Doing so requires that you identify which techniques are used and how they apply to the particular problem. The next example demonstrates how to do so.

**Proposition 34** *If $\mathbf{x}^1, \ldots, \mathbf{x}^k$ are $k$ linearly independent $n$-vectors and $\mathbf{x}$ is an $n$-vector for which there do not exist real numbers $t_1, \ldots, t_k$ such that $\mathbf{x} = t_1 \mathbf{x}^1 + \cdots + t_k \mathbf{x}^k$, then the $n$-vectors $\mathbf{x}^1, \ldots, \mathbf{x}^k, \mathbf{x}$ are linearly independent.*

**Proof of Proposition 34.**    (For reference purposes, each sentence of the proof is written on a separate line.)

**S1:**    To show that $\mathbf{x}^1, \ldots, \mathbf{x}^k, \mathbf{x}$ are linearly independent, let $s_1, \ldots, s_k, s$ be scalars for which $s_1 \mathbf{x}^1 + \cdots + s_k \mathbf{x}^k + s\mathbf{x} = \mathbf{0}$.

**S2:**    It is first shown that $s = 0$, for suppose not.

**S3:**    Then $s_1, \ldots, s_k$ satisfy $\mathbf{x} = -\frac{1}{s}\left(s_1 \mathbf{x}^1 + \cdots + s_k \mathbf{x}^k\right)$, but this contradicts the hypothesis that there do not exist real numbers $t_1, \ldots, t_k$ such that $\mathbf{x} = t_1 \mathbf{x}^1 + \cdots + t_k \mathbf{x}^k$.

**S4:**    To see that $s_1 = \cdots = s_k = 0$, note now that, because $s = 0$, $s_1 \mathbf{x}^1 + \cdots + s_k \mathbf{x}^k = \mathbf{0}$.

**S5:** But then the hypothesis that $\mathbf{x}^1, \ldots, \mathbf{x}^k$ are linearly independent ensures that $s_1 = \cdots = s_k = 0$.

The proof is now complete. □

**Analysis of Proof.** An interpretation of statements $S1$ through $S5$ follows.

**Interpretation of S1:** *To show that* $\mathbf{x}^1, \ldots, \mathbf{x}^k, \mathbf{x}$ *are linearly independent,*
*let* $s_1, \ldots, s_k, s$ *be scalars for which* $s_1\mathbf{x}^1 + \cdots + s_k\mathbf{x}^k + s\mathbf{x} = \mathbf{0}$.
The reason you may find this sentence challenging to understand is that
the author has not mentioned which technique is used to begin the proof. To
determine this, ask yourself which technique *you* would use to get started.
The keyword "no" appears in the hypothesis, so you might consider using the
contradiction or contrapositive methods (see Chapters 9 and 10). If this is
what the author is doing, then it should be assumed that $B$ is not true; that is,
that the $n$-vectors $\mathbf{x}^1, \ldots, \mathbf{x}^k, \mathbf{x}$ are not linearly independent. Rereading $S1$,
you can see that this is not the case. In fact, the author is using the forward-
backward method (without telling you). Working backward, a key question
associated with the conclusion is, "How can I show that some vectors (namely,
$\mathbf{x}^1, \ldots, \mathbf{x}^k, \mathbf{x}$) are linearly independent?" The author then uses Definition 29
to answer this question, so it must be shown that

**B1:** For all scalars $s_1, \ldots, s_k, s$ with $s_1\mathbf{x}^1 + \cdots + s_k\mathbf{x}^k + s\mathbf{x} = \mathbf{0}$,
it follows that $s_1 = \cdots = s_k = s = 0$.

Recognizing the keywords "for all" in the backward statement $B1$, the author
uses the choose method (see Chapter 5) to choose

**A1:** Scalars $s_1, \ldots, s_k, s$ with $s_1\mathbf{x}^1 + \cdots + s_k\mathbf{x}^k + s\mathbf{x} = \mathbf{0}$,

for which it must be shown that

**B2:** $s_1 = \cdots = s_k = s = 0$.

Indeed, the words "$\ldots$ let $s_1, \ldots, s_k, s$ be scalars for which $s_1\mathbf{x}^1 + \cdots + s_k\mathbf{x}^k +$
$s\mathbf{x} = \mathbf{0}$" in $S1$ indicate that the author is using the choose method. Accordingly, the author must now show that $B2$ is true. Rereading the proof, can
you see where and how the author does so?

**Interpretation of S2:** *It is first shown that* $s = 0$, *for suppose not.*
Here, the author is going to show that $s = 0$ (and will subsequently also
have to show that $s_1 = \cdots = s_k = 0$). The words, "$\ldots$ for suppose not" in
$S2$ indicate that the contradiction method (see Chapter 9) is used to do so.
Accordingly, the author assumes that

**A2:** $s \neq 0$.

From what the author has written subsequently, can you determine what the
contradiction is?

**Interpretation of S3:** *Then $s_1, \ldots, s_k$ satisfy $\mathbf{x} = -\frac{1}{s}\left(s_1\mathbf{x}^1 + \cdots + s_k\mathbf{x}^k\right)$, but this contradicts the hypothesis that there do not exist real numbers $t_1, \ldots, t_k$ such that $\mathbf{x} = t_1\mathbf{x}^1 + \cdots + t_k\mathbf{x}^k$.*

The author is working forward from $A1$ and $A2$ to reach a contradiction. Specifically, the author first solves the equation in $A1$ for $s\mathbf{x}$ by subtracting $s_1\mathbf{x}^1 + \cdots + s_k\mathbf{x}^k$ from both sides (see (c) and (d) in Table B.1) to obtain

**A3:** $s\mathbf{x} = -\left(s_1\mathbf{x}^1 + \cdots + s_k\mathbf{x}^k\right)$.

Then, the author divides both sides of the equality in $A3$ through by the scalar $s$, using the fact that $s \neq 0$ from $A2$, to obtain:

**A4:** $\mathbf{x} = -\frac{1}{s}\left(s_1\mathbf{x}^1 + \cdots + s_k\mathbf{x}^k\right)$.

At this point, the author claims that $A4$ is a contradiction of the hypothesis that there do not exist real numbers $t_1, \ldots, t_k$ such that $\mathbf{x} = t_1\mathbf{x}^1 + \cdots + t_k\mathbf{x}^k$. This is indeed the case because, in $A4$, the author has shown that there are real numbers $t_1 = -s_1/s, \ldots, t_k = -s_k/s$ that satisfy $\mathbf{x} = t_1\mathbf{x}^1 + \cdots + t_k\mathbf{x}^k$.

**Interpretation of S4:** *To see that $s_1 = \cdots = s_k = 0$, note now that, because $s = 0$, $s_1\mathbf{x}^1 + \cdots + s_k\mathbf{x}^k = \mathbf{0}$.*

Recall from $B2$ that it must also be shown that $s_1 = \cdots = s_k = 0$, which is what the author starts to do in $S4$ and finishes in $S5$. Here, the author works forward from $A1$ using the fact that it has just been shown that $s = 0$ and properties (i) and (c) from Table B.1 to state that

**A5:** $s_1\mathbf{x}^1 + \cdots + s_k\mathbf{x}^k = \mathbf{0}$.

**Interpretation of S5:** But then the hypothesis that $\mathbf{x}^1, \ldots, \mathbf{x}^k$ are linearly independent ensures that $s_1 = \cdots = s_k = 0$.

It is here that the author completes the choose method by showing, as required in $B2$, that $s_1 = \cdots = s_k = 0$. Can you determine which techniques the author uses to do so? First, the author works forward from the hypothesis that $\mathbf{x}^1, \ldots, \mathbf{x}^k$ are linearly independent using Definition 29, so

**A6:** For all scalars $t_1, \ldots, t_k$ for which $t_1\mathbf{x}^1 + \cdots + t_k\mathbf{x}^k = \mathbf{0}$, it follows that $t_1 = \cdots = t_k = 0$.

Recognizing the keywords " for all" in the forward statement $A6$, the author uses specialization (see Chapter 6). That is, the author specializes $A6$ with $t_1 = s_1, \ldots, t_k = s_k$. According to specialization, it is necessary to verify that these special values for $t_1, \ldots, t_k$ satisfy the certain property in $A6$; namely, $s_1\mathbf{x}^1 + \cdots + s_k\mathbf{x}^k = \mathbf{0}$. The author, however, leaves it to you to notice that this is true (see $A5$). The result of specialization in this case, as stated by the author in $S5$, is that

**A7:** $s_1 = \cdots = s_k = 0$.

This completes the choose method and the entire proof.

## B.2   EXAMPLES FROM MATRICES

An $(m \times n)$ **matrix** $A$ (read as "an $m$ by $n$ matrix $A$") is a rectangular table of real numbers organized in $m$ rows and $n$ columns. The **dimensions** of $A$ are the values of $m$ and $n$. The number in row $i$ $(1 \leq i \leq m)$ and column $j$ $(1 \leq j \leq n)$ of $A$ is called **element** $A_{ij}$ of the matrix. The following is an example of a $(2 \times 3)$ matrix:

$$A = \begin{bmatrix} 2 & -3 & 1 \\ 0 & 4 & -2 \end{bmatrix}.$$

The set of all $(m \times n)$ matrices is denoted here by $R^{m \times n}$.

You can think of an $(m \times n)$ matrix $A$ as an ordered collection of either its $n$ columns (in which column $j$ is denoted here as $A_{*j}$) or its $m$ rows (in which row $i$ is denoted here as $A_{i*}$), as follows:

$$A = \overbrace{\begin{bmatrix} \uparrow & \uparrow & \cdots & \uparrow \\ A_{*1} & A_{*2} & \cdots & A_{*n} \\ \downarrow & \downarrow & \cdots & \downarrow \end{bmatrix}}^{\text{columns}} = \left.\begin{bmatrix} \leftarrow & A_{1*} & \rightarrow \\ \leftarrow & A_{2*} & \rightarrow \\ \vdots & \vdots & \vdots \\ \leftarrow & A_{m*} & \rightarrow \end{bmatrix}\right\} \text{rows.}$$

You can think of an $n$-vector $\mathbf{x} = (x_1, \ldots, x_n)$ as a **column vector**—that is, as the following $(n \times 1)$ matrix:

$$\mathbf{x} = \begin{bmatrix} x_1 \\ x_2 \\ \vdots \\ x_n \end{bmatrix}.$$

Alternatively, you can think of an $n$-vector $\mathbf{x} = (x_1, \ldots, x_n)$ as a **row vector**—that is, as the following $(1 \times n)$ matrix denoted by $\mathbf{x}^t$ to differentiate the row vector from the column vector:

$$\mathbf{x}^t = [x_1, x_2, \ldots, x_n].$$

Several special matrices arise repeatedly. One is the **zero matrix**, denoted by $0$, in which each element is 0 and whose dimensions are understood from the context. Another special matrix is the **identity matrix**, consisting of $n$ rows and $n$ columns of zeros, except for ones along the diagonal (from the upper left to the lower right):

$$I = \begin{bmatrix} 1 & 0 & \cdots & 0 \\ 0 & 1 & \cdots & 0 \\ \vdots & \vdots & \ddots & \vdots \\ 0 & 0 & \cdots & 1 \end{bmatrix}.$$

It is possible to compare two matrices of the same dimensions for equality, as follows.

**Definition 30** *Two $(m \times n)$ matrices $A$ and $B$ are* **equal***, written $A = B$, if and only if, for all $i = 1, \ldots, m$ and $j = 1, \ldots, n$, $A_{ij} = B_{ij}$.*

You can create new matrices from existing matrices. For example, given $(m \times n)$ matrices $A$ and $B$ and a real number $t$:

$$
\begin{aligned}
A \pm B &= \text{the } (m \times n) \text{ matrix in which } (A \pm B)_{ij} &= A_{ij} \pm B_{ij} \\
tA &= \text{the } (m \times n) \text{ matrix in which } (tA)_{ij} &= tA_{ij} \\
A^t &= \text{the } (n \times m) \text{ matrix in which } (A^t)_{ji} &= A_{ij}.
\end{aligned}
$$

The foregoing matrix $A^t$ is called the **transpose of the matrix** $A$.

One other operation that arises in many applications is **matrix multiplication**. The result of multiplying the matrices $A$ and $B$ is the matrix $AB$ in which the element in row $i$ and column $j$ is the dot product of row $i$ of $A$ and column $j$ of $B$; that is,

$$(AB)_{ij} = A_{i*} \bullet B_{*j}.$$

In order to be able to compute the foregoing dot product, the dimension of the vector $A_{i*}$ must be the same as that of the vector $B_{*j}$; that is, the number of columns of $A$ must be the same as the number of rows of $B$. In summary, it is only possible to multiply a matrix $A$ by a matrix $B$ if $A \in R^{m \times p}$ and $B \in R^{p \times n}$, in which case, $AB \in R^{m \times n}$ and

$$(AB)_{ij} = A_{i*} \bullet B_{*j} \text{ for all } i = 1, \ldots, m \text{ and } j = 1, \ldots, n.$$

The foregoing matrix operations satisfy the properties in Table B.2, many of which you are asked to prove in the exercises at the end of this appendix.

One of the most important problems that arises in applications of linear algebra is that of solving $n$ linear equations in $n$ unknowns, which is stated in matrix notation as follows:

**Problem of Solving $Ax = b$:** Given an $(n \times n)$ matrix $A$ and a (column) $n$-vector $\mathbf{b}$, find a (column) $n$-vector $\mathbf{x}$ such that $Ax = \mathbf{b}$.

For a given $n$-vector $\mathbf{b}$, there may be no $n$-vector $\mathbf{x}$ for which $Ax = \mathbf{b}$, exactly one $n$-vector $\mathbf{x}$ for which $Ax = \mathbf{b}$, or infinitely many $n$-vectors $\mathbf{x}$ for which $Ax = \mathbf{b}$. As you will see shortly, the following property on the matrix $A$ ensures that there is exactly one solution to this problem.

**Definition 31** *An $(n \times n)$ matrix $A$ is* **invertible** *(also called* **nonsingular***) if and only if there is an $(n \times n)$ matrix $C$ such that $AC = CA = I$ (the $(n \times n)$ identity matrix).*

(a) $A + B = B + A$  (b) $(A + B) + C = A + (B + C)$

(c) $A + 0 = 0 + A = A$  (d) $A - A = 0$

(e) $0 - A = -A$  (f) $s(A + B) = sA + sB$

(g) $s(A - B) = sA - sB$  (h) $(s + t)A = sA + tA$

(i) $(s - t)A = sA - tA$  (j) $s(tA) = (st)A$

(k) $0A = 0$  (l) $s0 = 0$

(m) $A(B \pm C) = AB \pm AC$  (n) $(A \pm B)C = AC \pm BC$

(o) $(AB)C = A(BC)$  (p) $s(AB) = (sA)B = A(sB)$

(q) $IA = A$ and $BI = B$  (r) $0A = 0$ and $B0 = 0$

*Table B.2* Properties of operations for the real numbers $s$ and $t$ and the matrices $A, B, C, 0, I$ whose dimensions allow the operations to be performed.

### How to Do a Proof

The following proposition provides a condition under which it is possible to solve uniquely a system of $n$ linear equations in $n$ unknowns and illustrates how to do a proof involving matrices. Pay particular attention to how the form of the statement under consideration suggests the technique to use.

**Proposition 35** *If $A$ is an $(n \times n)$ invertible matrix, then, for every $n$-vector* **b**, *there is a unique $n$-vector* **x** *such that $Ax = b$.*

**Analysis of Proof.** Recognizing "for every" as the first keywords in the conclusion, it is reasonable to begin with the choose method (see Chapter 5). Accordingly, you should choose

**A1:** An $n$-vector **b**,

for which you must show that

**B1:** There is a unique $n$-vector **x** such that $Ax = b$.

Recognizing the keywords "unique" in $B1$, you should now use the backward uniqueness method (see Section 11.1). Accordingly, the first step is to construct the desired object in $B1$—namely, an $n$-vector **x** such that $Ax = b$.

Turning to the forward process to do so, because $A$ is invertible, by Definition 31 you know that

**A2:** There is an $(n \times n)$ matrix $C$ such that $AC = CA = I$.

This matrix $C$ is used to construct the desired $n$-vector $\mathbf{b}$ in $B1$. To see how, look at the property you want $\mathbf{b}$ to satisfy in $B1$—namely, that

$$A\mathbf{x} = \mathbf{b}. \tag{B.1}$$

You can "undo" the matrix $A$ on the left of (B.1) by multiplying both sides on the left by the matrix $C$ from $A2$. Doing so, you obtain

$$
\begin{aligned}
C(A\mathbf{x}) &= C\mathbf{b} \\
(CA)\mathbf{x} &= C\mathbf{b} \quad \text{(property (o) in Table B.2)} \\
I\mathbf{x} &= C\mathbf{b} \quad \text{(property of } C \text{ in } A2) \\
\mathbf{x} &= C\mathbf{b} \quad \text{(property (q) in Table B.2).}
\end{aligned} \tag{B.2}
$$

From (B.2), you have now constructed

**A3:** The $n$-vector $\mathbf{x} = C\mathbf{b}$.

According to the construction method, it must be shown that the value of $\mathbf{x}$ in $A3$ satisfies the property of $A\mathbf{x} = \mathbf{b}$ in $B1$. This is true because

$$
\begin{aligned}
A\mathbf{x} &= A(C\mathbf{b}) \quad \text{(constructed value of } \mathbf{x} \text{ in } A3) \\
&= (AC)\mathbf{b} \quad \text{(property (o) in Table B.2)} \\
&= I\mathbf{b} \quad \text{(property of } C \text{ in } A2) \\
&= \mathbf{b} \quad \text{(property (q) in Table B.2).}
\end{aligned}
$$

It remains from $B1$ to address the uniqueness of $\mathbf{x}$. Here, the direct uniqueness method is used. Accordingly, you should now assume that

**A4:** There is an $n$-vector $\mathbf{y}$ such that $A\mathbf{y} = \mathbf{b}$.

To establish the uniqueness of $\mathbf{x}$, you must use $A3$ and $A4$ to show that

**B2:** $\mathbf{x} = \mathbf{y}$.

To do so, from $A3$ and $A4$, you know that $A\mathbf{x} = \mathbf{b}$ and $A\mathbf{y} = \mathbf{b}$, so

**A5:** $A\mathbf{x} = A\mathbf{y}$.

Using the matrix $C$ to "undo" $A$ in $A5$ by multiplying $C$ on both sides of $A5$, you have

$$
\begin{aligned}
C(A\mathbf{x}) &= C(A\mathbf{y}) \\
(CA)\mathbf{x} &= (CA)\mathbf{y} \quad \text{(property (o) in Table B.2)} \\
I\mathbf{x} &= I\mathbf{y} \quad \text{(property of } C \text{ in } A2) \\
\mathbf{x} &= \mathbf{y} \quad \text{(property (q) in Table B.2)}
\end{aligned}
$$

This final equality establishes $B2$, and so the proof is now complete.

In the proof that follows, note that the techniques are not referenced by their names and no explanation is given for how the vector $\mathbf{x}$ is constructed.

Also, the steps using the properties of matrix operations are performed without justification.

**Proof of Proposition 35.** It is first shown that there is an $n$-vector $\mathbf{x}$ such that $A\mathbf{x} = \mathbf{b}$. To that end, from the hypothesis that $A$ is invertible, by definition, there is an $(n \times n)$ matrix $C$ such that $AC = CA = I$. Constructing $\mathbf{x} = C\mathbf{b}$, it is easy to see that $A\mathbf{x} = A(C\mathbf{b}) = (CA)\mathbf{b} = I\mathbf{b} = \mathbf{b}$. To see that this value of $\mathbf{x}$ is unique, suppose that the $n$-vector $\mathbf{y}$ also satisfies $A\mathbf{y} = \mathbf{b}$. You would then have

$$
\begin{aligned}
C(A\mathbf{x}) &= C(A\mathbf{y}) \\
(CA)\mathbf{x} &= (CA)\mathbf{y} \\
I\mathbf{x} &= I\mathbf{y} \\
\mathbf{x} &= \mathbf{y}.
\end{aligned}
$$

The fact that $\mathbf{x} = \mathbf{y}$ shows that the value of $\mathbf{x} = C\mathbf{b}$ is unique and completes the proof. ☐

A summary of how to do proofs is given at the end of this appendix. Now it is time to see how to read a proof.

### How to Read a Proof

It is now proved that, if every system of linear equations associated with an $(n \times n)$ matrix $A$ is uniquely solvable, then the columns of $A$ are linearly independent.

**Proposition 36** *If $A$ is an $(n \times n)$ matrix such that, for every $n$-vector $\mathbf{b}$, there is a unique $n$-vector $\mathbf{x}$ such that $A\mathbf{x} = \mathbf{b}$, then the columns of $A$ are linearly independent $n$-vectors.*

**Proof of Proposition 36.** (For reference purposes, each sentence of the proof is written on a separate line.)

**S1:** To see that the columns of $A$ are linearly independent, let $t_1, \ldots, t_n$ be real numbers such that $A_{*1}t_1 + \cdots + A_{*n}t_n = \mathbf{0}$.

**S2:** Letting $\mathbf{t}$ be the column vector whose components are $t_1, \ldots, t_n$, you then have from the hypothesis that $\mathbf{x} = \mathbf{t}$ is the only solution to $A\mathbf{x} = \mathbf{0}$.

**S3:** However, because $A\mathbf{0} = \mathbf{0}$, it must be that $\mathbf{t} = \mathbf{0}$; that is, $t_1 = 0, \ldots, t_n = 0$.

The proof is now complete. ☐

**Analysis of Proof.** An interpretation of statements $S1$ through $S3$ follows.

**Interpretation of S1:** *To see that the columns of $A$ are linearly independent, let $t_1, \ldots, t_n$ be real numbers such that $A_{*1}t_1 + \cdots + A_{*n}t_n = \mathbf{0}$.*

The reason you may find this sentence challenging to understand is that the author has not mentioned which technique is used to begin the proof. To determine this, ask yourself which technique *you* would use to get started. The keyword "for every" appears in the hypothesis, so you might consider using specialization. The author does use specialization, but not to begin the proof. Rather, the author is using the forward-backward method (without telling you). Working backward, a key question associated with the conclusion is, "How can I show that some vectors (namely, $A_{*1}, \ldots, A_{*n}$) are linearly independent?" The author then uses Definition 29 to answer this question, so it must be shown that

> **B1:** For all real numbers $t_1, \ldots, t_n$ for which $A_{*1}t_1 + \cdots + A_{*n}t_n = \mathbf{0}$, it follows that $t_1 = \cdots = t_n = 0$.

Recognizing the keywords "for all" in the backward statement $B1$, the author uses the choose method to choose

> **A1:** Real numbers $t_1, \ldots, t_n$ for which $A_{*1}t_1 + \cdots + A_{*n}t_n = \mathbf{0}$,

for which it must be shown that

> **B2:** $t_1 = \cdots = t_n = 0$.

Indeed, the words "... let $t_1, \ldots, t_n$ be real numbers such that ..." in $S1$ indicate that the choose method is used.

**Interpretation of S2:** *Letting $\mathbf{t}$ be the column vector whose components are $t_1, \ldots, t_n$, you then have from the hypothesis that $\mathbf{x} = \mathbf{t}$ is the only solution to $A\mathbf{x} = \mathbf{0}$.*

The words "... from the hypothesis" in $S2$ indicate that the author is now working forward. Can you explain how the author does so? The answer is that the author uses specialization (see Chapter 6) because the keywords "for every" appear in the hypothesis in the statement that, for every $n$-vector $\mathbf{b}$, there is a unique $n$-vector $\mathbf{x}$ such that, $A\mathbf{x} = \mathbf{b}$. Accordingly, the author specializes this for-all statement to the specific value $\mathbf{b} = \mathbf{0}$, which is an $n$-vector, to obtain the statement:

> **A2:** There is a unique $n$-vector $\mathbf{x}$ ($= \mathbf{t}$) such that $A\mathbf{x} = \mathbf{0}$.

**Interpretation of S3:** *However, because $A\mathbf{0} = \mathbf{0}$, it must be that $\mathbf{t} = \mathbf{0}$; that is, $t_1 = 0, \ldots, t_n = 0$.*

The author recognizes the keyword "unique" in the forward statement $A2$ and so uses the forward uniqueness method (see Section 11.1). Accordingly, the author identifies two $n$-vectors that solve the system of linear equations—namely, $\mathbf{x} = \mathbf{0}$ (because $A\mathbf{0} = \mathbf{0}$ from property (r) in Table B.2) and $\mathbf{x} = \mathbf{t}$ (see $A2$). Because from $A2$ there is a unique solution to this system of linear equations, by the forward uniqueness method, it must be the case that the two solutions **t** and **0** are the same; that is,

   **A3:** $\mathbf{t} = \mathbf{0}$.

Working forward from $A3$, using the definition of what it means for the two $n$-vectors **t** and **0** to be equal, it follows that

   **A4:** $t_1 = 0, \ldots, t_n = 0$.

The proof is now complete because $A4$ is the same as $B2$.

## Summary

Creating proofs is not a precise science. Some general suggestions are provided here. When trying to prove that "$A$ implies $B$," consciously choose a technique based on keywords that appear in $A$ and $B$. For example, if the quantifiers "there is" and "for all" appear, then consider using the corresponding construction, choose, induction, and specialization methods. If no keywords are apparent, then it is probably best to proceed with the forward-backward method. Remember that, as you proceed through a proof, different techniques are needed as the form of the statement under consideration changes. If you are unsuccessful at completing a proof, there are several avenues to pursue before giving up. You might try asking yourself why $B$ cannot be false, thus leading you to the contradiction (or contrapositive) method. If you are really stuck, it can sometimes be advantageous to leave the problem for a while because, when you return, you might see a new approach. Undoubtedly, you will learn many tricks of your own as you solve more and more problems.

Reading proofs can be challenging because the author does not always refer to the techniques by name, several steps may be combined in a single sentence with little or no justification, and the steps of a proof are not necessarily presented in the order in which they were performed when the proof was done. To read a proof, you have to reconstruct the author's thought processes. Doing so requires that you identify which techniques are used and how they apply to the particular problem. Begin by trying to determine which technique is used to start the proof. Then try to follow the methodology associated with that technique. Watch for quantifiers to appear, for then it is likely that the corresponding construction, choose, induction, and specialization methods are used. The inability to follow a particular step of a written proof is often because of the lack of sufficient detail. To fill in the gaps, learn to ask yourself how you would proceed to do the proof. Then try to see if the written proof matches your thought process.

## Exercises

**Note:** Solutions to those exercises marked with a $B$ are in the back of this book. Solutions to those exercises marked with a $W$ are located on the web at http://www.wiley.com/college/solow/.

**Note:** All proofs should contain an analysis of proof and a condensed version. Definitions for all mathematical terms are provided in the glossary at the end of the book.

$^B$**B.1**    Suppose you are trying to prove each of the properties in Table B.1.

    a. "How can I show that two $n$-vectors are equal?" is a valid key question for all of the properties in Table B.1 except three. Identify those three properties and state an associated key question that is valid for those three properties.

    b. What is a key answer to the key question in quotation marks in part (a)? Apply this key answer to the specific problem of trying to prove that, for $n$-vectors $\mathbf{x}$ and $\mathbf{y}$, $\mathbf{x} + \mathbf{y} = \mathbf{y} + \mathbf{x}$, and thus create a new statement $B1$ in the backward process.

    c. Based on your answer to part (b), which proof technique would you use next and why? Illustrate how this technique would be applied to the statement $B1$ in part (b).

$^B$**B.2**    Prove property (a) in Table B.1 that, if $\mathbf{x}$ and $\mathbf{y}$ are $n$-vectors, then $\mathbf{x} + \mathbf{y} = \mathbf{y} + \mathbf{x}$.

**B.3**    Prove property (b) in Table B.1 that, if $\mathbf{x}$, $\mathbf{y}$, and $\mathbf{z}$ are $n$-vectors, then $(\mathbf{x} + \mathbf{y}) + \mathbf{z} = \mathbf{x} + (\mathbf{y} + \mathbf{z})$.

$^B$**B.4**    Write an analysis of proof that corresponds to the following condensed proof. Indicate which techniques are used and how they are applied. Fill in the details of any missing steps where appropriate.

> **Proposition.** If $\mathbf{x}$ and $\mathbf{y}$ are $n$-vectors and $t$ is a real number, then $t(\mathbf{x} + \mathbf{y}) = t\mathbf{x} + t\mathbf{y}$. (Property (f) in Table B.1.)
>
> **Proof.** Let $i$ be an integer with $1 \le i \le n$. Then,
>
> $$[t(\mathbf{x} + \mathbf{y})]_i = t(\mathbf{x} + \mathbf{y})_i = t(x_i + y_i) = tx_i + ty_i = (t\mathbf{x} + t\mathbf{y})_i.$$
>
> It now follows that $t(\mathbf{x} + \mathbf{y}) = t\mathbf{x} + t\mathbf{y}$, and so the proof is complete.  $\square$

**B.5**    Write an analysis of proof that corresponds to the following condensed proof. Indicate which techniques are used and how they are applied. Fill in the details of any missing steps where appropriate.

**Proposition.** If $\mathbf{x}^1, \ldots, \mathbf{x}^k$ are linearly independent $n$-vectors and $\mathbf{x}$ is an $n$-vector for which there are real numbers $t_1, \ldots, t_k$ such that $\mathbf{x} = t_1\mathbf{x}^1 + \cdots + t_k\mathbf{x}^k$, then the numbers $t_1, \ldots, t_k$ are unique.

**Proof.** To see that the numbers $t_1, \ldots, t_k$ are unique, suppose that the numbers $s_1, \ldots, s_k$ also satisfy $\mathbf{x} = s_1\mathbf{x}^1 + \cdots + s_k\mathbf{x}^k$. You would then have

$$\mathbf{x} = s_1\mathbf{x}^1 + \cdots + s_k\mathbf{x}^k = t_1\mathbf{x}^1 + \cdots + t_k\mathbf{x}^k.$$

But then $(s_1 - t_1)\mathbf{x}^1 + \cdots + (s_k - t_k)\mathbf{x}^k = \mathbf{0}$. However, because $\mathbf{x}^1, \ldots, \mathbf{x}^k$ are linearly independent, it now follows that $s_1 = t_1, \ldots, s_k = t_k$. It has thus been established that $t_1, \ldots, t_k$ are the unique numbers such that $\mathbf{x} = t_1\mathbf{x}^1 + \cdots + t_k\mathbf{x}^k$, completing the proof. $\square$

**B.6**    Prove that, if $\mathbf{x}$ and $\mathbf{y}$ are nonzero $n$-vectors for which $\mathbf{x} \bullet \mathbf{y} = 0$, then $\mathbf{x}$ and $\mathbf{y}$ are linearly independent.

[B]**B.7**    Write what it means for the $n$-vectors $\mathbf{x}^1, \ldots, \mathbf{x}^k$ to be **linearly dependent**—that is, not linearly independent.

[B]**B.8**    Use the definition in Exercise B.7 to prove that, if $\mathbf{x}^1, \ldots, \mathbf{x}^k$ are $n$-vectors, then $\mathbf{x}^1, \ldots, \mathbf{x}^k, \mathbf{0}$ are linearly dependent.

**B.9**    Use the definition in Exercise B.7 to prove that, if $\mathbf{x}^1, \ldots, \mathbf{x}^k$ are linearly dependent $n$-vectors, then for any $n$-vector $\mathbf{x}$, the $n$-vectors $\mathbf{x}^1, \ldots, \mathbf{x}^k, \mathbf{x}$ are linearly dependent.

[W]**B.10**    Suppose you are trying to prove all of the properties in Table B.2.

a. What is a key question that is common to all of these properties?

b. What is a key answer to your question in part (a)? Apply this answer to the specific problem of trying to prove that, for $(m \times n)$ matrices $A$ and $B$, $A + B = B + A$ and thus create a new statement $B1$ in the backward process.

c. Based on your answer to part (b), which proof technique would you use next and why? Illustrate how this technique would be applied to the statement $B1$ in part (b).

[B]**B.11**    Prove property (a) in Table B.2 that, if $A$ and $B$ are $(m \times n)$ matrices, then $A + B = B + A$.

**B.12**    Prove property (b) in Table B.2 that, if $A$, $B$, and $C$ are $(m \times n)$ matrices, then $(A + B) + C = A + (B + C)$.

**B.13**    Prove that, if $A$ is an $(n \times n)$ invertible matrix, then there is a unique $(n \times n)$ matrix $C$ such that $AC = CA = I$.

$^B$**B.14**    Write an analysis of proof that corresponds to the following condensed proof. Indicate which techniques are used and how they are applied. Fill in the details of any missing steps where appropriate.

> **Proposition.** If $A$, $B$, and $C$ are matrices for which you can perform the following operations, then $A(B + C) = AB + AC$. (Property (m) in Table B.2.)
>
> **Proof.** Let $i$ and $j$ be integers with $1 \le i \le m$ and $1 \le j \le n$. Then,
>
> $$
> \begin{aligned}
> [A(B+C)]_{ij} &= A_{i*} \bullet (B+C)_{*j} = (A_{i*} \bullet B_{*j}) + (A_{i*} \bullet C_{*j}) \\
> &= (AB)_{ij} + (AC)_{ij} = (AB + AC)_{ij}.
> \end{aligned}
> $$
>
> It now follows that $A(B + C) = AB + AC$, and so the proof is complete.   $\square$

**B.15**    Write an analysis of proof that corresponds to the following condensed proof. Indicate which techniques are used and how they are applied. Fill in the details of any missing steps where appropriate.

> **Proposition.**   If $A$ and $C$ are $(n \times n)$ matrices such that $AC = I$ and $C$ is invertible, then $CA = I$.
>
> **Proof.** Because $C$ is invertible, by definition there is an $(n \times n)$ matrix $D$ such that $CD = DC = I$. From the hypothesis that $AC = I$, you have that
>
> $$(AC)D = ID \quad \text{or} \quad A(CD) = D \quad \text{or} \quad AI = D \quad \text{or} \quad A = D.$$
>
> The conclusion follows on multiplying both sides of the last equality on the left by $C$ and using the fact that $CD = I$.   $\square$

**B.16**    Identify the error in the following proof.

> **Proposition.** If $A$ is an $(n \times n)$ matrix with the property that, for every $n$-vector $\mathbf{b}$ there is a unique $n$-vector $\mathbf{x}$ such that $A\mathbf{x} = \mathbf{b}$, then $A$ is invertible.
>
> **Proof.** According to the definition of an invertible matrix, it is necessary to show that there is an $(n \times n)$ matrix $C$ such that $AC = CA = I$. To that end, from the hypothesis, for each $j = 1, \ldots, n$, let column $j$ of $C$ be the unique solution $\mathbf{x}$ to the system of linear equations $A\mathbf{x} = I_{*j}$. It then follows that $AC = I$ because for each column $j$, $(AC)_{*j} = AC_{*j} = I_{*j}$. The proof is now complete.   $\square$

# Appendix C
# Examples of Proofs from Modern Algebra

*Modern algebra* is the study of numbers and systems of objects that have properties similar to those of numbers and their operations. The objective of this appendix is not to teach modern algebra, but rather to provide examples of how the various techniques you have learned are used in doing proofs in this subject. You will also see how to read and understand such written proofs as they might appear in a textbook or other mathematical literature. It is assumed that the reader is familiar with the basic properties of numbers; however, the material in this appendix is completely self-contained.

## C.1  EXAMPLES FROM THE INTEGERS

The set of **integers** is denoted by $Z = \{\ldots, -2, -1, 0, 1, 2, \ldots\}$. From your previous experience with the integers, you already know about positive and negative integers, the basic algebraic operations of addition, subtraction, multiplication and division, as well as how to compare such numbers using $<, >, =, \leq$, and $\geq$. It is also assumed that you are familiar with all of the basic properties these operations and comparisons satisfy (for example, $m + n = n + m$, $(m + n)p = mp + np$, and if $m \leq n$ then $-m \geq -n$).

It is also useful to compare the set of integers to other sets of numbers, such as the **rationals**, denoted by $Q = \{p/q : p, q \in Z$ and $q \neq 0\}$. For example, one difference between the integers and the rationals is that, when you divide two rational numbers, the result is a rational number (assuming the divisor is not 0). In contrast, when you divide two integers, the result is not necessarily an integer. For example, $5/3$ is not an integer because, when 5 is divided by 3, there is a remainder of 2. However, sometimes the result of dividing two integers does result in an integer; for example, 3 divides 18 evenly because 18 = 6(3), thus leaving no remainder. This property is stated in the following definition.

**Definition 32** *An integer $a$ **divides** an integer $b$, written $a|b$, if and only if there is an integer $c$ such that $b = ca$.*

Another property that differentiates the integers from the rationals is that any nonempty set of positive integers always contains a smallest element. For example, the smallest element of the following set is 4:

$$\{a \in Z : a > 0 \text{ and } a^2 > 10\}.$$

In contrast, the following set of rationals has no smallest element:

$$\{r \in Q : r > 0 \text{ and } r^2 > 10\}.$$

This differentiating property of the integers is an axiom (see Chapter 3) that is assumed to be true and is stated formally as follows.

**The Least Integer Principle:** Every nonempty set of positive integers contains a least element.

The Least Integer Principle is a statement that you can assume is true when doing a proof. Consequently, because the keyword (*for*) *every* appears in the Least Integer Principle, you can use specialization (see Chapter 6) when working forward from this statement. To do so, identify a specific set of positive integers, say, $M$, and show that $M \neq \emptyset$. Specialization then allows you to state that $M$ has a least element. The process of applying specialization in this way is referred to as "using the Least Integer Principle."

### How to Do a Proof

You will now see how the foregoing concepts are used in proving that long-hand division works. That is, when you want to divide any integer $b$ by any integer $a \geq 1$, you find the largest number of times $q$ you can divide $a$ into

$b$ without having $qa$ exceed $b$, thus leaving a remainder of $r$, with $0 \leq r < b$. That is, you would define

$$q = \max\{k \in Z : ka \leq b\},$$
$$r = b - aq = b - a\max\{k \in Z : ka \leq b\}. \qquad \text{(C.1)}$$

One primary issue of concern is how do you know that the set in (C.1) has a largest element? The answer is to use the Least Integer Principle. To do so, it is necessary to convert the problem of wanting to find the maximum of a set of integers into an equivalent problem of wanting to find the minimum of a set of integers. To that end, the next proposition shows that the following set of integers has a least element (which is the remainder $r$):

$$M = \{w \in Z : w \geq 0, \text{ and there is an integer } k \text{ such that } w = b - ak\}.$$

In the associated proof, pay particular attention to how the form of the statement under consideration suggests the technique to use.

**Proposition 37** *If $a$ and $b$ are integers with $a \geq 1$, then $M = \{w \in Z : w \geq 0$, and there is an integer $k$ such that $w = b - ak\}$ has a least element.*

**Analysis of Proof.** Note that the conclusion contains the hidden keywords "there is" because you can rewrite the conclusion as follows:

**B1:** There is an integer $r$ such that $r$ is the least element of $M$.

Recognizing the keywords "there is" in the backward statement $B1$, you should consider using the construction method (see Chapter 4) to produce the desired integer $r$. One way to do so is to turn to the forward process and use previous knowledge (see Section 3.2) in the form of the Least Integer Principle, which states that

**A1:** Every nonempty set of positive integers contains a least element.

Recognizing the keyword "every" in the forward statement $A1$, you should consider using specialization (see Chapter 6). If you could specialize $A1$ to the set $M$ in the conclusion, then the result would be that $M$ has a least element, which is precisely $B1$.

To specialize $A1$ to the set $M$ in the conclusion, you must verify that $M$ satisfies the certain property in $A1$ of being a nonempty set of positive integers; that is, you must show that

**B2:** $M = \{w \in Z : w \geq 0$, and there is an integer $k$ with $w = b - ak\}$ is a nonempty set of positive integers.

To show that $M$ is not empty, you must show that

**B3:** There is an integer $k$ such that $w = b - ak \geq 0$.

Recognizing the keywords "there is" in the backward statement $B3$, you should use the construction method to produce the desired value of $k$. So, what value of $k$ should you use? Clearly $k$ must be $\geq 0$, so what about using $k = 0$? In this case you would have $w = b - ak = b - a(0) = b$, which works if $b \geq 0$. But what if $b < 0$? In this case, constructing $k = b$ results in $w = b - ak = b - ab = b(1 - a)$. Recalling from the hypothesis of the proposition that $a \geq 1$, you have $b < 0$ and $1 - a \leq 0$, so $w = b(1 - a) \geq 0$, as desired. In summary, the following proof by cases (see Section 12.1) is appropriate, depending on whether $b \geq 0$ or not.

**Case 1:** $b \geq 0$.
In this case construct $k = 0$, for then $w = b - ak = b \geq 0$, and so $B3$ is true.

**Case 2:** $b < 0$.
In this case construct $k = b$, for then $w = b - ak = b(1 - a) \geq 0$ because $a \geq 1$, and so again $B3$ is true.

From $B2$, it remains to show that $M$ is a set of positive integers. The set $M = \{w \in Z : w \geq 0,$ and there is an integer $k$ with $w = b - ak\}$ will contain positive integers except in the case when $0 \in M$. However, in this case 0 is the least element of $M$, and so the conclusion is true and the proof is complete. Here again, you should recognize a proof by cases, depending on whether $0 \in M$ or not, as summarized in the following condensed proof, which begins with this last proof by cases.

**Proof of Proposition 37.** In the event that $0 \in M$, then 0 is the least element of $M$, and so the proof would be done. You can therefore assume that $0 \notin M$. In this case the Least Integer Principle ensures the existence of a least element in $M$, provided that $M \neq \emptyset$; thus, it is necessary to show that

$$M = \{w \in Z : w \geq 0, \text{ and there is an integer } k \text{ with } w = b - ak\} \neq \emptyset.$$

However, you can see that $M$ is not empty because, if $b \geq 0$, then $w = b \in M$; whereas if $b < 0$, then $w = b - ab = b(1 - a) \geq 0$ is an element of $M$. Thus, in either case, $M \neq \emptyset$, and so the proof is complete. $\square$

A summary of suggestions on how to do a proof is given at the end of this appendix. Now you will see how to read a proof.

## How to Read a Proof

Reading proofs can be challenging because the author does not always refer to the techniques by name, several steps may be combined in a single sentence with little or no justification, and the steps of a proof are not necessarily presented in the order in which they were performed when the proof was done (as you just saw in the proof of Proposition 34). To read a proof, you have to reconstruct the author's thought processes. Doing so requires that you identify which techniques are used and how they apply to the particular problem. In the next proposition, it is proved that long-hand division works.

**Proposition 38 (The Division Algorithm)** *If* $a, b \in Z$ *with* $a \geq 1$, *then there are unique integers* $q$ *and* $r$ *with* $0 \leq r < a$ *such that* $b = aq + r$.

**Proof of Proposition 38.**    (For reference purposes, each sentence of the proof is written on a separate line.)

**S1:**    From Proposition 34, the set $M = \{w \in Z : w \geq 0$, and there is an integer $k$ with $w = b - ak\}$ has a least element, say, $r$.

**S2:**    Because $r \in M$, by definition of $M$, $r \geq 0$, and there is an integer $q$ such that $r = b - aq$; that is, $b = aq + r$.

**S3:**    To see that $r < a$, suppose not.

**S4:**    Then $w = b - a(q + 1) \in M$ because $w = b - a(q + 1) = b - aq - a = r - a \geq 0$.

**S5:**    However, you now have the contradiction that $w < r$ because $a > 0$ and so $w = b - a(q + 1) = b - aq - a < b - aq = r$.

**S6:**    To address the uniqueness of $q$ and $r$, suppose that $m$ and $n$ are also integers for which $b = am + n$ and $0 \leq n < a$.

**S7:**    Then $b = aq + r = am + n$, so $a(q - m) = n - r$, and also, from $0 \leq r < a$ and $0 \leq n < a$, you have that $-a < n - r < a$.

**S8:**    It now follows that $-a < a(q - m) < a$, and so $q = m$.

**S9:**    Finally, because $q = m$, you have $n - r = 0$, so $n = r$.

The proof is now complete.    $\square$

**Analysis of Proof.**    An interpretation of statements $S1$ through $S9$ follows.

**Interpretation of S1:** *From Proposition 34, the set* $M = \{w \in Z : w \geq 0$, *and there is an integer* $k$ *with* $w = b - ak\}$ *has a least element, say,* $r$.

The reason you may find this sentence challenging to understand is that the author has not mentioned which technique is used to begin the proof. To determine this, ask yourself which technique *you* would use to get started. Recognizing the keywords "there are" and "unique" in the conclusion should lead you (and the author) to use a backward uniqueness method (see Section 11.1). Can you figure out whether the author has used the direct or indirect uniqueness method? In either case, the first step is to use the construction method to produce the integers $q$ and $r$. Can you determine how the author does so? Reading $S1$, perhaps you can see that the author uses the previous knowledge (see Section 3.2) of Proposition 34 to

**A1:**  Construct $r$ as the least element of the set $M = \{w \in Z : w \geq 0$, and there is an integer $k$ with $w = b - ak\}$.

The author must still construct $q$ and show that $r$ and $q$ have the desired properties in the conclusion of the proposition.

**Interpretation of S2:** *Because $r \in M$, by definition of $M$, $r \geq 0$, and there is an integer $q$ such that $r = b - aq$; that is, $b = aq + r$.*

Here the author works forward from the fact that $r \in M$, which by definition of $M$ means

**A2:** $r \geq 0$, and there is an integer $q$ such that $r = b - aq$.

From $A2$, you can see that the author has also constructed the value of $q$. According to the construction method, it is necessary to show that $r$ and $q$ have the desired properties in the conclusion; namely,

**B1:** $0 \leq r < a$ and $b = aq + r$.

From $A2$, you can see that $r \geq 0$ and $q$ satisfies $r = b - aq$, which you can rewrite as $b = aq + r$. Looking at $B1$, it remains to show that

**B2:** $r < a$.

This is precisely what $S3$ is about.

**Interpretation of S3:** *To see that $r < a$, suppose not.*

The words "suppose not" in $S3$ indicate that the author is going to show $B2$ by using the contradiction method (see Chapter 9). Accordingly, you can assume that

**A3 (NOT B2):** $r \geq a$.

To determine what the contradiction is, read ahead in the proof.

**Interpretation of S4:** *Then $w = b - a(q+1) \in M$ because $w = b - a(q+1) = b - aq - a = r - a \geq 0$*

This sentence is challenging to understand unless you keep in mind that the author is trying to reach a contradiction. Evidently, the value of $w = b - a(q + 1)$ in $S4$ is helpful in reaching that contradiction. However, unless you have read ahead to determine the contradiction, for now all you can do is verify that the author is correct in stating that

**A4:** $w = b - a(q + 1) \in M$.

To see that this value of $w$ is in $M$, by definition of $M$ it must be shown that $w \geq 0$ and there is an integer $k$ such that $w = b - ak$. From $A4$, it is easy to see that $k = q + 1$, so the author does not mention this. However, the author does note in $S4$ that $w \geq 0$, which is true because

$$
\begin{aligned}
w &= b - a(q + 1) && \text{(from } A4\text{)} \\
&= b - aq - a && \text{(algebra)} \\
&= r - a && \text{(from } A2\text{)} \\
&\geq 0 && \text{(from } A3\text{)}.
\end{aligned}
$$

It remains to understand how the value of $w$ in $A4$ is used to reach a contradiction.

**Interpretation of S5:** *However, you now have the contradiction that $w < r$ because $a > 0$ and so $w = b - a(q + 1) = b - aq - a < b - aq = r$.*

Here, the author reaches the contradiction that $w < r$. Can you see why this is a contradiction? The answer is that, if $w < r$, then $w$ would be a smaller element of $M$ (see $A4$) than $r$, which is the smallest element of $M$ (see $A1$). In $S5$ the author provides the following justification that $w < r$:

$$
\begin{aligned}
w &= b - a(q + 1) & \text{(from } A4) \\
&= b - aq - a & \text{(algebra)} \\
&< b - aq & (a > 0 \text{ from the hypothesis)} \\
&= r & \text{(from } A2).
\end{aligned}
$$

This completes the construction of the integers $q$ and $r$. To finish the proof, the author must now address the issue of the uniqueness of these two numbers.

**Interpretation of S6:** *To address the uniqueness of $q$ and $r$, suppose that $m$ and $n$ are also integers for which $b = am + n$ and $0 \leq n < a$.*

The author is using the direct uniqueness method (see Section 11.1). Accordingly, in addition to the already-constructed integers $q$ and $r$ that have been shown to satisfy the following properties:

**A5:**  $0 \leq r < a$
**A6:**  $b = aq + r,$

the author correctly assumes that $m$ and $n$ are also integers that satisfy

**A7:**  $0 \leq n < a$
**A8:**  $b = am + n.$

According to the direct uniqueness method, the author must show that

**B3:**  $m = q$ and $n = r.$

Can you see where and how the author establishes that $B3$ is true?

**Interpretation of S7:** *Then $b = aq + r = am + n$, so $a(q - m) = n - r$, and also, from $0 \leq r < a$ and $0 \leq n < a$, you have that $-a < n - r < a$.*

The author is working forward from $A6$ and $A8$ to claim correctly that

**A9:**  $aq + r = am + n.$

Subtracting $am$ and $r$ from both sides of $A9$ results in

**A10:**  $a(q - m) = n - r.$

Likewise, the author works forward from $A5$ and $A7$ by algebra to state correctly that

**A11:**  $-a < n - r < a.$

The foregoing forward steps should be helpful in showing that $B3$ is true.

**Interpretation of S8:** *It now follows that* $-a < a(q-m) < a$, *and so* $q = m$.

The author continues to work forward and, in fact, establishes that $q = m$, as needed in $B3$. Specifically, the author replaces $n - r$ in $A11$ with $a(q - m)$ from $A10$ to claim that

**A12:** $-a < a(q - m) < a$.

Without telling you, the author then divides $A12$ through by $a > 0$ (see the hypothesis) to obtain

**A13:** $-1 < q - m < 1$.

Finally, the author realizes, without telling you, that the only integer strictly between $-1$ and $1$ is $0$; thus,

**A14:** $q - m = 0$; that is, $q = m$.

To establish $B3$, it still remains to show that $r = n$.

**Interpretation of S9:** *Finally, because* $q = m$, *you have* $n - r = 0$, *so* $n = r$.

It is here that the author establishes that $r = n$ and thus that $B3$ is true. Can you determine how the author reaches the conclusion that $r = n$? The answer is that the author works forward by substituting $q - m = 0$ from $A14$ in $A10$ to obtain

**A15:** $a(0) = n - r$; that is, $0 = n - r$; therefore $r = n$.

Having established that $B3$ is true, the direct uniqueness method and hence the whole proof are now complete.

A summary of how to read proofs is given at the end of this appendix.

## C.2    EXAMPLES FROM GROUPS

One accomplishment of modern algebra is the development of a mathematical framework that enables one to study not only numbers but also many other mathematical objects—such as functions (see Appendix A) and vectors and matrices (see Appendix B)—that have operations with properties similar to those of numbers. The approach for doing so is to replace the set $Z$ of integers with a set $G$ of general objects (which could be functions, vectors, or matrices, for example). It is also necessary to have some way to combine two objects in $G$ to create a new object in $G$. To that end, let $\odot$ be such an operation; that is, if $a, b \in G$, then $a \odot b$ is the element in $G$ obtained by combining $a$ and $b$ in some unspecified way. For each specific set $G$, it is necessary to specify how to combine two elements using the operation $\odot$, as in the following examples.

**Example 1:** To study the set of integers under addition, set $G = Z$ and $\odot = +$. You already know how to combine two integers using addition.

**Example 2:** To study the set of rational numbers except for 0 under multiplication, set $G = Q - \{0\}$ and $\odot = *$. You already know how to combine two rational numbers using multiplication.

**Example 3:** To study $(n \times n)$ invertible matrices (see Appendix B) under matrix multiplication, set $G = \{A : A$ is an $(n \times n)$ invertible matrix$\}$ and $\odot = *$. You already know how to multiply two $(n \times n)$ matrices.

The operation $\odot$ has no special properties other than combining two elements of $G$. Thus, for a set $G$ with $a, b \in G$ and a general operation $\odot$, it may or may not be the case that $a \odot b = b \odot a$. For instance, this *commutative* property of $\odot$ holds for Examples 1 and 2 but not for Example 3. One property that does hold in all three examples is the following:

$$\text{For all } a, b, c \in G, \ (a \odot b) \odot c = a \odot (b \odot c). \tag{C.2}$$

Also, if you examine Examples 1, 2, and 3 carefully, you will see that in each case there is a special element in $G$ that when combined with any other element in $G$ results in the original element. That is,

$$\text{There is an element } e \in G \text{ such that for all } a \in G, \ a \odot e = e \odot a = a. \tag{C.3}$$

In Example 1, $e = 0$; in Example 2, $e = 1$; and in Example 3, $e = I$, the $(n \times n)$ identity matrix.

Finally, each of the foregoing three examples also has the property that each element in the set has a unique inverse element that when combined with the original element results in $e$; that is,

$$\text{For all } a \in G, \text{ there is an } a^{-1} \in G \text{ such that } a \odot a^{-1} = a^{-1} \odot a = e. \tag{C.4}$$

In Example 1, $a^{-1} = -a$; in Example 2, $a^{-1} = 1/a$; and in Example 3, $a^{-1}$ is the inverse of the matrix $a$.

Putting together the properties in (C.2), (C.3), and (C.4), a part of modern algebra involves studying the following mathematical structure.

**Definition 33** *Let $G$ be a set and $\odot$ be an operation on $G$ with the property that, for all $a, b \in G$, $a \odot b \in G$. The pair $(G, \odot)$ is a* **group** *if and only if the following properties hold:*

1. *(Associative Law) For all $a, b, c \in G$, $(a \odot b) \odot c = a \odot (b \odot c)$.*

2. *(Existence of an Identity Element) There is an element $e \in G$ such that, for all $a \in G$, $a \odot e = e \odot a = a$. ($e$ is called the* **identity element**.)

3. *(Existence of an Inverse Element) For each $a \in G$, there is an element $a^{-1} \in G$ such that $a \odot a^{-1} = a^{-1} \odot a = e$. (The element $a^{-1}$ is called the* **inverse** *of $a$.)*

(a) The identity element $e \in G$ is unique.
(b) There is a unique inverse element $a^{-1} \in G$.
(c) (*Left Cancellation Law*) If $a \odot b = a \odot c$, then $b = c$.
(d) (*Right Cancellation Law*) If $b \odot a = c \odot a$, then $b = c$.
(e) (*Inverse of the Inverse*) $(a^{-1})^{-1} = a$.
(f) (*Inverse of a Product*) $(a \odot b)^{-1} = b^{-1} \odot a^{-1}$.

*Table C.1*  Properties of a group $(G, \odot)$ for all elements $a, b, c \in G$.

Whenever $(G, \odot)$ is a group, you know that the three properties in Definition 33 hold. You can therefore work forward by specializing the for-all statements in each property to appropriate elements of $G$. This is done here in many proofs without actually referring to the specialization method. In addition to the properties in the definition, a group $(G, \odot)$ satisfies many other properties, such as those listed in Table C.1, some of which you are asked to prove in the exercises at the end of this appendix.

Another concept related to a group $(G, \odot)$ is raising an element $a \in G$ to an integer power; that is, for an integer $n \geq 0$:

$$a^0 = e, \quad a^1 = a, \quad a^2 = a \odot a, \ldots, \quad a^n = \underbrace{(((a \odot a) \odot a) \odot \cdots \odot a)}_{n \text{ times}} \quad (\text{C.5})$$

$$a^{-n} = (a^{-1})^n. \quad (\text{C.6})$$

## How to Do a Proof

The following proposition is used to illustrate how to do a proof involving a group. As always, observe how keywords in the statements under consideration suggest what proof technique to use.

**Proposition 39**  *If $(G, \odot)$ is a group and $a \in G$, then, for all integers $m$ and $n$, $a^m \odot a^n = a^{m+n}$.*

**Analysis of Proof.**    Recognizing the keywords "for all" in the conclusion, you might consider using the choose method. However, because the objects are integers, you can also consider using induction (see Section 11.2). Here, both methods are used. That is, you can choose

   **A1:** An integer $m$,

for which you must show that

   **B1:** For all integers $n$, $a^m \odot a^n = a^{m+n}$.

To use induction now, the statement $B1$ should be in the form, "For all integers $n \geq$ some initial integer, $n_0$, $P(n)$ is true." What is missing in $B1$ is the initial integer, $n_0$. To introduce this initial integer, you can rewrite $B1$ as follows:

**B2:** For all integers $n \geq 0$, $a^m \odot a^n = a^{m+n}$, and
for all integers $n \leq 0$, $a^m \odot a^n = a^{m+n}$.

Both of the statements in $B2$ are now proved by induction, in which

**P(n):** $a^m \odot a^n = a^{m+n}$.

According to the induction method, the first step is to show that $P(n)$ is true for $n = 0$; that is, that $a^m \odot a^0 = a^{m+0} = a^m$. But this is clear because

$$
\begin{aligned}
a^m \odot a^0 &= a^m \odot e && \text{[definition of } a^0 \text{ in (C.5)]} \\
&= a^m && \text{[property (2) of } e \text{ in Definition 33].}
\end{aligned}
$$

The next step of induction is to assume that $P(n)$ is true, so assume that

**P(n):** $a^m \odot a^n = a^{m+n}$.

You must then show that the statement is true for $n + 1$; that is, you must show that

**P(n + 1):** $a^m \odot a^{n+1} = a^{m+n+1}$.

So, start with the left side of $P(n+1)$ and try to rewrite it so as to use the induction hypothesis $P(n)$, as follows (using the fact that $n > 0$):

$$
\begin{aligned}
a^m \odot a^{n+1} &= a^m \odot (a^n \odot a) && \text{[definition of } a^{n+1} \text{ for } n > 0 \text{ in (C.5)]} \\
&= (a^m \odot a^n) \odot a && \text{[property (1) in Definition 33]} \\
&= a^{m+n} \odot a && \text{[induction hypothesis } P(n)\text{].}
\end{aligned}
$$

Rewriting $a^{m+n} \odot a$ to obtain $a^{m+n+1}$ now depends on whether $m + n \geq 0$, so you should proceed by cases.

**Case 1:** $m + n \geq 0$.    (It will be shown that $a^{m+n} \odot a = a^{m+n+1}$.)
In this case, by (C.5), it follows that $a^{m+n} \odot a = a^{m+n+1}$ and so $P(n+1)$ is true.

**Case 2:** $m + n < 0$.    (It will be shown that $a^{m+n} \odot a = a^{m+n+1}$.)
In this case you also have that $P(n+1)$ is true because

$$
\begin{aligned}
a^{m+n} \odot a &= (a^{m+n} \odot e) \odot a && \text{[property (2) in Definition 33]} \\
&= [(a^{m+n} \odot a) \odot a^{-1}] \odot a && \text{[properties (1) and (3) in Def. 33]} \\
&= (a^{m+n+1} \odot a^{-1}) \odot a && \text{[from (C.6)]} \\
&= a^{m+n+1} \odot (a^{-1} \odot a) && \text{[property (1) in Definition 33]} \\
&= a^{m+n+1} \odot e && \text{[property (3) in Definition 33]} \\
&= a^{m+n+1} && \text{[property (2) in Definition 33].}
\end{aligned}
$$

Turning now to the second statement in $B2$, it has already been shown that $P(n)$ is true for $n = 0$. However, because you now want to show that $P(n)$ is true for all integers $n \leq 0$, you should assume that $P(n)$ is true and show that $P(n - 1)$ is also true. Thus, you should assume that

**P(n):** $a^m \odot a^n = a^{m+n}$.

You must show that

**P(n − 1):** $a^m \odot a^{n-1} = a^{m+n-1}$.

So, start with the left side of $P(n-1)$ and try to rewrite it so as to use the induction hypothesis $P(n)$, as follows (using the fact that $n < 0$):

$$
\begin{aligned}
a^m \odot a^{n-1} &= a^m \odot (a^{n-1} \odot e) && \text{[property (2) in Definition 33]} \\
&= a^m \odot (a^{n-1} \odot a) \odot a^{-1}) && \text{[properties (1) and (3) in Def. 33]} \\
&= a^m \odot (a^n \odot a^{-1}) && \text{[definition of } a^n \text{ in (C.6)]} \\
&= (a^m \odot a^n) \odot a^{-1} && \text{[property (1) in Definition 33]} \\
&= a^{m+n} \odot a^{-1} && \text{[induction hypothesis } P(n)].
\end{aligned}
$$

Rewriting $a^{m+n} \odot a^{-1}$ to obtain $a^{m+n-1}$ now depends on whether $m+n \le 0$, so you should proceed by cases.

**Case 1:** $m + n \le 0$.    (It will be shown that $a^{m+n} \odot a^{-1} = a^{m+n-1}$.)
   In this case, by (C.6), $a^{m+n} \odot a^{-1} = a^{m+n-1}$ and so $P(n+1)$ is true.

**Case 2:** $m + n > 0$.    (It will be shown that $a^{m+n} \odot a^{-1} = a^{m+n-1}$.)
   In this case you have that

$$
\begin{aligned}
a^{m+n} \odot a^{-1} &= (a^{m+n} \odot e) \odot a^{-1} && \text{[property (2) in Definition 33]} \\
&= (a^{m+n} \odot a^{-1}) \odot a) \odot a^{-1} && \text{[prop. (3) and (1) in Def. 33]} \\
&= (a^{m+n-1} \odot a) \odot a^{-1} && \text{[from (C.5)]} \\
&= a^{m+n-1} \odot (a \odot a^{-1}) && \text{[property (1) in Definition 33]} \\
&= a^{m+n-1} \odot e && \text{[property (3) in Definition 33]} \\
&= a^{m+n-1} && \text{[property (2) in Definition 33].}
\end{aligned}
$$

Thus, in this case, $P(n+1)$ is also true. The proof is now complete because you have used induction to establish that $B2$ is true.

**Proof of Proposition 39.**   Let $m$ be an integer. (The word "let" indicates that the choose method is used.) It will first be shown by induction that, for every integer $n \ge 0$, $a^m \odot a^n = a^{m+n}$. For $n = 0$, it is clear that $a^m \odot a^0 = a^m \odot e = a^m$. Assume now that $a^m \odot a^n = a^{m+n}$. Then for $n+1$,

$$
\begin{aligned}
a^m \odot a^{n+1} &= a^m \odot (a^n \odot a) \\
&= (a^m \odot a^n) \odot a \\
&= a^{m+n} \odot a \\
&= \begin{cases} a^{m+n+1} & \text{if } m + n \ge 0 \\ (a^{m+n+1} \odot a^{-1}) \odot a & \text{if } m + n < 0 \end{cases} \\
&= a^{m+n+1}
\end{aligned}
$$

It remains to show by induction that, for every integer $n \le 0$, $a^m \odot a^n = a^{m+n}$. The result for $n = 0$ has already been shown, so assume now that

$a^m \odot a^n = a^{m+n}$. Then for $n - 1$, you have

$$
\begin{aligned}
a^m \odot a^{n-1} &= a^m \odot (a^n \odot a^{-1}) \\
&= (a^m \odot a^n) \odot a^{-1} \\
&= a^{m+n} \odot a^{-1} \\
&= \begin{cases} a^{m+n-1} & \text{if } m+n < 0 \\ (a^{m+n-1} \odot a) \odot a^{-1} & \text{if } m+n \geq 0 \end{cases} \\
&= a^{m+n-1}
\end{aligned}
$$

The proof is now complete. $\square$

**How to Read a Proof**

The next proposition illustrates how to read a proof and uses the following axiom:

**Least Integer Principle:** Every nonempty set of positive integers contains a least element.

**Proposition 40** *If $a$ is an element of a group $(G, \odot)$ and $r \neq s$ are integers for which $a^r = a^s$, then there is a smallest integer $n > 0$ such that $a^n = e$.*

**Proof of Proposition 40.** (For reference purposes, each sentence of the proof is written on a separate line.)

**S1:** The result is established by showing that the set $M = \{\text{integers } k > 0 : a^k = e\}$ is not empty because then the Least Integer Principle ensures the existence of the integer $n$.

**S2:** To see that $M \neq \emptyset$, assume without loss of generality that $s > r$.

**S3:** Then because $a^r = a^s$, it follows that $e = a^s \odot a^{-r}$.

**S4:** Then from Proposition 36, you have that $a^s \odot a^{-r} = a^{s-r} = e$, and so $M \neq \emptyset$.

The proof is now complete. $\square$

**Analysis of Proof.** An interpretation of statements $S1$ through $S4$ follows.

**Interpretation of S1:** *The result is established by showing that the set $M = \{\text{integers } k > 0 : a^k = e\}$ is not empty because then the Least Integer Principle ensures the existence of the integer $n$.*
 The reason you may find this sentence challenging to understand is that the author has not mentioned which technique is used to begin the proof. To determine this, ask yourself which technique *you* would use to get started. Recognizing the keywords "there is" in the conclusion, you should use the

construction method (see Chapter 4) to produce the desired integer $n$. From $S1$, you can see that the author uses the Least Integer Principle to construct the integer $n$ as the smallest element of the set $M = \{\text{integers } k > 0 : a^k = e\}$. Then, recognizing the keywords "for all" in the Least Integer Principle, the author specializes this statement to the specific set $M$; however, to do so, it is necessary to show that $M$ is a nonempty set of positive integers. Noting that, by definition, $M$ consists of positive integers, you need only show that

**B1:** $M = \{\text{integers } k > 0 : a^k = e\} \neq \emptyset$.

**Interpretation of S2:** *To see that $M \neq \emptyset$, assume without loss of generality that $s > r$.*

The words "without loss of generality" in $S2$ indicate that the author is using a proof by cases (see Section 12.1), but can you determine why and how? To do so, ask yourself how you would prove that $M \neq \emptyset$. By definition of $M$, you must show that

**B2:** There is an integer $k > 0$ such that $a^k = e$.

Recognizing the keywords "there is" in $B2$ should lead you (and the author) to use the construction method to produce the desired integer $k$. Read forward in the proof to find the value of $k$ the author constructs. You can see in $S4$ that the value is $k = s - r$. Assuming this is the case, according to the construction method applied to $B2$, the author must show that $k = s - r > 0$ and $a^k = e$. Indeed, in $S2$, the author assumes, "without loss of generality," that $s > r$, so $k = s - r > 0$, as needed. The only question is, how can the author assume that $s > r$? The answer is that the author works forward from the hypothesis that $r \neq s$ to realize that

**A1:** Either $s > r$ or $r > s$.

Recognizing the keywords "either/or" in $A1$, the author proceeds with a proof by cases. Here, only the case $s > r$ is shown; the case $r > s$ is left for you to do on your own. Thus, the author is justified in assuming that $s > r$.

According to the construction method applied to $B2$, it remains to show that, for $k = s - r$, $a^k = e$.

**Interpretation of S3:** *Then because $a^r = a^s$, it follows that $e = a^s \odot a^{-r}$.*

In trying to establish that $a^k = a^{s-r} = e$, the author works forward from the hypothesis that $a^r = a^s$ using the properties of a group, as follows:

$$
\begin{aligned}
a^r &= a^s && [\text{hypothesis}] \\
a^r \odot a^{-r} &= a^s \odot a^{-r} && [\text{combine both sides with } a^{-r}] \\
e &= a^s \odot a^{-r} && [\text{Property (3) of Definition 33}].
\end{aligned}
$$

To complete the proof, it remains to show that $a^s \odot a^{-r} = a^{s-r}$.

**Interpretation of S4:** *Then from Proposition 36, you have that $a^s \odot a^{-r} = a^{s-r} = e$, and so $M \neq \emptyset$.*

The author obtains the desired conclusion that $a^s \odot a^{-r} = a^{s-r}$ by using previous knowledge (see Section 3.2). Specifically, the author specializes Proposition 36 to the integers $m = s$ and $n = -r$ to conclude correctly that

> **A2:** $a^s \odot a^{-r} = a^{s-r}$.

The author then mentions in $S4$ that the proof is now complete, which is correct because it has been shown that $a^{s-r} \in M$, and so the set $M \neq \emptyset$, thus allowing the author to apply the Least Integer Principle to $M$ to construct the desired integer $n$ in the conclusion of the proposition.

## Summary

Creating proofs is not a precise science. Some general suggestions are provided here. When trying to prove that "$A$ implies $B$," consciously choose a technique based on keywords that appear in $A$ and $B$. For example, if the quantifiers "there is" and "for all" appear, then consider using the corresponding construction, choose, induction, and specialization methods. If no keywords are apparent, then it is probably best to proceed with the forward-backward method. Remember that, as you proceed through a proof, different techniques are needed as the form of the statement under consideration changes. If you are unsuccessful at completing a proof, there are several avenues to pursue before giving up. You might try asking yourself why $B$ cannot be false, thus leading you to the contradiction (or contrapositive) method. If you are really stuck, it can sometimes be advantageous to leave the problem for a while because, when you return, you might see a new approach. Undoubtedly, you will learn many tricks of your own as you solve more and more problems.

Reading proofs can be challenging because the author does not always refer to the techniques by name, several steps may be combined in a single sentence with little or no justification, and the steps of a proof are not necessarily presented in the order in which they were performed when the proof was done. To read a proof, you have to reconstruct the author's thought processes. Doing so requires that you identify which techniques are used and how they apply to the particular problem. Begin by trying to determine which technique is used to start the proof. Then try to follow the methodology associated with that technique. Watch for quantifiers to appear, for then it is likely that the corresponding construction, choose, induction, and specialization methods are used. The inability to follow a particular step of a written proof is often because of the lack of sufficient detail. To fill in the gaps, learn to ask yourself how you would proceed to do the proof. Then try to see if the written proof matches your thought process.

## Exercises

**Note:** Solutions to those exercises marked with a $B$ are in the back of this book. Solutions to those exercises marked with a $W$ are located on the web at http://www.wiley.com/college/solow/.

**Note:** All proofs should contain an analysis of proof and a condensed version. Definitions for all mathematical terms are provided in the glossary at the end of the book.

$^B$**C.1**    Prove that, if $a$ and $b$ are integers for which $a|b$, then $a|(-b)$.

**C.2**    Prove that, if $a$ and $b$ are integers for which $a|b$ and $b|a$, then $a = \pm b$.

$^W$**C.3**    Prove that, if $a, b$, and $c$ are integers for which $a|b$ and $a|c$, then $a|(b+c)$.

$^B$**C.4**    Write an analysis of proof that corresponds to the following condensed proof. Indicate which techniques are used and how they are applied. Fill in the details of any missing steps where appropriate.

> **Definition.** An integer $p > 1$ is **prime** if and only if the only positive integers that divide $p$ are 1 and $p$.

> **Proposition.** For any integer $a > 1$, there are primes $p_1, \ldots, p_k$ such that $a = p_1 p_2 \cdots p_k$.

> **Proof.** Suppose not; then there is an integer $a > 1$ such that $a$ cannot be written as a product of primes. Let $b$ be the first such integer, which exists by the Least Integer Principle. Because $b$ is not prime, there are positive integers $b_1$ and $b_2$ with $1 < b_1, b_2 < b$ such that $b = b_1 b_2$. Because $1 < b_1, b_2 < b$, there are primes $q_1, \ldots, q_m$ and $r_1, \ldots, r_n$ such that $b_1 = q_1 q_2 \cdots q_m$ and $b_2 = r_1 r_2 \cdots r_n$. But then $b = b_1 b_2 = (q_1 q_2 \cdots q_m)(r_1 r_2 \cdots r_n)$, which contradicts the statement that $b$ cannot be written as a product of primes and thus completes the proof.  $\square$

**C.5**    Write an analysis of proof that corresponds to the following condensed proof. Indicate which techniques are used and how they are applied. Fill in the details of any missing steps where appropriate.

> **Proposition.** If $a$ and $b$ are integers not both 0, then there is a least positive integer $d$ such that $d = ax + by$, where $x, y \in Z$ and $d|a$ and $d|b$.

> **Proof.** Let $M = \{ax + by > 0 : x, y \in Z\}$, and note that $a^2 + b^2 \in M$ because $a$ and $b$ are not both 0, so $M \neq \emptyset$. By the Least Integer Principle, $M$ contains a smallest integer, say, $d = am + bn > 0$. It remains to show that $d|a$ and $d|b$. By the Division Algorithm (Proposition 35), there are integers $q$ and

$r$ with $0 \leq r < d$ such that $a = dq + r = (am + bn)q + r$. Thus, $r = a(1 - mq) + b(-nq) \in M$, but, because $d$ is the smallest positive element of $M$, it must be that $r = 0$. But then $a = dq$ and hence $d|a$. A similar argument shows that $d|b$. ☐

$^B$**C.6** Prove property (a) in Table C.1 that, if $e$ is the identity element of a group $(G, \odot)$, then $e$ is the only element in $G$ with the property that, for all elements $a \in G$, $a \odot e = e \odot a = a$.

**C.7** Prove property (b) in Table C.1 that, if $a^{-1}$ is the inverse of the element $a$ of a group $(G, \odot)$, then $a^{-1}$ is the only element in $G$ with the property that $a \odot a^{-1} = a^{-1} \odot a = e$.

$^W$**C.8** Prove property (c) in Table C.1 that, if $a, b$, and $c$ are elements of a group $(G, \odot)$ with the property that $a \odot b = a \odot c$, then $b = c$.

**C.9** Prove property (d) in Table C.1 that, if $a, b$, and $c$ are elements of a group $(G, \odot)$ with the property that $b \odot a = c \odot a$, then $b = c$.

$^W$**C.10** Prove property (e) in Table C.1 that, if $a$ is an element of a group $(G, \odot)$, then $(a^{-1})^{-1} = a$.

**C.11** Prove property (f) in Table C.1 that, if $a$ and $b$ are elements of a group $(G, \odot)$, then $(a \odot b)^{-1} = b^{-1} \odot a^{-1}$.

$^B$**C.12** Write an analysis of proof that corresponds to the following condensed proof. Indicate which techniques are used and how they are applied. Fill in the details of any missing steps where appropriate.

> **Proposition.** If $(G, \odot)$ is a group and $x \in G$, then $H = \{x^k : k \in Z\}$ together with the operation $\odot$ from $G$ is a group.
>
> **Proof.** Note first that the operation $\odot$ in $H$ combines two elements of $H$ and produces an element of $H$. This is because for $a, b \in H$, there are integers $i, j$ such that $a = x^i$ and $b = x^j$. So, by Proposition 36, $a \odot b = x^i \odot x^j = x^{i+j} \in H$. The remaining properties of a group are now established for $(H, \odot)$.
>
> Definition 33 is used to establish that $H$ is a group. To that end, let $a, b, c \in H$. Then $(a \odot b) \odot c = a \odot (b \odot c)$ because $a, b, c \in G$ and property (1) of $(G, \odot)$ being a group.
>
> The identity element of $H$ is the identity element of $G$, say, $e = x^0 \in H$. This is because, for $a \in H$, you have $a \odot e = e \odot a = a$ from the corresponding property of the group $(G, \odot)$.
>
> The inverse of $x^k \in H$ is $x^{-k} \in H$ because, by Proposition 36,
>
> $$x^k \odot x^{-k} = x^0 = e \quad \text{and} \quad x^{-k} \odot x^k = x^0 = e.$$
>
> The proof is now complete. ☐

**C.13**    Write an analysis of proof that corresponds to the following condensed proof. Indicate which techniques are used and how they are applied. Fill in the details of any missing steps where appropriate.

> **Proposition.** Let $(G, \odot)$ be a group with identity element $e$. For all elements $a, b \in G$, $a \odot b = b \odot a$ if and only if, for all elements $a, b \in G$, $(a \odot b)^{-1} = a^{-1} \odot b^{-1}$.

> **Proof.** Suppose first that, for all $x, y \in G$, $x \odot y = y \odot x$ and let $a, b \in G$. It will be shown that $(a \odot b)^{-1} = a^{-1} \odot b^{-1}$. However, you have that
>
> $$(a \odot b)^{-1} = b^{-1} \odot a^{-1} = a^{-1} \odot b^{-1}.$$
>
> Now suppose that, for all $x, y \in G$, $(x \odot y)^{-1} = x^{-1} \odot y^{-1}$ and let $a, b \in G$. It will be shown that $a \odot b = b \odot a$. However, because $(a \odot b)^{-1} = a^{-1} \odot b^{-1}$ and $(a \odot b)^{-1} = b^{-1} \odot a^{-1}$, you have that
>
> $$\begin{aligned} a^{-1} \odot b^{-1} &= b^{-1} \odot a^{-1} \\ b \odot (a^{-1} \odot b^{-1}) &= b \odot (b^{-1} \odot a^{-1}) \\ b \odot (a^{-1} \odot b^{-1}) &= a^{-1} \\ a \odot [b \odot (a^{-1} \odot b^{-1})] &= a \odot a^{-1} \\ (a \odot b) \odot (a^{-1} \odot b^{-1}) &= e \\ [(a \odot b) \odot (a^{-1} \odot b^{-1})] \odot b &= e \odot b \\ (a \odot b) \odot a^{-1} &= b \\ [(a \odot b) \odot a^{-1}] \odot a &= b \odot a \\ a \odot b &= b \odot a. \end{aligned}$$
>
> The proof is now complete.   $\square$

# Appendix D
# Examples of Proofs from Real Analysis

*Real analysis* is the study of properties of the real numbers. The objective of this appendix is not to teach real analysis, but rather to provide examples of how the various techniques you have learned are used in doing proofs in this subject. You will also see how to read and understand such written proofs as they might appear in a textbook or other mathematical literature. It is assumed that the reader is familiar with the basic properties of the real numbers; however, the material in this appendix is completely self-contained.

## D.1  EXAMPLES FROM THE REAL NUMBERS

The set of **real numbers** (also called the **reals**) is denoted by $R$ and consists of those numbers that can be expressed in decimal format, possibly with an infinite number of positions to the right of the decimal point. From your previous experience with the real numbers, you already know about positive

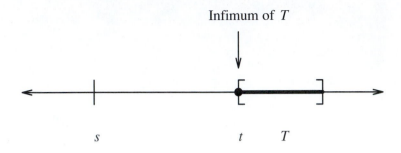

*Fig. D.1*   Lower bounds and the infimum $t$ for a set $T$.

and negative values and the basic algebraic operations of addition, subtraction, multiplication, and division, as well as how to compare such numbers using $<, >, =, \le$, and $\ge$. It is also assumed that you are familiar with all of the basic properties these operations and comparisons satisfy (for example, $x + y = y + x$, $(x + y)z = (xz) + (yz)$, if $x \le y$ then $-x \ge -y$, and, for every real number $x$, either $x < 0$, $x = 0$, or $x > 0$).

It is also useful to compare the set of real numbers to other sets of numbers, such as the **integers**, denoted by $Z = \{\ldots, -1, 0, 1, \ldots\}$, and the **rationals**, denoted by $Q = \{p/q : p, q \in Z$ and $q \ne 0\}$. For example, one difference between the reals and the integers is that, when you divide two real numbers, the result is a real number (assuming the divisor is not 0). In contrast, when you divide two integers, the result is not necessarily an integer (for example, $4/3$ is not an integer).

There are also differences between the reals and the rationals, but they are more subtle. For example, as is proved subsequently here, there are real numbers, such as $\sqrt{2}$, that are not rational. Another property that differentiates the reals from the rationals is based on the following definition.

**Definition 34** *A real number $t$ is a* **lower bound for a set $T$ of real numbers** *if and only if, for every element $x \in T$, $t \le x$.*

Some sets of real numbers, such as $\{x \in R : x > 0\}$, have a lower bound, whereas other sets, such as $\{x \in R : x < 2\}$, do not. If a set $T$ has a lower bound, say, $t$, then $T$ has infinitely many lower bounds because any real number $s < t$ is also a lower bound for $T$ (see Figure D.1). Of particular interest in this case is the *largest* lower bound, a concept that is formalized as follows.

**Definition 35** *A real number $t$ is the* **infimum of a set $T$ of real numbers** *(also called the* **greatest lower bound***) if and only if (1) $t$ is a lower bound for $T$ and (2) for any lower bound $s$ for $T$, $s \le t$.*

A property that differentiates the reals from the rationals arises by trying to identify when a set $T$ of real numbers has an infimum. Working with numerous examples should lead you to the conclusion that, as long as the set $T$ is not empty and has a lower bound, $T$ has an infimum. Although no proof exists for this fact, it is an axiom (see Chapter 3) that is assumed to be true and is stated formally as follows.

**The Infimum Property of** $R$**:** Every nonempty set of real numbers that has a lower bound has an infimum.

The Infimum Property of $R$ is a statement that you can assume is true when doing a proof. Consequently, because the keyword (*for*) *every* appears in the Infimum Property, you can use specialization when working forward from this statement. To do so, identify a specific nonempty set of real numbers, say, $M$, and show that $M$ has a lower bound. Then specialization allows you to state that $M$ has an infimum. The process of applying specialization in this way is referred to as "using the Infimum Property."

### How to Do a Proof

You will now see how the foregoing concepts are used in doing a proof. In the example that follows, pay particular attention to how the form of the statement under consideration suggests the technique to use.

**Proposition 41** *If* $T = \{s \in R : s > 0 \text{ and } s^2 > 2\}$, $x$ *is a real number with* $0 < x$ *and* $x^2 < 2$, *and* $n$ *is a positive integer with* $\frac{1}{n} < \frac{2-x^2}{2x+1}$, *then* $x + \frac{1}{n}$ *is a lower bound for* $T$.

**Analysis of Proof.** Not seeing any keywords—such as *there is, for all, no, not,* and so on—in the hypothesis or conclusion of the proposition, it is reasonable to begin with the forward-backward method. A key question associated with the conclusion is, "How can I show that a real number (namely, $x + \frac{1}{n}$) is a lower bound for a set (namely, $T$)?" Using Definition 34, one answer is to show that

**B1:** For all elements $s \in T$, $s \geq x + \frac{1}{n}$.

Recognizing the keywords "for all" in $B1$ and after identifying the objects (elements $s$), the certain property ($s \in T$), and the something that happens ($s \geq x + \frac{1}{n}$), you can use the choose method (see Chapter 5) to choose

**A1:** An element $s \in T$,

for which you must show that

**B2:** $s \geq x + \frac{1}{n}$.

Now work forward from $A1$ and the hypothesis to show that $B2$ is true. From $A1$ you know that $s \in T$, so by the defining property of $T$,

**A2:** $s > 0$ and $s^2 > 2$.

If you can show that $2 > \left(x + \frac{1}{n}\right)^2$, then you would have $s^2 > 2 > \left(x + \frac{1}{n}\right)^2$, and because $s > 0$ (see $A2$), it would follow that $s > x + \frac{1}{n}$, as desired in $B2$. In summary, you should now try to show that

**B3:** $2 > \left(x + \frac{1}{n}\right)^2$.

You can obtain $B3$ by working forward from the hypothesis that $\frac{1}{n} < \frac{2-x^2}{2x+1}$ using algebra, as follows:

$$\left(x + \frac{1}{n}\right)^2 \;=\; x^2 + \frac{2x}{n} + \frac{1}{n^2} \quad \text{(expand the square)} \tag{D.1}$$

$$=\; x^2 + \frac{1}{n}\left(2x + \frac{1}{n}\right) \quad \text{(factor out } \tfrac{1}{n}) \tag{D.2}$$

$$\leq\; x^2 + \frac{1}{n}\left(2x + 1\right) \quad (1 \leq n) \tag{D.3}$$

$$<\; x^2 + (2 - x^2) \quad \text{(from the hypothesis } \tfrac{1}{n} < \tfrac{2-x^2}{2x+1}) \tag{D.4}$$

$$=\; 2. \tag{D.5}$$

The choose method and the whole proof are now complete.

**Proof of Proposition 41.**   To see that $x + \frac{1}{n}$ is a lower bound for the set $T$, let $s \in T$. (The word "let" here indicates that the choose method is used.) It will be shown that $x + \frac{1}{n} < s$. Because $s \in T$, $s > 0$ and $s^2 > 2$. It now follows from the algebra in equations (D.1) through (D.5) that $s^2 > 2 > \left(x + \frac{1}{n}\right)^2$. Because $s > 0$, you have that $s > x + \frac{1}{n}$, which shows that $x + \frac{1}{n}$ is a lower bound for $T$, thus completing the proof.  □

A summary of suggestions on how to do a proof is given at the end of this appendix. Now you will see how to read a proof.

### How to Read a Proof

Reading proofs can be challenging because the author does not always refer to the techniques by name, several steps may be combined in a single sentence with little or no justification, and the steps of a proof are not necessarily presented in the order in which they were performed when the proof was done. To read a proof, you have to reconstruct the author's thought processes. Doing so requires that you identify which techniques are used and how they apply to the particular problem. The next example demonstrates how to do so and establishes formally that $\sqrt{2}$ is a real number that is not rational.

**Proposition 42** *If the infimum property of real numbers holds; that is, if every nonempty set of real numbers that has a lower bound has an infimum, then there is a real number $x$ such that $x^2 = 2$ and $x$ is not rational.*

**Proof of Proposition 42.**     (For reference purposes, each sentence of the proof is written on a separate line.)

**S1:**     The desired value of $x$ is the infimum of the set $T = \{s \in R : s > 0$ and $s^2 > 2\}$.

**S2:**     The set $T$ has an infimum because $2 \in T$, and so $T \neq \emptyset$ and also $0$ is a lower bound for $T$.

**S3:**     It must now be shown that $x^2 = 2$.

**S4:**     Now if $x^2 > 2$, then a value of the positive integer $n$ for which

$$\frac{1}{n} < \frac{x^2 - 2}{2x}$$

can be shown to satisfy $x - \frac{1}{n} \in T$, and because $x - \frac{1}{n} < x$, it would follow that $x$ is not a lower bound for $T$, which is a contradiction.

**S5:**     On the other hand, if $x^2 < 2$, then a value of the positive integer $n$ for which

$$\frac{1}{n} < \frac{2 - x^2}{2x + 1}$$

can be shown to satisfy $x + \frac{1}{n}$ is a lower bound for $T$, and because $x < x + \frac{1}{n}$, it would follow that $x$ is not the greatest lower bound for $T$, which is a contradiction.

**S6:**     Having ruled out $x^2 > 2$ and $x^2 < 2$, it must be that $x^2 = 2$.

**S7:**     The fact that $x$ is not rational is proved in Proposition 13 in Section 9.2 and is not repeated here.

The proof is now complete.     □

**Analysis of Proof.**     An interpretation of statements $S1$ through $S7$ follows.

**Interpretation of S1:**     *The desired value of $x$ is the infimum of the set $T = \{s \in R : s > 0$ and $s^2 > 2\}$.*

The reason you may find this sentence challenging to understand is that the author has not mentioned which technique is used to begin the proof. To determine this, ask yourself which technique *you* would use to get started. Recognizing the keywords "there is" in the conclusion of the proposition should lead you (and the author) to use the construction method (see Chapter 4). Indeed, as stated in $S1$, the author constructs the value of

**A1:** $x =$ infimum of the set $T = \{s \in R : s > 0 \text{ and } s^2 > 2\}$.

Before showing that this value of $x$ is correct, as required by the construction method, you should ask yourself if this set $T$ does, in fact, have an infimum; otherwise, it is not possible to construct $x$ in this way. The author asked the same question and establishes in $S2$ that $T$ does have an infimum.

**Interpretation of S2:** *The set $T$ has an infimum because $2 \in T$, and so $T \neq \emptyset$ and also 0 is a lower bound for $T$.*

Here the author claims that the set $T$ in $S1$ has an infimum. Once again, the author has failed to mention what proof techniques are used and has also skipped several steps, which you must recreate. Specifically, the author recognizes the keywords "(for) every" in the hypothesis, "... every nonempty set of real numbers that has a lower bound has an infimum" and therefore uses specialization (see Chapter 6). To specialize this for-all statement to the set $T$ in $S1$, it is necessary to show that $T$ satisfies the certain property of being nonempty and having a lower bound. Indeed, in $S2$ the author claims without justification that $2 \in T$ (which you can verify because $2 > 0$ and $2^2 = 4 > 2$), and so $T \neq \emptyset$. In $S2$ the author also states without justification that 0 is a lower bound for $T$ (which you are asked to verify in the exercises). The result of specializing the hypothesis is that $T$ does have an infimum, so the construction of $x$ in $S1$ is valid.

**Interpretation of S3:** *It must now be shown that $x^2 = 2$.*

Following the construction method in $S1$, the author must show that the value of $x$ satisfies the desired properties, so it must be shown that

**B1:** $x^2 = 2$ and $x$ is not rational.

Reading the rest of the proof, can you determine how the author does so?

**Interpretation of S4:** *Now if $x^2 > 2$, then a value of the positive integer $n$ for which*

$$\frac{1}{n} < \frac{x^2 - 2}{2x}$$

*can be shown to satisfy $x - \frac{1}{n} \in T$, and because $x - \frac{1}{n} < x$, it would follow that $x$ is not a lower bound for $T$, which is a contradiction.*

This sentence is challenging to understand unless you keep in mind from $B1$ that the author is trying to show that $x^2 = 2$ and read ahead to $S6$. In $S6$ the author notes that the cases $x^2 > 2$ and $x^2 < 2$ have been ruled out, so the only remaining possibility is that $x^2 = 2$, as desired. In other words, without telling you, the author is working forward using the property of real numbers to note that

**A2:** Either $x^2 = 2$ or $x^2 > 2$ or $x^2 < 2$.

To establish that $x^2 = 2$ the author proceeds in $S4$ to rule out the possibility that $x^2 > 2$. Then, in $S5$ the author rules out $x^2 < 2$. It remains to determine how the author rules out these two cases.

The answer is that the author uses the contradiction method (see Chapter 9). Specifically, to rule out the case when $x^2 > 2$, the author uses the contradiction method in $S4$ and so assumes that $x^2 > 2$. Can you determine what the contradiction is? The answer is that at the end of $S4$ the author claims to have shown that $x$ is not a lower bound for $T$, which, if true, is a contradiction. This is because, according to $A1$, $x$ is the infimum of $T$, and so, by Definition 35, $x$ is a lower bound for $T$.

In order for the author to show that $x$ is not a lower bound for $T$, using the rules for writing the NOT of Definition 34 (see Chapter 8), the author must show that

**B2:** There is an element $s \in T$ such that $s < x$.

Recognizing the keywords "there is" in $B2$, the author uses the construction method to produce an element $s \in T$ such that $s < x$. Specifically, in $S4$ the author constructs $s = x - \frac{1}{n}$, where $n$ is a positive integer for which $\frac{1}{n} < \frac{x^2 - 2}{2x}$. According to the construction method, the author should now show that this value of $s$ satisfies the certain property ($s \in T$) and the something that happens ($s < x$) in $B2$. Unfortunately, the author leaves these details for you to verify. While it is perhaps easy to see that $s = x - \frac{1}{n} < x$ because $n > 0$, it is not so easy to verify that $s = x - \frac{1}{n} \in T$. Can you use the fact that $\frac{1}{n} < \frac{x^2 - 2}{2x}$ to show that $s = x - \frac{1}{n} \in T$?

**Interpretation of S5:** *On the other hand, if $x^2 < 2$, then a value of the positive integer $n$ for which*

$$\frac{1}{n} < \frac{2 - x^2}{2x + 1}$$

*can be shown to satisfy $x + \frac{1}{n}$ is a lower bound for $T$, and because $x < x + \frac{1}{n}$, it would follow that $x$ is not the greatest lower bound for $T$, which is a contradiction.*

Here, the author rules out the case that $x^2 < 2$ by assuming that $x^2 < 2$ and reaching a contradiction. Specifically, in $S5$ the author claims to have shown that $x$ is not the greatest lower bound for $T$, which, if true, contradicts the fact that, by construction in $A1$, $x$ is the greatest lower bound for $T$.

In order for the author to show that $x$ is not the greatest lower bound for $T$, using the rules for writing the NOT of Definition 35 (see Chapter 8), the author must show that

**B3:** Either $x$ is not a lower bound for $T$ or there is a lower bound $s$ for $T$ such that $s > x$.

Recognizing the keywords "either/or" in the backward statement $B3$, the author uses a proof by elimination (see Section 12.1). Accordingly, the author assumes that

**A3:** $x$ is a lower bound for $T$,

and so it must now be shown that

**B4:** There is a lower bound $s$ for $T$ such that $s > x$.

Recognizing the keywords "there is" in $B4$, the author uses the construction method to produce this value of $s$. Specifically, in $S5$ the author constructs $s = x + \frac{1}{n}$, where $n$ is a positive integer for which $\frac{1}{n} < \frac{2-x^2}{2x+1}$. According to the construction method, the author should now show that this value of $s$ satisfies the certain property ($s$ is a lower bound for $T$) and the something that happens ($s > x$) in $B4$. Unfortunately, the author leaves these details for you to verify. While it is perhaps easy to see that $s = x + \frac{1}{n} > x$ because $n > 0$, it is not so easy to verify that $s = x + \frac{1}{n}$ is a lower bound for $T$. This fact is proved in the preceding Proposition 38 in this appendix.

**Interpretation of S6:** *Having ruled out $x^2 > 2$ and $x^2 < 2$, it must be that $x^2 = 2$.*

The author is now claiming that $x^2 = 2$, as required in $B1$ to complete the construction method. The author is justified in making this claim because in $S4$ the case $x^2 > 2$ has been ruled out by contradiction, whereas in $S5$ the case $x^2 < 2$ has been ruled out by contradiction.

**Interpretation of S7:** *The fact that $x$ is not rational is proved in Proposition 13 in Section 9.2 and is not repeated here.*

The author observes from $B1$ that it must still be shown that the constructed value of $x$ is not rational but notes that this has already been proved previously, thus completing the proof.

A summary of how to read proofs is given at the end of this appendix.

## D.2    EXAMPLES FROM SEQUENCES

In solving many problems, it is necessary to work with an infinite collection of real numbers. For example, if you start with $x_1 = 2$ and repeatedly compute

$$x_{k+1} = \frac{1}{2}\left(x_k + \frac{2}{x_k}\right), \quad \text{for } k = 1, 2, \ldots \tag{D.6}$$

you will generate an infinite number of real numbers that, in fact, are getting closer and closer to the value of $\sqrt{2}$. These concepts are now formalized, and you will see how to read and do proofs using these concepts, starting with the following definition.

**Definition 36** *A* **sequence of real numbers** *is a function* $x : N \to R$, *where* $N = \{1, 2, \ldots\}$, *and* $R$ *is the set of real numbers. For* $k \in N$, *the notation* $x_k$ *is used instead of* $x(k)$, *and the sequence is written as follows:*

$$X = (x_1, x_2, \ldots) \quad \text{or} \quad X = (x_k : k \in N) \quad \text{or} \quad X = (x_k).$$

For example, the collection of numbers defined by equation (D.6) is a sequence, as is $X = \left(\frac{1}{1}, \frac{1}{2}, \ldots, \frac{1}{k}, \ldots\right)$.

It is also possible to create new sequences from existing sequences. For example, given the sequences $X = (x_1, x_2, \ldots)$ and $Y = (y_1, y_2, \ldots)$,

$$X \pm Y = (x_1 \pm y_1, x_2 \pm y_2, \ldots).$$

Of interest is whether the values in a given sequence are *converging*—that is, getting closer and closer to some specific number. In that regard, there are two separate convergence questions pertaining to a sequence $X = (x_1, x_2, \ldots)$:

1. Given a specific number $x$, does the sequence $X$ converge to $x$?

2. Is there any real number $x$ to which the sequence converges and, if so, to which real number?

Some sequences, such as the one defined by (D.6), converge, and others, such as the sequence $X = (1, 2, \ldots, k, \ldots)$, do not converge. The objective now is to provide a formal definition of what it means for a sequence to converge to a given real number.

Intuitively, if a sequence $X$ converges to the given real number $x$, then the further along you are in the sequence, the closer the numbers in the sequence should be to $x$. The *distance* from a point $x_k$ in the sequence to the number $x$ is measured using the absolute value; that is, $|x - x_k|$. The set of all points whose distance from $x$ is less than some fixed amount $\epsilon > 0$ is central to convergence and is defined formally as follows (see Figure D.2):

**Definition 37** *Given a real number* $x$ *and a distance* $\epsilon > 0$, *the* **neighborhood of radius** $\epsilon$ **around** $x$ *is* $N_\epsilon(x) = \{y \in R : |x - y| < \epsilon\}$.

fNote from Figure D.2 that, as $\epsilon$ gets closer to 0, the $\epsilon$-neighborhoods around $x$ are "shrinking" to $x$. This observation gives rise to the following notion of a sequence $X$ converging to $x$ by requiring that

> No matter how close $\epsilon$ is to 0, "most" of the points in the sequence lie inside $N_\epsilon(x)$.

Mathematicians have determined that "most" means all of the points in the sequence after some point, say, $x_j$. Putting together the pieces, the formal definition of convergence, which contains three quantifiers, follows.

**Definition 38** *A given sequence* $X = (x_1, x_2, \ldots)$ **converges to a real number** $x$, *written* $(x_k) \to x$, *if and only if, for every real number* $\epsilon > 0$, *there is an integer* $j \in N$ *such that, for all* $k \in N$ *with* $k > j$, $x_k \in N_\epsilon(x)$ *that is,* $|x_k - x| < \epsilon$.

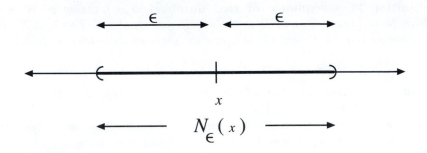

*Fig. D.2*   The neighborhood of radius $\epsilon$ around the real number $x$.

## How to Do a Proof

The following proposition is used to illustrate how to do a proof involving a sequence. As always, observe how keywords in the statements under consideration suggest which proof technique to use.

**Proposition 43**   *The sequence $X = \left(1 - \frac{1}{k^2}\right)$ converges to 1.*

**Analysis of Proof.**   Not seeing any keywords—such as *there is, for all, no, not*, and so on—in the proposition, it is reasonable to begin with the forward-backward method. A key question associated with the conclusion is, "How can I show that a sequence [namely, $X = \left(1 - \frac{1}{k^2}\right)$] converges to a specific number (namely, 1)?" Using Definition 38, one answer is to show that

**B1:**   For every real number $\epsilon > 0$, there is an integer $j \in N$ such that, for all $k \in N$ with $k > j$, $\left|1 - \frac{1}{k^2} - 1\right| = \frac{1}{k^2} < \epsilon$.

Recognizing the keywords "for all" as the first quantifier in the backward statement $B1$, you should now use the choose method to choose

**A1:**   A real number $\epsilon > 0$,

for which it must be shown that

**B2:**   There is an integer $j \in N$ such that, for all $k \in N$ with $k > j$, $\frac{1}{k^2} < \epsilon$.

Recognizing the keywords "there is" as the first quantifier in $B2$, you should now consider using the construction method. You can turn to the forward process in an attempt to do so; however, another approach is to assume that you have *already* constructed the desired value of $j \in N$. To complete the construction method, you would then have to show that your value of $j$ satisfies the something that happens in $B2$; namely, that

**B3:**   For all $k \in N$ with $k > j$, $\frac{1}{k^2} < \epsilon$.

The idea now is to try to prove $B3$ and in so doing to discover what value of $j \in N$ allows you to do so. Proceeding with the backward process and recognizing the keywords "for all" in $B3$, you should now use the choose method to choose

**A2:** An integer $k \in N$ with $k > j$,

for which you must show that

**B4:** $\frac{1}{k^2} < \epsilon$.

Now, from $A2$, you know that $k > j \geq 1$, so $k^2 > j^2$ and thus

**A3:** $\frac{1}{k^2} < \frac{1}{j^2}$.

You can obtain $B4$ from $A3$ provided that

$$\frac{1}{j^2} < \epsilon. \tag{D.7}$$

Indeed, this tells you that you need to construct $j$ so that equation (D.7) holds. Solving the inequality (D.7) for $j$ using the fact that $\epsilon > 0$ leads to

$$j > \frac{1}{\sqrt{\epsilon}}. \tag{D.8}$$

In other words, constructing $j$ to be any integer satisfying equation (D.8) enables you to show that $B4$ is true, thus completing the proof.

**Proof of Proposition 43.** To show that the sequence $X = \left(1 - \frac{1}{k^2}\right)$ converges to 1, let $\epsilon > 0$. (The word "let" here indicates that the choose method is used.) Now let $j$ be any integer with $j > \frac{1}{\sqrt{\epsilon}}$, which is possible because $\epsilon > 0$. (This time the word "let" means that the construction method is used.) Now, for $k > j$ (here, the choose method is used), it follows that

$$\left|1 - \frac{1}{k^2} - 1\right| = \frac{1}{k^2} \quad \text{(algebra)}$$

$$< \frac{1}{j^2} \quad \text{(because } k > j \geq 1\text{)}$$

$$< \epsilon \quad \text{(because } j > \frac{1}{\sqrt{\epsilon}} \text{)}.$$

The proof is now complete. $\square$

A summary of how to do proofs is given at the end of this appendix.

## How to Read a Proof

The next proposition is used to illustrate how to read a proof and establishes that, if two sequences converge, then their sum converges.

**Proposition 44** *If $X = (x_1, x_2, \ldots)$ and $Y = (y_1, y_2, \ldots)$ are two sequences of real numbers that converge to the given real numbers $x$ and $y$, respectively, then the sequence $X + Y = (x_1 + y_1, x_2 + y_2, \ldots)$ converges to $x + y$.*

**Proof of Proposition 44.**    (For reference purposes, each sentence of the proof is written on a separate line.)

**S1:**    To show that $X + Y$ converges to $x + y$, let $\epsilon > 0$.

**S2:**    Because $X$ converges to $x$, by definition there is an integer $j_1 \in N$ such that, for all $k \in N$ with $k > j_1$, $|x_k - x| < \epsilon/2$.

**S3:**    Likewise, because $Y$ converges to $y$, there is an integer $j_2 \in N$ such that, for all $k \in N$ with $k > j_2$, $|y_k - y| < \epsilon/2$.

**S4:**    Now, defining $j = \max\{j_1, j_2\}$, you have from the convergence of $X$ and $Y$ that, for $k > j$,

$$
\begin{aligned}
|(x_k + y_k) - (x + y)| \;&=\; |(x_k - x) + (y_k - y)| \\
&\leq\; |x_k - x| + |y_k - y| \\
&<\; \tfrac{\epsilon}{2} + \tfrac{\epsilon}{2} \\
&=\; \epsilon.
\end{aligned}
$$

The proof is now complete.    $\square$

**Analysis of Proof.**    An interpretation of statements $S1$ through $S4$ follows.

**Interpretation of S1:** *To show that $X + Y$ converges to $x + y$, let $\epsilon > 0$.*
   The reason you may find this sentence challenging to understand is that the author has not mentioned which technique is used to begin the proof. To determine this, ask yourself which technique *you* would use to get started. Not seeing any keywords—such as *there is, for all, no, not,* and so on—in the hypothesis or conclusion of the proposition, it is reasonable to begin with the forward-backward method (which is what this author has done, without telling you). Working backward, a key question associated with the conclusion is, "How can I show that a sequence (namely, $X + Y$) converges to a real number (namely, $x + y$)?" The author then uses Definition 38 to answer this question, so it must be shown that

**B1:** For every real number $\epsilon > 0$, there is an integer $j \in N$ such that, for all $k \in N$ with $k > j$, $|(x_k + y_k) - (x + y)| < \epsilon$.

The three quantifiers in $B1$ should be processed from left to right (see Chapter 7). Because the first keywords in $B1$ are "for every," the author uses the choose method to choose

**A1:** A real number $\epsilon > 0$,

for which it must be shown that

**B2:** There is an integer $j \in N$ such that, for all $k \in N$ with $k > j$, $|(x_k + y_k) - (x + y)| < \epsilon$.

Indeed, the words "let $\epsilon > 0$" in $S1$ indicate that the author is using the choose method. Accordingly, the author must now show that $B2$ is true. Which technique would you use to do so?

The first quantifier in $B2$ is "there is," so the construction method should be used to produce the desired integer $j \in N$. Recognizing the need for the construction method, the author turns in $S2$ to the forward process. (Can you identify where in the proof the author constructs the value of $j$?)

**Interpretation of S2:** *Because $X$ converges to $x$, by definition there is an integer $j_1 \in N$ such that, for all $k \in N$ with $k > j_1$, $|x_k - x| < \epsilon/2$.*

The author is working forward from the hypothesis that $X$ converges to $x$. Specifically, using Definition 38 with $\bar{\epsilon}$ in place of $\epsilon$ to avoid overlapping notation, you can state that

**A2:** For every real number $\bar{\epsilon} > 0$, there is an integer $j \in N$ such that, for all $k \in N$ with $k > j$, $|x_k - x| < \bar{\epsilon}$.

Recognizing the keywords "for every" in $A2$, you should consider using specialization, as the author does. The only question is what specific value of $\bar{\epsilon}$ should you use for the specialization? Perhaps you should use the value of $\bar{\epsilon} = \epsilon$ that was chosen in $A1$? Rereading $S2$, you can see that the author uses the value $\bar{\epsilon} = \epsilon/2$ instead of $\bar{\epsilon} = \epsilon$. At this point it is not clear why the author has used this particular value. In any event, before specializing $A2$, it is necessary to verify that the specific value of $\bar{\epsilon} = \epsilon/2$ satisfies the certain property in $A2$ of being $> 0$, which is true because $\epsilon > 0$ (see $A1$). The result of specialization, as the author states in $S2$, is that,

**A3:** There is an integer $j_1 \in N$ such that, for all $k \in N$ with $k > j_1$, $|x_k - x| < \epsilon/2$.

**Interpretation of S3:** *Likewise, because $Y$ converges to $y$, there is an integer $j_2 \in N$ such that, for all $k \in N$ with $k > j_2$, $|y_k - y| < \epsilon/2$.*

The author works forward from the hypothesis that the sequence $Y$ converges to $y$ in the same way as is done for the sequence $X$ in $S2$, so

**A4:** There is an integer $j_2 \in N$ such that, for all $k \in N$ with $k > j_2$, $|y_k - y| < \epsilon/2$.

**Interpretation of S4:** *Now, defining $j = \max\{j_1, j_2\}$, you have from the convergence of $X$ and $Y$ that, for $k > j$, ....*

Recall from $B2$ that the construction method is being used to produce a value for the desired integer $j$. It is here in $S4$ that the author finally uses the values of $j_1$ from $A3$ and $j_2$ from $A4$ to construct $j = \max\{j_1, j_2\}$. According to the construction method, it remains to show that this value of $j$ satisfies the something that happens in $B2$; namely, that

**B3:** For all $k \in N$ with $k > j$, $|(x_k + y_k) - (x + y)| < \epsilon$.

Recognizing the keywords "for all" in $B3$, without telling you, the author uses the choose method to choose

**A5:** An integer $k \in N$ with $k > j$,

for which it must be shown that

**B4:** $|(x_k + y_k) - (x + y)| < \epsilon$.

Indeed, in $S4$, the author shows that $B4$ is true, but can you justify each of the following steps?

$$|(x_k + y_k) - (x + y)| \quad = \quad |(x_k - x) + (y_k - y)| \tag{D.9}$$

$$\leq \quad |x_k - x| + |y_k - y| \tag{D.10}$$

$$< \quad \frac{\epsilon}{2} + \frac{\epsilon}{2} \tag{D.11}$$

$$= \quad \epsilon. \tag{D.12}$$

It should be clear that the equality in (D.9) results from using algebra to rearrange the terms. Likewise, the inequality in (D.10) follows from algebra because $|a + b| \leq |a| + |b|$. The inequality in (D.11), however, is not so clear. The inequalities

$$|x_k - x| < \frac{\epsilon}{2} \quad \text{and} \quad |y_k - y| < \frac{\epsilon}{2}$$

are obtained from $A3$ and $A4$, respectively, by specializing their for-all statements. Specifically, the author specializes the following for-all statement from $A3$ using the specific value of $k$ chosen in $A5$:

**A6:** For all $k \in N$ with $k > j_1$, $|x_k - x| < \frac{\epsilon}{2}$.

To apply specialization to $A6$, the author must show that the value of $k$ from $A5$ satisfies the certain property in $A6$ (of being in $N$) and also that $k > j_1$. It is clear that $k \in N$; however, can you see why $k > j_1$? The answer is that $k > j$ (see $A5$) and $j$ is constructed as $\max\{j_1, j_2\}$; that is, $k > j = \max\{j_1, j_2\} \geq j_1$. Thus, you can specialize $A6$ to the value of $k$ from $A5$, resulting in

**A7:** $|x_k - x| < \frac{\epsilon}{2}$.

A similar specialization argument applied to the for-all statement in $A4$ results in

**A8:** $|y_k - y| < \frac{\epsilon}{2}$.

Combining $A7$ and $A8$ results in the inequality in equation (D.11).

Finally, the last inequality in equation (D.12) follows by simple algebra. Perhaps now it is also clear why the author specialized the for-all statements in the definition of convergence of the sequences $X$ and $Y$ to $\bar{\epsilon} = \epsilon/2$ instead of $\bar{\epsilon} = \epsilon$; namely, so that in equation (D.12), $\frac{\epsilon}{2} + \frac{\epsilon}{2} = \epsilon$, as required in $B1$.

## Summary

Creating proofs is not a precise science. Some general suggestions are provided here. When trying to prove that "$A$ implies $B$," consciously choose a technique based on keywords that appear in $A$ and $B$. For example, if the quantifiers "there is" and "for all" appear, then consider using the corresponding construction, choose, induction, and specialization methods. If no keywords are apparent, then it is probably best to proceed with the forward-backward method. Remember that, as you proceed through a proof, different techniques are needed as the form of the statement under consideration changes. If you are unsuccessful at completing a proof, there are several avenues to pursue before giving up. You might try asking yourself why $B$ cannot be false, thus leading you to the contradiction (or contrapositive) method. If you are really stuck, it can sometimes be advantageous to leave the problem for a while because, when you return, you might see a new approach. Undoubtedly, you will learn many tricks of your own as you solve more and more problems.

Reading proofs can be challenging because the author does not always refer to the techniques by name, several steps may be combined in a single sentence with little or no justification, and the steps of a proof are not necessarily presented in the order in which they were performed when the proof was done. To read a proof, you have to reconstruct the author's thought processes. Doing so requires that you identify which techniques are used and how they apply to the particular problem. Begin by trying to determine which technique is used to start the proof. Then try to follow the methodology associated with that technique. Watch for quantifiers to appear, for then it is likely that the corresponding construction, choose, induction, and specialization methods are used. The inability to follow a particular step of a written proof is often because of the lack of sufficient detail. To fill in the gaps, learn to ask yourself how you would proceed to do the proof. Then try to see if the written proof matches your thought process.

## Exercises

**Note:** Solutions to those exercises marked with a $B$ are in the back of this book. Solutions to those exercises marked with a $W$ are located on the web at http://www.wiley.com/college/solow/.

**Note:** All proofs should contain an analysis of proof and a condensed version. Definitions for all mathematical terms are provided in the glossary at the end of the book.

$^B$**D.1**    Let $X = (x_1, x_2, \ldots)$ and $Y = (y_1, y_2, \ldots)$ be sequences of real numbers that converge to the real numbers $x$ and $y$, respectively.

   a. State a common key question associated with trying to prove that each of the following statements is true: $X - Y$ converges to $x - y$, $X \bullet Y = (x_1 y_1, x_2 y_2, \ldots,)$ converges to $xy$, and $X/Y = (x_1/y_1, x_2/y_2, \ldots,)$ converges to $x/y$, provided that all denominators are not 0.

   b. Provide a specific answer to your question in part (a) for trying to prove that $X \bullet Y = (x_1 y_1, x_2 y_2, \ldots,)$ converges to $xy$ to create a new statement in the backward process.

   c. Based on your answer to part (b), which proof technique would you use next and why?

$^B$**D.2**    Recall from Proposition 39 that $\sqrt{2}$ is shown to be the infimum of the set $T = \{s \in R : s > 0 \text{ and } s^2 > 2\}$.

   a. What set should you use instead of $T$ if you want to prove the existence of the $n^{\text{th}}$ root of 2, where $n$ is a positive integer greater than 1?

   b. Prove that your set in part (a) has an infimum.

**D.3**    Recall from Proposition 39 that $\sqrt{2}$ is shown to be the infimum of the set $T = \{s \in R : s > 0 \text{ and } s^2 > 2\}$.

   a. What set should you use instead of $T$ if you want to prove the existence of the $\sqrt{a}$, where $a$ is a positive real number?

   b. Prove that your set in part (a) has an infimum.

**D.4**    Prove that, if a set $T$ of real numbers has an infimum, then that infimum is unique.

$^W$**D.5**    Complete the proof of Proposition 39 by proving that, if $T = \{s \in R : s > 0 \text{ and } s^2 > 2\}$, then 0 is a lower bound for $T$.

$^B$**D.6**    Write an analysis of proof that corresponds to the following condensed proof. Indicate which techniques are used and how they are applied. Fill in the details of any missing steps where appropriate.

**Proposition.** If $t$ is the infimum of a set $T$ of real numbers, then, for every real number $\epsilon > 0$, there is an element $x \in T$ such that $x < t + \epsilon$.

**Proof.** Let $\epsilon > 0$, for which it must be shown that there is an element $x \in T$ such that $x < t + \epsilon$. To that end, observe that, because $t$ is the greatest lower bound for $T$ and $t + \epsilon > t$, it follows that $t + \epsilon$ is not a lower bound for $T$. This, in turn, means that there is an element $x \in T$ such that $x < t + \epsilon$ and so the proof is complete. $\square$

**D.7**   Write an analysis of proof that corresponds to the following condensed proof. Indicate which techniques are used and how they are applied. Fill in the details of any missing steps where appropriate.

**Proposition.** If $T$ is a set of real numbers and $t$ is a lower bound for $T$ with the property that, for every real number $\epsilon > 0$, there is an element $x \in T$ such that $x < t + \epsilon$, then $t$ is the infimum of $T$.

**Proof.** Because $t$ is a lower bound for $T$, by definition, it remains only to show that, for any lower bound $s$ for $T$, $s \leq t$. So, let $s$ be a lower bound for $T$. Now if $s > t$, then letting $\epsilon = s - t > 0$, it follows from the hypothesis that there is an element $x \in T$ such that $x < t + \epsilon = t + (s - t) = s$. In other words, $x < s$, which contradicts the fact that $s$ is a lower bound for $T$, and this completes the proof. $\square$

[B]**D.8**   Write the negation of Definition 38; that is, write what it means for a sequence $X = (x_1, x_2, \ldots)$ not to converge to the real number $x$.

**D.9**   A sequence $X = (x_1, x_2, \ldots)$ is **monotone increasing** if and only if, for each $i = 1, 2, \ldots$, $x_i < x_{i+1}$. Write what it means for a sequence not to be monotone increasing.

[B]**D.10**   Prove that the sequence $X = (\frac{1}{1}, \frac{1}{2}, \frac{1}{3}, \ldots,)$ converges to 0.

**D.11**   Use Exercise D.8 to prove that the sequence $X = (\frac{1}{1}, \frac{1}{2}, \frac{1}{3}, \ldots,)$ does not converge to 1. Start by using the forward-backward method.

**D.12**   Prove that, if $X = (x_1, x_2, \ldots)$ and $Y = (y_1, y_2, \ldots)$ are sequences that converge to the real numbers $x$ and $y$, respectively, then the sequence $X - Y = (x_1 - y_1, x_2 - y_2, \ldots)$ converges to $x - y$, as follows:

a. First prove that the sequence $-Y = (-y_1, -y_2, \ldots)$ converges to $-y$.

b. Now use Proposition 41 to prove that $X - Y$ converges to $x - y$.

$^W$**D.13**    Write an analysis of proof that corresponds to the following condensed proof. Indicate which techniques are used and how they are applied. Fill in the details of any missing steps where appropriate.

> **Proposition.** If $x$ and $y$ are real numbers such that, for every real number $\epsilon > 0$, $|x - y| < \epsilon$, then $x = y$.
>
> **Proof.** Suppose, to the contrary, that $x \neq y$. Letting $\epsilon = |x - y| > 0$, it follows from the hypothesis that $|x - y| < |x - y|$. This contradiction completes the proof. $\square$

**D.14**    Write an analysis of proof that corresponds to the following condensed proof. Indicate which techniques are used and how they are applied. Fill in the details of any missing steps where appropriate.

> **Proposition.** If $X = (x_1, x_2, \ldots)$ is a sequence of real numbers that converges to the real number $x$, then $x$ is the only real number to which the sequence $X$ converges.
>
> **Proof.** Suppose that the sequence $X$ also converges to the real number $y$. It will be shown that $x = y$ by showing that, for every real number $\epsilon > 0$, $|x - y| < \epsilon$. (See Exercise D.13.) To that end, let $\epsilon > 0$. Because $X$ converges to $x$, by definition, there is an integer $j_1 \in N$ such that, for all integers $k \in N$ with $k > j_1$, $|x_k - x| < \epsilon/2$. Likewise, because $X$ also converges to $y$, there is an integer $j_2 \in N$ such that, for all integers $k \in N$ with $k > j_2$, $|x_k - y| < \epsilon/2$. Choosing $k \in N$ with $k > \max\{j_1, j_2\}$, it then follows that
>
> $$
> \begin{aligned}
> |x - y| &= |x - x_k + x_k - y| \\
> &\leq |x - x_k| + |x_k - y| \\
> &< \tfrac{\epsilon}{2} + \tfrac{\epsilon}{2} \\
> &= \epsilon.
> \end{aligned}
> $$
>
> The proof is now complete. $\square$

# Solutions to
# Selected Exercises

## Solutions to Selected Exercises in Chapter 1

**1.1** (a), (c), and (e) are statements.

**1.3** a. Hypothesis: The right triangle $XYZ$ with sides of lengths $x$ and $y$ and hypotenuse of length $z$ has an area of $z^2/4$.

Conclusion: The triangle $XYZ$ is isosceles.

b. Hypothesis: $n$ is an even integer.

Conclusion: $n^2$ is an even integer.

c. Hypothesis: $a$, $b$, $c$, $d$, $e$, and $f$ are real numbers for which $ad - bc \neq 0$.

Conclusion: The two linear equations $ax + by = e$ and $cx + dy = f$ can be solved for $x$ and $y$.

**1.6** Jack's statement is true. This is because the hypothesis that Jack did not get his car fixed is false. Therefore, according to rows 3 and 4 of Table 1.1, the if/then statement is true, regardless of the truth of the conclusion.

**1.9** a. True because $A : 2 > 7$ is false (see rows 3 and 4 of Table 1.1).

b. True because $B : 1 < 2$ is true (see rows 1 and 3 of Table 1.1).

**1.11** If you want to prove that "$A$ implies $B$" is true and you know that $B$ is false, then $A$ should also be false. The reason is that, if $A$ is false, then it does not matter whether $B$ is true or false because Table 1.1 ensures that "$A$ implies $B$" is true. On the other hand, if $A$ is true and $B$ is false, then "$A$ implies $B$" would be false.

**1.13**                 (T = true, F = false)

| $A$ | $B$ | $C$ | $B \Rightarrow C$ | $A \Rightarrow (B \Rightarrow C)$ |
|---|---|---|---|---|
| T | T | T | T | T |
| T | T | F | F | F |
| T | F | T | T | T |
| T | F | F | T | T |
| F | T | T | T | T |
| F | T | F | F | T |
| F | F | T | T | T |
| F | F | F | T | T |

**1.16** According to row 2 of Table 1.1, you must show that $A$ is true and $B$ is false.

**1.17** a. For $A$ to be true and $B$ to be false, it is necessary to find a real number $x > 0$ such that $\log_{10}(x) \leq 0$. For example, $x = 0.1 > 0$, while $\log_{10}(0.1) = -1 \leq 0$. Thus, $x = 0.1$ is a desired counterexample. (Any value of $x$ such that $0 < x \leq 1$ would provide a counterexample.)

  b. For $A$ to be true and $B$ to be false, it is necessary to find an integer $n > 0$ such that $n^3 < n!$. For example, $n = 6 > 0$, while $6^3 = 216 < 720 = 6!$. Thus, $n = 6$ is a desired counterexample. (Any integer $n \geq 6$ would provide a counterexample.)

## Solutions to Selected Exercises in Chapter 2

**2.3** a. This question is incorrect because it asks how to prove the hypothesis, not the conclusion. This key question is also incorrect because it uses specific notation from the problem.

  b. This question is incorrect because it asks how to prove the hypothesis, not the conclusion.

  c. This question is incorrect because it uses the specific notation given in the problem.

  d. This question is correct.

**2.9** Any answer to a key question for a statement $B$ that results in a new statement $B1$ must have the property that, if $B1$ is true, then $B$ is true. In this case, the answer that "**B1:** the integer is odd" does not mean that "**B:** the integer is prime." For example, the odd integer $9 = 3(3)$ is not prime.

**2.13 a.**  How can I show that two lines are parallel?

How can I show that two lines do not intersect?

How can I show that two lines tangent to a circle are parallel?

How can I show that two tangent lines passing through the endpoints of the diameter of a circle are parallel?

**b.**  How can I show that a function is a polynomial?

How can I show that the sum of two functions is a polynomial?

How can I show that the sum of two polynomials is a polynomial?

**2.17 a.**  Show that one number is $\leq$ the other number and vice versa.

Show that the ratio of the two (nonzero) numbers is 1.

Show that the two numbers are both equal to a third number.

**b.**  Show that the elements of the two sets are identical.

Show that each set is a subset of the other.

Show that both sets are equal to a third set.

**2.22**  (1) How can I show that the solution to a quadratic equation is positive?

(2) Show that the quadratic formula gives a positive solution.

(3) Show that the solution $-b/2a$ is positive.

**2.26**  The given statement is obtained by working forward from the conclusion rather than the hypothesis.

**2.29 a.**  $(x-2)(x-1) < 0$.

$x(x-3) < -2$.

$-x^2 + 3x - 2 > 0$.

**b.**  $x/z = 1/\sqrt{2}$.

Angle $X$ is a 45-degree angle.

$\cos(X) = 1/\sqrt{2}$.

**c.**  The circle has its center at $(3,2)$.

The circle has a radius of 5.

The circle crosses the $y$-axis at $(0,6)$ and $(0,-2)$.

$x^2 - 6x + 9 + y^2 - 4y + 4 = 25$.

**2.31**  (d) is not valid because "$x \neq 5$" is not stated in the hypothesis and so, if $x = 5$, it will not be possible to divide by $x - 5$.

**2.35**  Key Question:    How can I show that a real number is 0?

Key Answer:    Show that the number is less than or equal to 0 and that the number is greater than or equal to 0.

**2.40 Analysis of Proof.**  In this problem one has:

**A:**  The right triangle $XYZ$ is isosceles.

**B:**  The area of triangle $XYZ$ is $z^2/4$.

A key question for $B$ is, "How can I show that the area of a triangle is equal to a particular value?" One answer is to use the formula for computing the area of a triangle to show that

**B1:** $z^2/4 = xy/2$.

Working forward from the hypothesis that triangle $XYZ$ is isosceles,

**A1:** $x = y$, so
**A2:** $x - y = 0$.

Because $XYZ$ is a right triangle, from the Pythagorean theorem,

**A3:** $z^2 = x^2 + y^2$.

Squaring both sides of the equality in $A2$ and performing algebraic manipulations yields

**A4:** $(x - y)^2 = 0$.
**A5:** $x^2 - 2xy + y^2 = 0$.
**A6:** $x^2 + y^2 = 2xy$.

Substituting $A3$ in $A6$ yields

**A7:** $z^2 = 2xy$.

Dividing both sides by 4 finally yields the desired result:

**A8:** $z^2/4 = xy/2$.

**Proof.** From the hypothesis, $x = y$, or equivalently, $x - y = 0$. Performing algebraic manipulations yields $x^2 + y^2 = 2xy$. By the Pythagorean theorem, $z^2 = x^2 + y^2$, and, on substituting $z^2$ for $x^2 + y^2$, one obtains $z^2 = 2xy$, or, $z^2/4 = xy/2$. From the formula for the area of a right triangle, the area of $XYZ = xy/2$, so $z^2/4$ is the area of the triangle. ☐

## Solutions to Selected Exercises in Chapter 3

**3.1**  a.  Key Question:  How can I show that an integer (namely, $n^2$) is odd?

Abstract Answer:  Show that the integer equals two times some integer plus one.

Specific Answer  Show that $n^2 = 2k + 1$ for some integer $k$.

b.  Key Question:  How can I show that a real number (namely, $s/t$) is rational?

Abstract Answer:  Show that the real number is equal to the ratio of two integers in which the denominator is not zero.

Specific Answer:  Show that $s/t = p/q$, where $p$ and $q$ are integers and $q \neq 0$.

c. Key Question:        How can I show that two pairs of real numbers
                        (namely, $(x_1, y_1)$ and $(x_2, y_2)$) are equal?

   Abstract Answer:     Show that the first and second elements of one
                        pair of real numbers are equal to the corresponding
                        elements of the other pair.

   Specific Answer:     Show that $x_1 = x_2$ and $y_1 = y_2$.

**3.6**  (T = true, F = false)

a. Truth Table for the Converse of "$A$ Implies $B$."

| $A$ | $B$ | "$A$ Implies $B$" | "$B$ Implies $A$" |
|---|---|---|---|
| T | T | T | T |
| T | F | F | T |
| F | T | T | F |
| F | F | T | T |

b. Truth Table for the Inverse of "$A$ Implies $B$."

| $A$ | $B$ | $NOT\ A$ | $NOT\ B$ | "$NOT\ A$ Implies $NOT\ B$" |
|---|---|---|---|---|
| T | T | F | F | T |
| T | F | F | T | T |
| F | T | T | F | F |
| F | F | T | T | T |

The converse and inverse of "$A$ implies $B$" are equivalent. Both are
true except when $A$ is false and $B$ is true.

**3.9**  a. If $n$ is an odd integer, then $n^2$ is odd.

b. If $r$ is a real number such that $r^2 \neq 2$, then $r$ is rational.

c. If the quadrilateral $ABCD$ is a rectangle, then $ABCD$ is a parallelogram with one right angle.

**3.11 Analysis of Proof.**  Using the forward-backward method one is led
to the key question, "How can I show that one statement (namely, $A$)
implies another statement (namely, $C$)?" According to Table 1.1, the
answer is to assume that the statement to the left of the word "implies"
is true, and then reach the conclusion that the statement to the right of
the word "implies" is true. In this case, you assume

   **A1:** $A$ is true,

and try to reach the conclusion that

   **B1:** $C$ is true.

Working forward from the information given in the hypothesis, because "$A$ implies $B$" is true and $A$ is true, by row 1 in Table 1.1, it must be that

**A1:** $B$ is true.

Because $B$ is true, and "$B$ implies $C$" is true, it must also be that

**A2:** $C$ is true.

Hence the proof is complete.

**Proof.** To conclude that "$A$ implies $C$" is true, assume that $A$ is true. By the hypothesis, "$A$ implies $B$" is true, so $B$ must be true. Finally, because "$B$ implies $C$" is true, $C$ is true, thus completing the proof. $\square$

**3.18** Proposition 3 is used in the backward process to answer the key question, "How can I show that a triangle is isosceles?" Accordingly, it must be shown that the hypothesis of Proposition 3 is true for the current problem—that is, that $w = \sqrt{2uv}$. This is precisely what the author does by working forward from the hypothesis of the current proposition.

**3.22 Analysis of Proof.** A key question for this problem is, "How can I show that an integer (namely, $n^2$) is odd?" By definition, it must be shown that

**B1:** There is an integer $k$ such that $n^2 = 2k + 1$.

Turning to the forward process to determine the desired value of $k$ in $B1$, from the hypothesis that $n$ is odd, by definition,

**A1:** There is an integer $p$ such that $n = 2p + 1$.

Squaring both sides of the equality in $A1$ and applying algebra results in

**A2:** $n^2 = (2p + 1)^2 = 4p^2 + 2p + 1 = 2(2p^2 + p) + 1.$

The proof is completed on noting from $A2$ that $k = 2p^2 + p$ satisfies $B1$.

**Proof.** Because $n$ is odd, by definition, there is an integer $p$ such that $n = 2p + 1$. Squaring both sides of this equality and applying algebra, it follows that $n^2 = (2p + 1)^2 = 4p^2 + 2p + 1 = 2(2p^2 + p) + 1$. But this means that, for $k = 2p^2 + p$, $n^2 = 2k + 1$ and so $n^2$ is odd. $\square$

## Solutions to Selected Exercises in Chapter 4

**4.1**

|       | Object | Certain Property | Something Happens |
|-------|--------|------------------|-------------------|
| (a)   | two people | none | they have the same number of friends |
| (b)   | integer $x$ | none | $f(x) = 0$ |
| (c)   | a point $(x, y)$ | $x \geq 0$ and $y \geq 0$ | $y = m_1 x + b_1$ and $y = m_2 x + b_2$ |
| (d)   | angle $t'$ | $0 \leq t' \leq \pi$ | $\tan(t') > \tan(t)$ |
| (e)   | integers $m$ and $n$ | none | $am + bn = c$ |

**4.6**  The construction method arises in the proof of Proposition 2 because the statement,

> **B1:** $n^2$ can be expressed as two times some other integer

can be rewritten as follows to contain the quantifier "there is":

> **B2:** There is an integer $p$ such that $n^2 = 2p$.

The construction method is then used to produce the integer $p$ for which $n^2 = 2p$. Specifically, using the fact that $n$ is even, and hence that there is an integer $k$ such that $n = 2k$, the desired value for $p$ is constructed as $p = 2k^2$. This value for $p$ is correct because $n^2 = (2k)^2 = 4k^2 = 2(2k^2) = 2p$.

**4.7**  **Analysis of Proof.** The appearance of the keywords "there is" in the conclusion suggests using the construction method to find an integer $x$ for which $x^2 - 5x/2 + 3/2 = 0$. Factoring means that you want an integer $x$ such that $(x - 1)(x - 3/2) = 0$. Thus, the desired value is $x = 1$. On substituting this value of $x$ in $x^2 - 5x/2 + 3/2$ yields 0, so the quadratic equation is satisfied. This integer solution is unique because the only other solution is $x = 3/2$, which is not an integer.

**Proof.** Factoring $x^2 - 5x/2 + 3/2$ means that the only roots are $x = 1$ and $x = 3/2$. Thus, there exists an integer (namely, $x = 1$) such that $x^2 - 5x/2 + 3/2 = 0$. The integer is unique.  □

**4.10**  **Analysis of Proof.** One answer to the key question, "How can I show that a number (namely, $s + t$) is rational?" is to use the definition and show that

**B1:** There are integers $p$ and $q$ with $q \neq 0$ such that $s + t = \frac{p}{q}$.

Because of the keywords "there are" in the backward statement $B1$, the author recognizes the need for the construction method. Specifically, the author works forward from the hypothesis, by definition of a rational number, to claim that

**A1:** There are integers $a, b, c, d$ with $b, d \neq 0$ such that $s = \frac{a}{b}$ and $t = \frac{c}{d}$.

Adding the two fractions in $A1$ together results in

**A2:** $s + t = \frac{a}{b} + \frac{c}{d} = \frac{ad+bc}{bd}$.

From $A2$, the author constructs the values for the integers $p$ and $q$ in $B1$, as indicated when the author says, "Now set $p = ad + bc$ and $q = bd$." The remainder of this proof involves showing that these values of $p$ and $q$ satisfy the desired properties in $B1$. Specifically, $q = bd \neq 0$ because $b, d \neq 0$ (see $A1$) and $s + t = \frac{p}{q}$ from $A2$. The proof is now complete. $\square$

**4.14** The error in the proof occurs when the author says to divide by $p^2$ because it will not be possible to do so if $p = 0$, which can happen.

**4.18 Analysis of Proof.** The forward-backward method gives rise to the key question, "How can I show that an integer (namely, $a$) divides another integer (namely, $c$)?" By the definition, one answer is to show that

**B1:** There is an integer $k$ such that $c = ak$.

The appearance of the quantifier "there is" in $B1$ suggests turning to the forward process to construct the desired integer $k$.

From the hypothesis that $a|b$ and $b|c$, and by definition,

**A1:** There are integers $p$ and $q$ such that $b = ap$ and $c = bq$.

Therefore, it follows that

**A2:** $c = bq = (ap)q = a(pq)$,

and so the desired integer $k$ is $k = pq$.

**Proof.** Because $a|b$ and $b|c$, by definition, there are integers $p$ and $q$ for which $b = ap$ and $c = bq$. But then $c = bq = (ap)q = a(pq)$ and so $a|c$. $\square$

## Solutions to Selected Exercises in Chapter 5

**5.1**  a. Object:              real number $x$.
            Certain property:      none.
            Something happens:    $f(x) \leq f(x^*)$.

b. Object:                   element $x$.
   Certain property:       $x \in S$.
   Something happens:    $g(x) \geq f(x)$.

c. Object:                   element $x$.
   Certain property:       $x \in S$.
   Something happens:    $x \leq u$.

**5.5**   a. Choose real numbers $x'$, $y'$ with $x' < y'$.
      It will be shown that $f(x') < f(y')$.

   b. Choose an element $x' \in S$.
      It will be shown that $g(x') \geq f(x')$.

   c. Choose an element $x' \in S$.
      It will be shown that $x' \leq u$.

**5.8**   Key Question:    How can I show that a set (namely, $R$) is a subset of another set (namely, $T$)?

   Key Answer:      Show that every element of the first set is in the second set and so it must be shown that

   **B1:**             For every element $r \in R$, $r \in T$.

   **A1:**             Choose an element $r' \in R$ for which it must be shown that

   **B2:**             $r'$ is in $T$.

**5.13** a. Incorrect because $(x', y')$ should be chosen in $S$.

   b. Correct.

   c. Incorrect because $(x', y')$ should be chosen in $S$ (not in $T$). Also, you should show that $(x', y')$ is in $T$ (not in $S$).

   d. Incorrect because you should not choose specific values for $x$ and $y$ but rather general values for $x$ and $y$ for which $(x, y) \in S$.

   e. Correct. Note that this is just a notational modification of part (b).

**5.15** The choose method is used in the first sentence of the proof where it says, "Let $x$ be a real number." More specifically, the author asked the key question, "How can I show that a real number (namely, $x^*$) is a maximizer of a function (namely, $f(x) = ax^2 + bx + c$)?" Using the definition, one answer is to show that

   **B1:** For all real numbers $x$, $f(x^*) \geq f(x)$.

Recognizing the quantifier "for all" in $B1$, the author uses the choose method to choose

   **A1:** A real number $x$,

for which it must be shown that

   **B2:** $f(x^*) \geq f(x)$, that is, that $a(x^*)^2 + bx^* + c \geq ax^2 + bx + c$.

The author then reaches $B2$ by considering the following two cases.

**Case 1.** $x^* \geq x$. In this case, the author correctly notes that $x^* - x \geq 0$ and also that $a(x^* + x) + b \geq 0$ because $x^* = -b/(2a)$ and so

**A2:** $a(x^* + x) + b = -b/2 + ax + b = (2ax + b)/2.$

However, because $x^* = -b/(2a) \geq x$, multiplying the inequality through by $2a < 0$ and adding $b$ to both sides yields

**A3:** $2ax + b \geq 0.$

Thus, from $A2$ and $A3$, the author correctly concludes that

**A4:** $a(x^* + x) + b \geq 0.$

Multiplying $A4$ through by $x^* - x \geq 0$ and rewriting yields

**A5:** $a(x^*)^2 - ax^2 + bx^* - bx \geq 0.$

Bringing all $x$-terms to the right and adding $c$ to both sides yields $B2$, thus completing this case.

**Case 2.** $x^* < x$. In this case, the author leaves the following steps for you to create. Now $x^* - x < 0$ and also $a(x^* + x) + b \leq 0$ because

**A2:** $a(x^* + x) + b = -b/2 + ax + b = (2ax + b)/2.$

However, because $x^* = -b/(2a) < x$ and $a < 0$, it follows that

**A3:** $2ax + b < 0.$

Thus, from $A2$ and $A3$,

**A4:** $a(x^* + x) + b < 0.$

Multiplying $A4$ through by $x^* - x < 0$ and rewriting yields

**A5:** $a(x^*)^2 - ax^2 + bx^* - bx > 0.$

Bringing all $x$-terms to the right and adding $c$ to both sides yields $B2$, thus completing this case and the proof.

**5.17 Analysis of Proof.** Because the conclusion contains the keywords "for all," the choose method is used to choose

**A1:** Real numbers $x$ and $y$ with $x < y$,

for which it must be shown that

**B1:** $f(x) < f(y)$, that is, $mx + b < my + b.$

Work forward from the hypothesis that $m > 0$ to multiply both sides of the inequality $x < y$ in $A1$ by $m$ yielding

**A2:** $mx < my.$

Adding $b$ to both sides gives precisely $B1$.

**5.21 Analysis of Proof.** The forward-backward method gives rise to the key question, "How can I show that a function (namely, $x + 1$) is greater than or equal to another function (namely, $(x - 1)^2$) on a set (namely, $S$)?" The definition provides the answer that one must show that

   **B1:** For all $x \in S$, $x + 1 \geq (x - 1)^2$.

The appearance of the quantifier "for all" in the backward process suggests using the choose method to choose

   **A1:** An element $x \in S$,

for which it must be shown that

   **B2:** $x + 1 \geq (x - 1)^2$.

Bringing the term $x + 1$ to the right and performing algebraic manipulations, it must be shown that

   **B3:** $x(x - 3) \leq 0$.

However, working forward from $A1$ using the definition of $S$ yields

   **A2:** $0 \leq x \leq 3$.

But then $x \geq 0$ and $x - 3 \leq 0$ so $B3$ is true and the proof is complete.

**Proof.** Let $x \in S$, so $0 \leq x \leq 3$. It will be shown that $x + 1 \geq (x - 1)^2$. However, because $x \geq 0$ and $x \leq 3$, it follows that $x(x-3) = x^2 - 3x \leq 0$. Adding $x + 1$ to both sides and factoring leads to the desired conclusion that $x + 1 \geq (x - 1)^2$. $\square$

## Solutions to Selected Exercises in Chapter 6

**6.2** You can apply specialization to the statement,

   **A1:** For every object $X$ with the property $Q$, $T$ happens,

to reach the conclusion that $S$ happens for the particular object $Y$ if, when you replace $X$ with $Y$ in $A1$, the something that happens (namely, $T$) leads to the desired conclusion that $S$ happens. To apply specialization, you must show that the particular object $Y$ has the property $Q$ in $A1$, most likely by using the fact that $Y$ has property $P$.

   To use specialization in this problem you need to show two things: (1) that you can specialize $A1$ to the object $Y$ (which will be possible if, whenever property $P$ holds, so does property $Q$) and (2) that, if $T$ happens (plus any other information you can assume), then $S$ happens.

**6.3**    a. (1) Look for a specific real number, say $y$, with which to apply specialization and (2) conclude that $f(y) \leq f(x^*)$ as a new statement in the forward process.

b. (1) Look for a specific element, say $y$, with which to apply specialization, (2) show that $y \in S$, and (3) conclude that $g(y) \geq f(y)$ as a new statement in the forward process.

c. (1) Look for a specific element, say $y$, with which to apply specialization, (2) show that $y \in S$, and (3) conclude that $y \leq u$ as a new statement in the forward process.

**6.7**    a. You can reach the desired conclusion that $\sin(2X) = 2\sin(X)\cos(X)$ by specializing the statement, "For all angles $\alpha$ and $\beta$, $\sin(\alpha + \beta) = \sin(\alpha)\cos(\beta) + \cos(\alpha)\sin(\beta)$" to $\alpha = X$ and $\beta = X$, which you can do because $X$ is an angle. The result of this specialization is that $\sin(X + X) = \sin(X)\cos(X) + \cos(X)\sin(X)$, that is, $\sin(2X) = 2\sin(X)\cos(X)$.

b. You can reach the desired conclusion that $(A \cap B)^c = A^c \cup B^c$ by specializing the statement, "For any sets $S$ and $T$, $(S \cup T)^c = S^c \cap T^c$" to $S = A^c$ and $T = B^c$, which you can do because $A^c$ and $B^c$ are sets. The result of this specialization is that $(A^c \cup B^c)^c = (A^c)^c \cap (B^c)^c = A \cap B$. Applying the complement to both sides now leads to the desired conclusion that $A^c \cup B^c = (A \cap B)^c$.

**6.10** The author makes a mistake when saying that, "In particular, for the specific elements $x$ and $y$, and for the real number $t$, it follows that $tx + (1-t)y \in R$." In this sentence, the author is applying specialization to the statement

> **A1:**  For any two elements $u$ and $v$ in $R$, and for any real number $s$ with $0 \leq s \leq 1$, $su + (1 - s)v \in R$.

Specifically, the author is specializing A1 with $u = x$, $v = y$, and $s = t$. However, the author has failed to verify that these specific objects satisfy the certain properties in A1. In particular, to specialize A1 to $u = x$ and $v = y$, it is necessary to verify that $x \in R$ and $y \in R$. The author fails to do so (in fact it is not necessarily true that $x \in R$ and $y \in R$). Thus, the author is not justified in applying this specialization.

**6.13 Analysis of Proof.** The forward-backward method gives rise to the key question, "How can I show that a set (namely, $R$) is a subset of another set (namely, $T$)?" One answer is by the definition, so one must show that

> **B1:** For all elements $r \in R$, $r \in T$.

The appearance of the quantifier "for all" in the backward process suggests using the choose method. So choose

**A1:** An element $r' \in R$,

for which it must be shown that

**B2:** $r' \in T$.

Turning to the forward process, the hypothesis says that $R$ is a subset of $S$ and $S$ is a subset of $T$. By definition, this means, respectively, that

**A2:** For all elements $r \in R$, $r \in S$, and
**A3:** For all elements $s \in S$, $s \in T$.

Specializing $A2$ to $r = r'$ (which is in $R$ from $A1$), one has that

**A4:** $r' \in S$.

Specializing $A3$ to $r = r'$ (which is in $S$ from $A4$), one has that

**A5:** $r' \in T$.

The proof is now complete because $A5$ is the same as $B2$.

**Proof.** To show that $R \subseteq T$, it must be shown that, for all $r \in R$, $r \in T$. Let $r' \in R$. By hypothesis, $R \subseteq S$, so $r' \in S$. Also, by hypothesis, $S \subseteq T$, so $r' \in T$. ☐

**6.19 Analysis of Proof.** The forward-backward method gives rise to the key question, "How can I show that a set (namely, $S \cap T$) is convex?" One answer is by the definition, whereby it must be shown that

**B1:** For all $x, y \in S \cap T$, and for all $t$ with $0 \le t \le 1$,
$tx + (1 - t)y \in S \cap T$.

The appearance of the quantifier "for all" in the backward process suggests using the choose method to choose

**A1:** $x', y' \in S \cap T$, and $t'$ with $0 \le t' \le 1$,

for which it must be shown that

**B2:** $t'x' + (1 - t')y' \in S \cap T$.

Working forward from the hypothesis and $A1$, $B2$ will be established by showing that

**B3:** $t'x' + (1 - t')y'$ is in both $S$ and $T$.

Specifically, from the hypothesis that $S$ is convex, by definition, it follows that

**A2:** For all $x, y \in S$, and for all $0 \le t \le 1$, $tx + (1 - t)y \in S$.

Specializing $A2$ to $x = x'$, $y = y'$, and $t = t'$ (noting from $A1$ that $0 \le t' \le 1$) yields that

**A3:** $t'x' + (1 - t')y' \in S$.

A similar argument shows that $t'x' + (1 - t')y' \in T$, thus completing the proof.

**Proof.** To see that $S \cap T$ is convex, let $x', y' \in S \cap T$, and let $t'$ with $0 \le t' \le 1$. It will be established that $t'x' + (1 - t')y' \in S \cap T$. From the hypothesis that $S$ is convex, one has that $t'x' + (1 - t')y' \in S$. Similarly, $t'x' + (1 - t')y' \in T$. Thus it follows that $t'x' + (1 - t')y' \in S \cap T$, and so $S \cap T$ is convex. $\square$

**6.23 Analysis of Proof.** The author has used the forward-backward method to ask the key question, "How can I show that a function (namely, $g$) is greater than or equal to another function (namely, $f$) on a set (namely, $R$)?" The definition provides the answer that it is necessary to show that

**B1:** For every element $r \in R$, $g(r) \ge f(r)$.

Recognizing the quantifier "for every" in $B1$, the author uses the choose method to choose

**A1:** An element $x \in R$,

for which it must be shown that

**B2:** $g(x) \ge f(x)$.

To reach $B2$, the author turns to the forward process and works forward from the hypothesis that $R$ is a subset of $S$ by definition to claim that

**A2:** For every element $r \in R$, $r \in S$.

Recognizing the quantifier "for every" in $A2$, the author specializes $A2$ to the element $r = x \in R$ chosen in $A1$. This specialization results in

**A3:** $x \in S$.

Then the author works forward from the hypothesis that $g \ge f$ on $S$, which, by definition, means that

**A4:** For every element $s \in S$, $g(s) \ge f(s)$.

Recognizing the quantifier "for every" in $A4$, the author specializes $A4$ to the element $s = x \in S$ (see $A3$). The result of this specialization is

**A5:** $g(x) \ge f(x)$.

The proof is now complete because $A5$ is the same as $B2$. $\square$

## Solutions to Selected Exercises in Chapter 7

**7.1**  a. For the quantifier "there is":

   Object:                real number $y$.
   Certain property:      none.
   Something happens:     for every real number $x$, $f(x) \leq y$.

   For the quantifier "for every":

   Object:                real number $x$.
   Certain property:      none.
   Something happens:     $f(x) \leq y$.

   b. For the quantifier "there is":

   Object:                real number $M$.
   Certain property:      $M > 0$.
   Something happens:     $\forall$ element $x \in S$, $|x| < M$.

   For the quantifier "for all":

   Object:                element $x$.
   Certain property:      $x \in S$.
   Something happens:     $|x| < M$.

   c. For the first quantifier "for all":

   Object:                real number $\epsilon$.
   Certain property:      $\epsilon > 0$.
   Something happens:     there is a real number $\delta > 0$ such that, for all real numbers $y$ with $|x - y| < \delta$, $|f(x) - f(y)| < \epsilon$.

   For the quantifier "there is":

   Object:                real number $\delta$.
   Certain property:      $\delta > 0$.
   Something happens:     for all real numbers $y$ with $|x - y| < \delta$, $|f(x) - f(y)| < \epsilon$.

   For the second quantifier "for all":

   Object:                real number $y$.
   Certain property:      $|x - y| < \delta$.
   Something happens:     $|f(x) - f(y)| < \epsilon$.

   d. For the first quantifier "for all":

   Object:                real number $\epsilon$.
   Certain property:      $\epsilon > 0$.
   Something happens:     $\exists$ an integer $j \geq 1 \ni \forall$ integer $k$ with $k > j$, $|x_k - x| < \epsilon$.

   For the quantifier "there is":

   Object:                integer $j$.
   Certain property:      $j \geq 1$.
   Something happens:     $\forall$ integer $k$ with $k > j$, $|x_k - x| < \epsilon$.

   For the second quantifier "for all":

   Object:                integer $k$.
   Certain property:      $k > j$.
   Something happens:     $|x_k - x| < \epsilon$.

**7.3**  a. Both $S1$ and $S2$ are true. This is because, when you apply the choose method to each statement, in either case you will choose real numbers $x$ and $y$ with $0 \le x \le 1$ and $0 \le y \le 2$ for which you can then show that $2x^2 + y^2 \le 6$.

b. $S1$ and $S2$ are different—$S1$ is true and $S2$ is false. You can use the choose method to show that $S1$ is true. To see that $S2$ is false, consider $y = 1$ and $x = 2$. For these real numbers, $2x^2 + y^2 = 2(4) + 1 = 9 > 6$.

c. These two statements are the same when the properties $P$ and $Q$ do not depend on the objects $X$ and $Y$. This is the case in part (a) but not in part (b).

**7.5**  a. Recognizing the keywords "for all" as the first quantifier from the left, the first step in the backward process is to choose an object $X$ with a certain property $P$, for which it must be shown that there is an object $Y$ with property $Q$ such that something happens. Recognizing the keywords "there is," one then needs to construct an object $Y$ with property $Q$ such that something happens.

b. Recognizing the keywords "there is" as the first quantifier from the left, the first step in the backward process is to construct an object $X$ with property $P$. After constructing $X$, the choose method is used to show that, for the object $X$ you constructed, it is true that for all objects $Y$ with property $Q$ that something happens. Recognizing the keywords "for all," you would choose an object $Y$ with property $Q$ and show that something happens.

**7.6**  a. Recognizing the keywords "for all" as the first quantifier from the left, you should apply specialization to a particular object, say $X = Z$. To do so, you need to be sure that $Z$ satisfies the property $P$. If so, then you can conclude, as a new statement in the forward process, that

> **A1:** There is an object $Y$ with the certain property $Q$ such that something happens.

You can work forward from the object $Y$ and its certain property $Q$.

b. You know that there is an object $X$ such that

> **A1:** For every object $Y$ with the certain property $Q$, something happens.

Recognizing the keywords "for all" in the forward statement $A1$, you should apply specialization to a particular object, say $Y = Z$. To do so, you need to be sure that $Z$ satisfies the property $Q$. If so, then you can conclude, as a new statement in the forward process, that

**A2:** The something happens for $Z$.

**7.11 Analysis of Proof.** Recognizing the keywords "for every" in the conclusion, you should use the choose method to choose

**A1:** Real numbers $\epsilon > 0$ and $a > 0$,

for which you must show that

**B1:** There is an integer $n > 0$ such that $\frac{a}{n} < \epsilon$.

Recognizing the keywords "there is" in the backward statement $B1$, you should now use the construction method to produce an integer $n > 0$ for which you must then show that

**B2:** $\frac{a}{n} < \epsilon$.

To construct this integer $n$, work backward from the fact that you want $\frac{a}{n} < \epsilon$. Multiplying both sides of $B2$ by $n > 0$ and then dividing both sides by $\epsilon > 0$ (see $A1$), you can see that you need $\frac{a}{\epsilon} < n$. In other words, constructing $n$ to be any integer $> \frac{a}{\epsilon} > 0$, it follows that $B2$ is true and so the proof is complete.

**Proof.** Choose a real number $\epsilon > 0$. Now let $n > 0$ be an integer with $n > \frac{a}{\epsilon}$. It then follows that $\frac{a}{n} < \epsilon$ and so the proof is complete. $\square$

**7.12** a. Recognizing the keywords "for all" as the first quantifier from the left, the author uses the choose method (as indicated by the words "Let $x$ and $y$ be real numbers with $x < y$"). Accordingly, the author must now show that

**B1:** There is a rational number $r$ such that $x < r < y$.

b. The author is using previous knowledge of the statement in Exercise 7.11. Specifically, recognizing the keywords "for all" in the forward process, the author is specializing that statement to the particular value $\epsilon = y - x > 0$ and $a = 2 > 0$. The result of that specialization is that there is an integer $n > 0$ such that $\frac{2}{n} < \epsilon$, that is, $n\epsilon > 2$, as the author states in the second sentence of the proof.

c. The construction method is used because the author recognized the keywords "there is" in the backward statement $B1$ in the answer to part (a). Specifically, the author constructs the rational number $r = \frac{m}{n}$ and claims that this value satisfies the property that $x < r < y$ in $B1$.

d. The author is justified in claiming that the proof is complete because the author has constructed the rational number $r = \frac{m}{n}$ and, because

$n$ is a positive integer [see the answer to part (b)] and $nx < m < ny$, it follows that $x < \frac{m}{n} = r < y$. This means that $r$ satisfies the certain property in $B1$ and so the proof is complete.

**7.17 Analysis of Proof.** The keywords "for every" in the conclusion suggest using the choose method to choose

**A1:** A real number $x' > 2$,

for which it must be shown that

**B1:** There is a real number $y < 0$ such that $x' = 2y/(1 + y)$.

The keywords "there is" in $B1$ suggest using the construction method to construct the desired $y$. Working backward form the fact that $y$ must satisfy $x' = 2y/(1 + y)$, it follows that $y$ must be constructed so that

**B2:** $x' + x'y = 2y$, or
**B3:** $y(2 - x') = x'$, or
**B4:** $y = x'/(2 - x')$.

To see that the value of $y$ in $B4$ is correct, it is easily seen that $x' = 2y/(1 + y)$. However, it must also be shown that $y < 0$, which it is, because $x' > 2$.

**Proof.** Let $x' > 2$ and construct $y = x'/(2 - x')$. Because $x' > 2$, $y < 0$. It is also easy to verify that $x' = 2y/(1 + y)$.  □

## Solutions to Selected Exercises in Chapter 8

**8.1**  a. The real number $x^*$ is not a maximum of the function $f$ means that there is a real number $x$ such that $f(x) > f(x^*)$.

b. Suppose that $f$ and $g$ are functions of one variable. Then $g$ is not $\geq f$ on the set $S$ of real numbers means that there exists an element $x \in S$ such that $g(x) < f(x)$.

c. The real number $u$ is not an upper bound for a set $S$ of real numbers means that there exists an element $x \in S$ such that $x > u$.

**8.4**  a. There does not exist an element $x \in S$ such that $x \notin T$.

b. It is not true that, for every angle $t$ between $0$ and $\pi/2$, $\sin(t) > \cos(t)$ or $\sin(t) < \cos(t)$.

c. There does not exist an object with the certain property such that the something does not happen.

d. It is not true that, for every object with the certain property, the something does not happen.

**8.6**  a. Work forward from $NOT\ B$.
Work backward from $(NOT\ C)\ OR\ (NOT\ D)$.

b. Work forward from *NOT B*.
Work backward from $(NOT\ C)\ AND\ (NOT\ D)$.

c. Work forward from $(NOT\ C)\ OR\ (NOT\ D)$.
Work backward from *NOT A*.

d. Work forward from $(NOT\ C)\ AND\ (NOT\ D)$.
Work backward from *NOT A*.

**8.9**   a. $x = 3$ is a counterexample because $x^2 = 9 > 3 = x$. (Any value of a real number $x$ for which $x^2 > x$ provides a counterexample.)

b. $n = 4$ is a counterexample because $n^2 = 16 < 24 = n!$. (Any value of an integer $n$ for which $n^2 < n!$ provides a counterexample.)

c. $a = 2$, $b = 3$, and $c = 5$ constitute a counterexample because $a|(b+c)$ [that is, $2|(3+5)$] and yet $a = 2$ does not divide $b = 3$ (nor does $a = 2$ divide $c = 5$). (Any value of the integers $a$, $b$, and $c$ for which $a|(b+c)$ and either $a$ does not divide $b$ or $a$ does not divide $c$ provides a counterexample.)

## Solutions to Selected Exercises in Chapter 9

**9.1**   a. Assume that $l$, $m$, and $n$ are three consecutive integers and that 24 divides $l^2 + m^2 + n^2 + 1$.

b. Assume that $n > 2$ is an integer and $x^n + y^n = z^n$ has an integer solution for $x$, $y$, and $z$.

c. Assume that $f$ and $g$ are two functions such that $g \geq f$, $f$ is unbounded above, and $g$ is not unbounded above.

**9.4**   a. There are not a finite number of primes.

b. The positive integer $p$ cannot be divided by any positive integer other than 1 and $p$.

c. The lines $l$ and $l'$ do not intersect.

**9.6**   a. Recognizing the keywords "there is," you should use the construction method to construct an element $s \in S$ for which you must show that $s$ is also in $T$.

b. Recognizing the quantifier "for all" as the first keywords from the left, you should use the choose method to choose

**A1:**  An element $s \in S$,

for which you must show that

**B1:**  There is no element $t \in T$ such that $s > t$.

Recognizing the keyword "no" in the backward statement $B1$, you should proceed with the contradiction method and assume that

**A2:**  There is an element $t \in T$ such that $s > t$.

You must then work forward from $A1$ and $A2$ to reach a contradiction.

c. Recognizing the keyword "no," you should first use the contradiction method, whereby you should assume that

> **A1:** There is a real number $M > 0$ such that, for all elements $x \in S$, $|x| < M$.

To reach a contradiction, you will probably have to apply specialization to the forward statement, "for all elements $x \in S$, $|x| < M$." Alternatively, you can pass the $NOT$ through the nested quantifiers and hence you must show that

> **B1:** $\forall M > 0$, $\exists x \in S$ such that $|x| \geq M$.

Recognizing the quantifier "for all" as the first keywords in the backward statement $B1$, you can now use the choose method to choose

> **A1:** A real number $M > 0$,

for which you must show that

> **B2:** There is an element $x \in S$ such that $|x| \geq M$.

Recognizing the keywords "there is" in the backward statement $B2$, you should then use the construction method to produce the desired element $x \in S$ for which $|x| \geq M$.

**9.8** **Analysis of Proof.** To use the contradiction method, assume:

> **A:**    $n$ is an integer for which $n^2$ is even.
> **A1 (NOT B):**    $n$ is not even, that is, $n$ is odd.

Work forward from these assumptions using the definition of an odd integer to reach the contradiction that $n^2$ is odd. Applying a definition to work forward from $NOT$ $B$ yields

> **A2:** There exists an integer $k$ such that $n = 2k + 1$.

Squaring both sides of the equality in $A2$ and performing simple algebraic manipulations leads to

> **A3:**  $n^2 = (2k + 1)^2 = 4k^2 + 4k + 1$, so
> **A4:**  $n^2 = 2(2k^2 + 2k) + 1$,

which says that $n^2 = 2p + 1$, where $p = 2k^2 + 2k$. Thus $n^2$ is odd, and this contradiction to the hypothesis that $n^2$ is even completes the proof.

**Proof.** Assume, to the contrary, that $n$ is odd and $n^2$ is even. Hence, there is an integer $k$ such that $n = 2k + 1$. Consequently,

$$n^2 = (2k + 1)^2 = 2(2k^2 + 2k) + 1,$$

and so $n^2$ is odd, contradicting the initial assumption. ☐

**9.10 Analysis of Proof.** Proceed by assuming that there is a chord of a circle that is longer than its diameter. Using this assumption and the properties of a circle, you must arrive at a contradiction.

Let $AC$ be the chord of the circle that is longer than the diameter of the circle (see the figure below). Let $AB$ be a diameter of the circle. This construction is valid because, by definition, a diameter is a line passing through the center terminating at the perimeter of the circle.

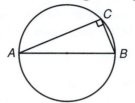

It follows that angle $ACB$ has 90 degrees because $ACB$ is an angle inscribed in a semicircle. Hence the triangle $ABC$ is a right triangle in which $AB$ is the hypotenuse. Then the desired contradiction is that the hypotenuse of a right triangle is shorter than one side of the triangle, which is impossible.

**Proof.** Assume that there does exist a chord, say $AC$, of a circle that is longer than a diameter. Construct a diameter that has one of its ends coinciding with one end of the chord $AC$. Joining the other ends produces a right triangle in which the diameter $AB$ is the hypotenuse. But then the hypotenuse is shorter than one leg of the right triangle, which is a contradiction. ☐

**9.14 Analysis of Proof.** To use contradiction, assume that no two people have the same number of friends; that is, everybody has a different number of friends. Because there are $n$ people, each of whom has a different number of friends, you can number the people in an increasing sequence according to the number of friends each person has. In other words,

> person number 1 has no friends,
> person number 2 has 1 friend,
> person number 3 has 2 friends,
> $\vdots$
> person number $n$ has $n - 1$ friends.

By doing so, there is a contradiction that the last person is friends with all the other $n - 1$ people, including the first one, who has no friends.

**Proof.** Assume, to the contrary, that no two people have the same number of friends. Number the people at the party in such a way that, for each $i = 1, \ldots, n$, person $i$ has $i - 1$ friends. It then follows that the person with $n - 1$ friends is a friend of the person who has no friends, which is a contradiction.  □

**9.22** The contradiction is that $b$ is both odd (as stated in the hypothesis) and even (as has been shown because $b = -2a$ is even).

**9.26 Analysis of Proof.** The proof is done by contradiction because the author is assuming

      **NOT B:** $xz - y^2 > 0$.

A contradiction is reached by showing that the square of a number is negative. Specifically, the author shows that

      **B1:** $(az - cx)^2 < 0$.

The contradiction in $B1$ is obtained by working forward from *NOT B* and the hypothesis. Specifically, from *NOT B* and the hypothesis that $ac - b^2 > 0$, it follows that

      **A1:** $xz > y^2$,
      **A2:** $ac > b^2$.

Multiplying corresponding sides of the inequalities in $A1$ and $A2$ yields:

      **A3:** $(ac)(xz) > b^2 y^2$.

Then, adding $2by$ to both sides of the hypothesis $az - 2by + cx = 0$ and squaring both sides results in:

      **A4:** $(az + cx)^2 = 4b^2 y^2$.

From $A3$, it follows that $4b^2 y^2 < 4(ac)(xz)$ and so, from $A4$,

      **A5:** $(az + cx)^2 < 4(ac)(xz)$.

Expanding the left side of $A5$, subtracting $4(ac)(xz)$ from both sides, and then factoring yields the contradiction that $(az - cx)^2 < 0$.  □

## Solutions to Selected Exercises in Chapter 10

**10.1** a. Work forward from:    $n$ is an odd integer.
        Work backward from:  $n^2$ is an odd integer.

      b. Work forward from:    $S$ is a subset of $T$ and $T$ is bounded.
        Work backward from:  $S$ is bounded.

      c. Work forward from:    The integer $p > 1$ is not prime.
        Work backward from:  There is an integer $1 < n \leq \sqrt{p}$, such that $n | p$.

**10.4** Statement (b) is a result of the forward process because you can assume that there is a real number $t$ with $0 < t < \pi/4$ such that $\sin(t) = r\cos(t)$. The answer in part (b) results on squaring both sides and replacing $\cos^2(t)$ with $1 - \sin^2(t)$.

**10.7** The correct key question is in (d). To understand why, recall that with the contrapositive method you work forward from *NOT B* and backward from *NOT A*. Thus, you should apply the key question to the statement, "The derivative of the function $f$ at the point $x$ is 0."

    a. Incorrect because the key question is applied to *NOT B*. Also, the question uses symbols and notation from the specific problem.

    b. Incorrect because the key question uses symbols and notation from the specific problem.

    c. Incorrect because the key question is applied to *NOT B*.

    d. Correct.

**10.9** The contradiction method is used in this proof because the author works forward from $A$ and *NOT B* to reach the contradiction that the integer $p > p$.

**10.12** With the contrapositive method, you will assume

      **NOT B:** $p$ is not prime.

You must then show that

      **NOT A:** There is an integer $m$ with $1 < m \le \sqrt{p}$ such that $m|p$.

Recognizing the keywords "there is" in the backward statement *NOT A*, you should use the construction method to construct an integer $m$ with the property that $1 < m \le \sqrt{p}$ such that $m|p$.

**10.15 Analysis of Proof.** By the contrapositive method, assume that

      **NOT B:** $p = q$.

It must be shown that

      **NOT A:** $\sqrt{pq} = (p + q)/2$.

However, from *NOT B* and the fact that $p, q > 0$, it follows that

      **A1:** $\sqrt{pq} = \sqrt{p^2} = p = (p + p)/2 = (p + q)/2$.

The proof is now complete because *NOT A* is true.

**Proof.** By the contrapositive method, assume that $p = q$. It must be shown that $\sqrt{pq} = (p + q)/2$. However, because $p, q > 0$, it follows that

$$\sqrt{pq} = \sqrt{p^2} = p = (p + p)/2 = (p + q)/2.$$

The proof is now complete. $\square$

**10.24 Analysis of Proof.** This is a proof by the contrapositive method because the author assumes that

      **NOT B:** $p$ is not prime

and eventually shows that

**NOT A:** There is an integer $m$ with $1 < m \leq \sqrt{p}$ such that $m|p$.

Recognizing the keywords "there is" in $NOT\ A$, the construction method is used to produce the value for the integer $m$. To do so, the author works forward from $NOT\ B$ to claim that, because $p$ is not prime,

**A1:** There is an integer $n$ with $1 < n < p$ such that $n|p$.

The author now considers two possibilities: $n \leq \sqrt{p}$ and $n > \sqrt{p}$. In the former case,

**A2:** $n \leq \sqrt{p}$,

and the author constructs

**A3:** $m = n$.

This value for $m$ in $A3$ is correct because, from $A1$ and $A2$, $1 < n = m$ and $m = n \leq \sqrt{p}$. Also, from $A1$, $n|p$ and $m = n$, so $m|p$.

In the latter case,

**A4:** $n > \sqrt{p}$.

To construct the value for $m$, the author works forward from $A1$ and the definition of $n|p$ to state that

**A5:** There is an integer $k$ such that $p = nk$.

The author then constructs

**A6:** $m = k$.

The author shows that this value for $m$ is correct by establishing that

**A7:** $1 < k \leq \sqrt{p}$.

To do so, the author argues by contradiction that $1 < k$. For otherwise, from $A5$, it would follow that $p = nk \leq n$, which cannot happen because, from $A1$, $n < p$. Finally, $k \leq \sqrt{p}$ because otherwise, $k > \sqrt{p}$ and then, since $n > \sqrt{p}$ from $A4$, it would follow that $nk > \sqrt{p}\sqrt{p} = p$, which cannot happen because, from $A5$, $p = nk$. Observe that the author omits noting that $k|p$, which is true because, from $A5$, $p = nk = nm$. The proof is now complete.  $\square$

**Solutions to Selected Exercises in Chapter 11**

**11.1 a.**    i)    Show that the lines $y = mx + b$ and $y = cx + d$ both pass through the points $(x_1, y_1)$ and $(x_2, y_2)$.

        ii)    Conclude that $y = mx + b$ and $y = cx + d$ are the same line.

**b.**    i)    Show that $(x_1, y_1)$ and $(x_2, y_2)$ are solutions to the equations $ax + by = 0$ and $cx + dy = 0$.

        ii)    Conclude that $(x_1, y_1) = (x_2, y_2)$.

**c.**    i)    Show that $(a + bi)(r + si) = (a + bi)(t + ui) = 1$.

        ii)    Conclude that $r + si = t + ui$.

**11.3 a.**    i)    First, construct a line, say $y = mx + b$, that goes through the two given points. Then assume that $y = cx + d$ also passes through those two points. You must now work forward to show that the two lines are the same—that is, that $m = c$ and $b = d$.

        ii)    First, construct a line, say $y = mx + b$, that goes through the two given points. Then assume that a different line, say $y = cx + d$, also passes through those two points. You must now work forward to reach a contradiction.

**b.**    i)    First, construct a solution, say $(x_1, y_1)$, to the system of equations $ax + by = 0$ and $cx + dy = 0$. Then assume that you have another solution to the equations, say $(x_2, y_2)$. You must now work forward to show that $(x_1, y_1) = (x_2, y_2)$.

        ii)    First, construct a solution, say $(x_1, y_1)$, to the two equations $ax + by = 0$ and $cx + dy = 0$. Then assume that you have a different solution, say $(x_2, y_2) \neq (x_1, y_1)$, to the two equations. You must now work forward to reach a contradiction.

**c.**    i)    First, construct a complex number, say $r + si$, that satisfies $(a+bi)(r+si) = 1$. Then assume that you have another complex number, say $t + ui$, that also satisfies $(a + bi)(t + ui) = 1$. You must now work forward to show that $r + si$ and $t + ui$ are the same—that is, that $r + si = t + ui$.

        ii)    First, construct a complex number, say $r+si$, that satisfies $(a+bi)(r + si) = 1$. Then assume that there is a different complex number, say $t+ui \neq r+si$, that also satisfies $(a+bi)(t+ui) = 1$. You must now work forward to reach a contradiction.

**11.8 Analysis of Proof.** The direct uniqueness method is used, whereby one must first construct the real number $y$. This is done in Exercise 7.17, so

    **A1:** $y < 0$ and $x = 2y/(1 + y)$.

It remains to show the uniqueness by assuming that $z$ is also a number with

    **A2:** $z < 0$ and $x = 2z/(1 + z)$.

Working forward via algebraic manipulations, it will be shown that

**B1:** $y = z$.

Specifically, combining $A1$ and $A2$ yields

**A3:** $x = 2y/(1 + y) = 2z/(1 + z)$.

Dividing both sides of $A3$ by 2 and clearing the denominators, you have that

**A4:** $y + yz = z + yz$.

Subtracting $yz$ from both sides of $A4$ yields the desired conclusion that $y = z$.

**Proof.** The existence of the real number $y$ is established in Exercise 7.17. To show that $y$ is unique, suppose that $y$ and $z$ satisfy $y < 0$, $z < 0$, $x = 2y/(1+y)$, and also $x = 2z/(1+z)$. But then $2y/(1+y) = 2z/(1+z)$ and so $y + yz = z + yz$, or, $y = z$, as desired. $\square$

**11.12 Analysis of Proof.** The issue of existence is addressed first. To construct the complex number $c + di$ that satisfies $(a + bi)(c + di) = 1$, multiply the two terms using complex arithmetic to obtain

$$ac - bd = 1, \quad \text{and}$$
$$bc + ad = 0.$$

Solving these two equations for the two unknowns $c$ and $d$ in terms of $a$ and $b$ leads you to construct $c = a/(a^2 + b^2)$ and $d = -b/(a^2 + b^2)$ (noting that the denominator is not 0 because, by the hypothesis, at least one of $a$ and $b$ is not 0). To see that this construction is correct, note that

$$
\begin{aligned}
\textbf{A1:} \ (a + bi)(c + di) &= (ac - bd) + (bc + ad)i \\
&= [(a^2 + b^2)/(a^2 + b^2)] + 0i \\
&= 1.
\end{aligned}
$$

To establish the uniqueness, suppose that $e + fi$ is also a complex number that satisfies

**A2:** $(e + fi)(a + bi) = 1$.

It will be shown that

**B1:** $c + di = e + fi$.

Working forward by multiplying both sides of the equality in $A1$ by $e + fi$ and using associativity yields

**A3:** $[(e + fi)(a + bi)](c + di) = (e + fi)$.

Because $(e + fi)(a + bi) = 1$ from $A2$, it follows from $A3$ that $c + di = e + fi$, and so $B1$ is true, completing the proof.

**Proof.** Because either $a \neq 0$ or $b \neq 0$, $a^2 + b^2 \neq 0$, and so it is possible to construct the complex number $c + di$ in which $c = a/(a^2 + b^2)$ and $d = -b/(a^2 + b^2)$, for then

$$(a + bi)(c + di) = (ac - bd) + (bc + ad)i = 1.$$

To see the uniqueness, assume that $e + fi$ also satisfies $(a + bi)(e + fi) = 1$. Multiplying the foregoing displayed equality through by $e + fi$ yields

$$[(e + fi)(a + bi)](c + di) = e + fi.$$

Using the fact that $(a + bi)(e + fi) = 1$, it follows that $c + di = e + fi$ and so the uniqueness is established. $\square$

## Solutions to Selected Exercises in Chapter 12

**12.1** a. Applicable.
   b. Not applicable because the statement contains the quantifier "there is" instead of "for all."
   c. Applicable.
   d. Applicable.

**12.5** a. Verify that $P(n)$ is true for the initial value of $n = n_0$. Then, assuming that $P(n)$ true, prove that $P(n-1)$ is true.
   b. Verify that $P(n)$ is true for some integer $n_0$. Assuming that $P(n)$ is true for $n$, prove that $P(n+1)$ is true and that $P(n-1)$ is also true.

**12.7 Proof.** First it must be shown that $P(n)$ is true for $n = 1$. Replacing $n$ by 1, it must be shown that $1(1!) = (1+1)! - 1$. But this is clear because $1(1!) = 1 = (1+1)! - 1$.

Now assume that $P(n)$ is true and use that fact to show that $P(n+1)$ is true. So assume

$$P(n): \ 1(1!) + \cdots + n(n!) = (n+1)! - 1.$$

It must be shown that

$$P(n+1): \ 1(1!) + \cdots + (n+1)(n+1)! = (n+2)! - 1.$$

Starting with the left side of $P(n+1)$ and using $P(n)$:

$$
\begin{aligned}
1(1!) + \cdots + n(n!) + (n+1)(n+1)! \\
&= [1(1!) + \cdots + n(n!)] + (n+1)(n+1)! \\
&= [(n+1)! - 1] + (n+1)(n+1)! \qquad (P(n) \text{ is used here.}) \\
&= (n+1)![1 + (n+1)] - 1 \\
&= (n+1)!(n+2) - 1 \\
&= (n+2)! - 1. \ \square
\end{aligned}
$$

**12.12 Proof.** The statement is true for $n = 1$ because the subsets of a set consisting of one element, say $x$, are $\{x\}$ and $\emptyset$; that is, there are $2^1 = 2$ subsets. Assume that, for a set with $n$ elements, the number of subsets is $2^n$. It will be shown that, for a set with $n+1$ elements, the number of subsets is $2^{n+1}$. For a set $S$ with $n+1$ elements, one can construct all the subsets by listing first all those subsets that include the first $n$ elements, and then, to each such subset, one can add the last element of $S$. By the induction hypothesis, there are $2^n$ subsets using the first $n$ elements. An additional $2^n$ subsets are created by adding the last element of $S$ to each of the subsets of $n$ elements. Thus the total number of subsets of $S$ is $2^n + 2^n = 2^{n+1}$, and so the statement is true for $n + 1$. $\square$

**12.15 Proof.** Let

$$S = 1 + 2 + \cdots + n.$$

Then

$$S = n + (n - 1) + \cdots + 1.$$

On adding the two foregoing equations, one obtains

$$2S = n(n + 1), \text{ that is, } S = n(n + 1)/2. \quad \square$$

**12.19** The author relates $P(n + 1)$ to $P(n)$ by expressing the product of the $n$ terms associated with the left side of the equality in $P(n + 1)$ to the product of the $n$ terms associated with left side of the equality in $P(n)$ and one additional term—specifically,

$$\underbrace{\prod_{k=2}^{n+1} \left(1 - \frac{1}{k^2}\right)}_{\text{left side of } P(n+1)} = \underbrace{\prod_{k=2}^{n} \left(1 - \frac{1}{k^2}\right)}_{\text{left side of } P(n)} \underbrace{\left(1 - \frac{1}{(n + 1)^2}\right)}_{\text{extra term}}$$

The induction hypothesis is used when the author subsequently replaces $\prod_{k=2}^{n} (1 - \frac{1}{k^2})$, which is the left side of $P(n)$, with $\frac{n+1}{2n}$, which is the right side of $P(n)$.

**12.24** The mistake occurs in the last sentence, where it states that, "Then, because all the colored horses in this (second) group are brown, the uncolored horse must also be brown." How do you know that there is a colored horse in the second group? In fact, when the original group of $n + 1$ horses consists of exactly 2 horses, the second group of $n$ horses does not contain a colored horse. The entire difficulty is caused by the fact that the statement should have been verified for the initial integer $n = 2$, not $n = 1$. This, of course, you will not be able to do.

## Solutions to Selected Exercises in Chapter 13

**13.2** a. This means that $C$ cannot be true. Thus, to complete the proof, you now need to do the second case by showing that "$D$ implies $B$."

b. This situation arises when the author recognizes that $x \geq 2$ and $x \leq 1$ is an impossibility and so goes on to finish the proof by assuming the other half of the either/or statement, namely that $x \leq 2$ and $x \geq 1$.

**13.3** a. To apply a proof by elimination to the statement, "If $A$, then $C$ OR $D$ OR $E$," you would assume that $A$ is true, $C$ is not true, and $D$ is not true; you must conclude that $E$ is true. (Alternatively, you can assume that $A$ is true and that any two of the three statements $C$, $D$, and $E$ are not true; you would then have to conclude that the remaining statement is true.)

b. To apply a proof by cases to the statement, "If $C$ OR $D$ OR $E$, then $B$," you must do all three of the following proofs: (1) "If $C$, then $B$," (2) "If $D$, then $B$," and (3) "If $E$, then $B$."

**13.5** The author uses a proof by cases when the following statement containing the keywords "either/or" is encountered in the forward process:

**A1:** Either $(x \leq 0$ and $x - 3 \geq 0)$ or $(x \geq 0$ and $x - 3 \leq 0)$.

Accordingly, the author considers the following two cases:

**Case 1:** $x \leq 0$ and $x - 3 \geq 0$. The author then observes that this cannot happen and so proceeds to Case 2.

**Case 2:** $x \geq 0$ and $x - 3 \leq 0$. The author then reaches the desired conclusion.

**13.10** a. If $x$ is a real number that satisfies $x^3 + 3x^2 - 9x - 27 \geq 0$, then $x \leq -3$ or $x \geq 3$.

b. **Analysis of Proof.** With this proof by elimination, you assume that

$$\textbf{A:} \quad x^3 + 3x^2 - 9x - 27 \geq 0 \text{ and}$$
$$\textbf{A1 (NOT C):} \quad x > -3.$$

It must be shown that

$$\textbf{B1 (D):} \quad x \geq 3, \text{ that is, } x - 3 \geq 0.$$

By factoring $A$, it follows that

$$\textbf{A2:} \quad x^3 + 3x^2 - 9x - 27 = (x - 3)(x + 3)^2 \geq 0.$$

From $A1$, because $x > -3$, $(x+3)^2$ is strictly positive. Thus, dividing both sides of $A2$ by $(x + 3)^2$ yields $B1$ and completes the proof.

**Proof.** Assume that $x^3 + 3x^2 - 9x - 27 \geq 0$ and $x > -3$. Then it follows that $x^3 + 3x^2 - 9x - 27 = (x-3)(x+3)^2 \geq 0$. Because $x > -3$, $(x+3)^2$ is positive, so $x - 3 \geq 0$, that is, $x \geq 3$. $\square$

**13.14 Analysis of Proof.** The appearance of the keywords "either/or" in the hypothesis suggest proceeding with a proof by cases.

**Case 1.** Assume that

**A1:** $a|b$.

It must be shown that

**B1:** $a|(bc)$.

$B1$ gives rise to the key question, "How can I show that an integer (namely, $a$) divides another integer (namely, $bc$)?" Applying the definition means you must show that

**B2:** There is an integer $k$ such that $bc = ka$.

Recognizing the keywords "there is" in $B2$, you should use the construction method to produce the desired integer $k$. Working forward from $A1$ by definition, you know that

**A2:** There is an integer $p$ such that $b = pa$.

Multiplying both sides of the equality in $A2$ by $c$ yields

**A3:** $bc = cpa$.

From $A3$, the desired value for $k$ in $B2$ is $cp$, thus completing this case.

**Case 2.** In this case, you should assume that

**A1:** $a|c$.

You must show that

**B1:** $a|(bc)$.

The remainder of the proof in this case is similar to that in Case 1 and is not repeated.

**Proof.** Assume, without loss of generality, that $a|b$. By definition, there is an integer $p$ such that $b = pa$. But then, $bc = (cp)a$, and so $a|(bc)$. $\square$

## Solutions to Selected Exercises in Chapter 14

**14.1** a. For all elements $s \in S$, $s \leq z$.

b. There is an element $s \in S$ such that $s \geq z$.

**14.4 Analysis of Proof.** Recognizing the keyword "min" in the backward process, you must show that the following quantified statement is true:

> **B1:** For all numbers $x$, $x(x-2) \geq -1$, that is, $x^2 - 2x + 1 \geq 0$.

Recognizing the keywords "for all" in the backward statement $B1$, the choose method is used to choose

> **A1:** A real number $y$,

for which it must be shown that

> **B2:** $y^2 - 2y + 1 \geq 0$.

Now $B2$ is true because $y^2 - 2y + 1 = (y-1)^2 \geq 0$ and so the proof is complete.

**Proof.** To show that, for all real numbers $x$, $x(x-2) \geq -1$, one only needs to show that, for all $x$, $x^2 - 2x + 1 \geq 0$. To that end, choose a real number $y$ and note that $y^2 - 2y + 1 = (y-1)^2 \geq 0$. $\square$

## Solutions to Selected Exercises in Chapter 15

**15.1** a. Contrapositive or contradiction method because the keyword "no" appears in the conclusion.

b. Induction method because the conclusion is true for every integer $n \geq 4$.

c. Forward-Backward method because there are no keywords in the hypothesis or conclusion.

d. Max/Min method because the conclusion contains the keyword "maximum."

e. Uniqueness method because the conclusion contains the keywords "one and only one."

**15.3** a. With the contradiction method, you would assume that $p$ and $q$ are odd integers and that $x$ is a rational number that satisfies the equation $x^2 + 2px + 2q = 0$. You would then work forward from this information to reach a contradiction.

b. To use induction, first show that the statement is true for $n = 4$, that is, that $4! > 4^2$. For the second part of induction, you would assume that $n! > n^2$ and try to prove that $(n+1)! > (n+1)^2$. To do so, you should relate $(n+1)!$ to $n!$ so that you can use the induction hypothesis. You can also use the fact that $n \geq 4$, if necessary.

c. Working backward, you are led to the key question, "How can I show that a function (namely, $f + g$) is convex?" You can use the definition to answer this question. To show that $f + g$ is convex, you should also work forward from the hypothesis that $f$ and $g$ are convex by using the definition of a convex function.

d. According to the max/min method, you should convert the conclusion to the following equivalent quantified statement:

> **B1:** For all real numbers $a$, $b$, and $c$ with $a^2 + b^2 + c^2 = 1$, $ab + bc + ac \leq 1$.

Recognizing the keywords "for all" in the backward statement $B1$, the choose method should be used next.

e. To apply the direct uniqueness method, you must first show that there is a line $M$ perpendicular to the given line $L$ through the point $P$. Next, you should assume that $N$ is also a line perpendicular to $L$ through the point $P$. You should then work forward to show that $M$ and $N$ are the same.

To apply the indirect uniqueness method, you must first show that there is a line $M$ perpendicular to the given line $L$ through the point $P$. Next, you should assume that $N$ is a line, different from $M$, perpendicular to $L$ through the point $P$. You should then work forward to reach a contradiction.

## Solutions to Selected Exercises in Appendix A

**A.1** One common key question is, "How can I show that two sets are equal?" An associated answer to this question—obtained from the definition of two sets being equal—is, "Show that the first set is a subset of the second set and that the second set is a subset of the first set."

**A.4** **Analysis of Proof.** The forward-backward method is used to begin the proof because there are no keywords in either the hypothesis or the conclusion. A key question associated with the conclusion is, "How can I show that a set (namely, $(A \cap B)^c$) is a subset of another set (namely, $A^c \cup B^c$)?" Using the definition of subset, one answer is to show that

> **B1:** For every element $x \in (A \cap B)^c$, $x \in A^c \cup B^c$.

Recognizing the keywords "for every" in $B1$, you should now use the choose method to choose

> **A1:** An element $x \in (A \cap B)^c$,

for which you must show that

> **B2:** $x \in A^c \cup B^c$.

Working forward from $A1$, by definition of complement, you can say that

**A2:** $x \notin A \cap B$.

Now, from the definition for the intersection of two sets, $A2$ is equivalent to the statement NOT $(x \in A$ AND $x \in B)$. Using the techniques in Chapter 8, you can rewrite this statement as follows:

**A3:** $x \notin A$ OR $x \notin B$.

Again using the definition of the complement of a set, $A3$ becomes

**A4:** $x \in A^c$ OR $x \in B^c$.

Finally, by the definition for the union of two sets, from $A4$ you have

**A5:** $x \in A^c \cup B^c$.

The proof is now complete because $A5$ is the same as $B2$.

**A.10** The function $f : A \rightarrow B$ is not surjective if and only if there is an element $y \in B$ such that, for every element $x \in A$, $f(x) \neq y$. The statement is obtained by using the rules in Chapter 8 to negate the definition of $f$ being surjective, as follows:

> NOT(For every element $y \in B$, there is an element $x \in A$ such that $f(x) = y$).

> There is an element $y \in B$ such that NOT(there is an element $x \in A$ such that $f(x) = y$).

> There is an element $y \in B$ such that, for every element $x \in A$, NOT($f(x) = y$).

> There is an element $y \in B$ such that, for every element $x \in A$, $f(x) \neq y$.

**A.12** Yes, the function $f : R \rightarrow R^+ = \{x \in R : x \geq 0\}$ defined by $f(x) = |x|$ is surjective because of the following proof.

**Analysis of Proof.** From the backward process, a key question associated with the conclusion is, "How can I show that a function (namely, $f(x) = |x|$) is surjective?" One answer is by the definition, so you must show that

> **B1:** For every element $y \in R^+$, there is an element $x \in R$ such that $f(x) = |x| = y$.

Recognizing the keywords "for every" in $B1$, you should use the choose method to choose

**A1:** An element $y \in R^+$,

for which you must show that

**B2:** There is an element $x \in R$ such that $|x| = y$.

Recognizing the keywords "there is" in the backward statement $B2$, you should now use the construction method to construct the value of $x \in R$ such that $|x| = y$. However, because from $A1$ $y \in R^+ = \{x \in R : x \geq 0\}$, you know that $y \geq 0$. Thus, you can

**A2:** Construct $x = y$.

According to the construction method, you must show that the value of $x$ in $A2$ satisfies the certain property ($x \in R$) and the something that happens ($|x| = y$) in $B2$. Now $x \in R$ because $x = y$ is a real number. Finally, because $x = y \geq 0$, it follows that $|x| = |y| = y$. The proof is now complete.

**Proof.** To show that $f : R \rightarrow R^+$ defined by $f(x) = |x|$ is surjective, let $y \in R^+$. (The word "let" here indicates that the choose method is used.) It must be shown that there is an element $x \in R$ such that $|x| = y$. To that end, let $x = y$ and $y$ is a real number. (The word "let" here indicates that the construction method is used.) But because $x = y \geq 0$, it follows that $|x| = |y| = y$. □

**A.16  Analysis of Proof.** The keyword "unique" in the conclusion suggests using the backward uniqueness method. Accordingly, it is first necessary to construct a fixed point $x_*$ of $f$. However, this fixed point is given in the hypothesis. It remains to show that there is at most one fixed point of $f$. Here, the author uses the indirect uniqueness method to do so and thus assumes that, in addition to the fixed point $x_*$ of $f$,

**A1:** $y_* \neq x_*$ is also a fixed point of $f$.

The author now works forward from $A1$, the hypothesis, and the fact that $x_*$ is a fixed point of $f$ to reach the contradiction that $\alpha \geq 1$. Specifically, the author specializes the hypothesis that, for all real numbers $x$ and $y$, $|f(y) - f(x)| \leq \alpha|y - x|$ to the values $y = y_*$ and $x = x_*$, to obtain

**A2:** $|f(y_*) - f(x_*)| \leq \alpha|y_* - x_*|$.

The author now works forward from the fact that $x_*$ and $y_*$ are fixed points of $f$ to obtain

$$\begin{aligned} |x_* - y_*| &= |f(x_*) - f(y_*)| & (x_* \text{ and } y_* \text{ are fixed points of } f) \\ &\leq \alpha|x_* - y_*| & (\text{from } A2). \end{aligned}$$

The author then divides both sides of the foregoing inequality by the number $|x_* - y_*|$, which is $> 0$ because $x_* \neq y_*$ (see $A1$), to obtain that

$\alpha \geq 1$. But this contradicts the hypothesis that $\alpha < 1$, and so the proof is complete.

## Solutions to Selected Exercises in Appendix B

**B.1**  a. A key question associated with the properties in (j), (k), and (l) in Table B.1 is, "How can I show that two real numbers are equal?" This is because the dot product of two $n$-vectors is a real number.

b. Using the definition, one answer to the key question, "How can I show that two $n$-vectors are equal?" is to show that all $n$ components of the two vectors are equal. Applying this answer to the $n$-vectors $\mathbf{x+y} = \mathbf{y+x}$ means you must show that, for every integer $i = 1, \ldots, n$, $(\mathbf{x} + \mathbf{y})_i = (\mathbf{y} + \mathbf{x})_i$; that is, you must show that

>  **B1:**  For every integer $i = 1, \ldots, n$, $x_i + y_i = y_i + x_i$.

c. Based on the answer in part (b), the next technique you should use is the choose method because the keywords "for all" appear in the backward statement $B1$. For that statement, you would choose

>  **A1:**  An integer $i$ with $1 \leq i \leq n$,

for which you must show that

>  **B2:**  $x_i + y_i = y_i + x_i$.

**B.2**  **Analysis of Proof.**  Following the approach in Exercise B.1, a key question associated with showing that $\mathbf{x} + \mathbf{y} = \mathbf{y} + \mathbf{x}$ is, "How can I show that two $n$-vectors are equal?" By definition, one answer is to show that all components are equal; that is,

>  **B1:**  For every integer $i = 1, \ldots, n$, $(\mathbf{x} + \mathbf{y})_i = (\mathbf{y} + \mathbf{x})_i$; that is, $x_i + y_i = y_i + x_i$.

Recognizing the keywords "for all" in the backward statement $B1$, you should proceed with the choose method and choose

>  **A1:**  An integer $i$ with $1 \leq i \leq n$,

for which you must show that

>  **B2:**  $x_i + y_i = y_i + x_i$.

However, $B2$ is true because of the commutative property of addition of real numbers and so the proof is complete.

**Proof.**  To see that $\mathbf{x+y} = \mathbf{y+x}$, let $i$ be an integer with $1 \leq i \leq n$. (The word "let" here indicates that the choose method is used.) But then you have that $(\mathbf{x} + \mathbf{y})_i = x_i + y_i = y_i + x_i = (\mathbf{y} + \mathbf{x})_i$ by the commutative property of addition of numbers, and so the proof is complete.  □.

**B.4  Analysis of Proof.** The forward-backward method is used to begin the proof because the hypothesis and conclusion do not contain any keywords. A key question associated with showing that $t(\mathbf{x}+\mathbf{y}) = t\mathbf{x}+t\mathbf{y}$ is, "How can I show that two $n$-vectors (namely, $t(\mathbf{x} + \mathbf{y})$ and $t\mathbf{x} + t\mathbf{y}$) are equal?" By definition, one answer is to show that all components are equal; that is,

> **B1:**  For every integer $i = 1, \ldots, n$, $[t(\mathbf{x} + \mathbf{y})]_i = (t\mathbf{x} + t\mathbf{y})_i$.

Recognizing the keywords "for all" in the backward statement $B1$, you should proceed with the choose method and choose

> **A1:**  An integer $i$ with $1 \leq i \leq n$,

for which you must show that

> **B2:**  $[t(\mathbf{x} + \mathbf{y})]_i = (t\mathbf{x} + t\mathbf{y})_i$.

However, $B2$ is true because $[t(\mathbf{x} + \mathbf{y})]_i = t(\mathbf{x} + \mathbf{y})_i = t(x_i + y_i) = tx_i + ty_i = (t\mathbf{x} + t\mathbf{y})_i$ and so the proof is complete.

**B.7**  The $n$-vectors $\mathbf{x}^1, \ldots, \mathbf{x}^k$ are **linearly dependent** if and only if there are real numbers $t_1, \ldots, t_k$, not all 0, such that $t_1\mathbf{x}^1 + \cdots + t_k\mathbf{x}^k = \mathbf{0}$. This statement is obtained by using the rules in Chapter 8 to negate the definition of the $n$-vectors $\mathbf{x}^1, \ldots, \mathbf{x}^k$ being linearly independent, as follows:

> NOT[For all real numbers $t_1, \ldots, t_k$ with $t_1\mathbf{x}^1 + \cdots + t_k\mathbf{x}^k = \mathbf{0}$, it follows that $t_1 = \cdots = t_k = 0$].
>
> There are real numbers $t_1, \ldots, t_k$ with $t_1\mathbf{x}^1 + \cdots + t_k\mathbf{x}^k = \mathbf{0}$ such that NOT[$t_1 = \cdots = t_k = 0$].
>
> There are real numbers $t_1, \ldots, t_k$ not all 0 such that $t_1\mathbf{x}^1 + \cdots + t_k\mathbf{x}^k = \mathbf{0}$.

**B.8  Analysis of Proof.** The forward-backward method is used to begin the proof because the hypothesis and conclusion do not contain keywords. A key question associated with the conclusion is, "How can I show that some $n$-vectors (namely, $\mathbf{x}^1, \ldots, \mathbf{x}^k, \mathbf{0}$) are linearly dependent?" Using the definition in Exercise B.7, one answer is to show that

> **B1:**  There are real numbers $t_1, \ldots, t_k, t_{k+1}$ not all 0 such that
> $t_1\mathbf{x}^1 + \cdots + t_k\mathbf{x}^k + t_{k+1}\mathbf{0} = \mathbf{0}$.

Recognizing the keywords "there are" in the backward statement $B1$, you should now use the construction method to produce the real numbers $t_1, \ldots, t_k, t_{k+1}$ in $B1$. To that end, you can

> **A1:**  Construct the real numbers $t_1 = 0, \ldots, t_k = 0, t_{k+1} = 1$.

According to the construction method, you must show that the values constructed in $A1$ satisfy the certain property of not all being 0 as well as the something that happens in $B1$. It is easy to see that not all the values in $A1$ are 0 because $t_{k+1} = 1$. Finally, the constructed values in $A1$ satisfy the something that happens in $B1$ because

$$t_1 \mathbf{x}^1 + \cdots + t_k \mathbf{x}^k + t_{k+1}\mathbf{0} = 0\mathbf{x}^1 + \cdots + 0\mathbf{x}^k + 1(\mathbf{0}) = \mathbf{0},$$

the last equality being true from the properties of vector operations in Table B.1. The proof is now complete.

**Proof.** To see that the $n$-vectors $\mathbf{x}^1, \ldots, \mathbf{x}^k, \mathbf{0}$ are linearly dependent, by the definition in Exercise B.7, it will be shown that there are real numbers $t_1, \ldots, t_k, t_{k+1}$ not all 0 such that $t_1 \mathbf{x}^1 + \cdots + t_k \mathbf{x}^k + t_{k+1}\mathbf{0} = \mathbf{0}$. Specifically, the real numbers $t_1 = 0, \ldots, t_k = 0, t_{k+1} = 1$ are not all 0 and satisfy

$$t_1 \mathbf{x}^1 + \cdots + t_k \mathbf{x}^k + t_{k+1}\mathbf{0} = 0\mathbf{x}^1 + \cdots + 0\mathbf{x}^k + 1(\mathbf{0}) = \mathbf{0}.$$

The proof is now complete. $\square$

**B.11  Analysis of Proof.** Following the approach in Exercise B.10, a key question associated with showing that $A + B = B + A$ is, "How can I show that two matrices are equal?" By definition, one answer is to show that they have the same dimensions (which they do because both $A + B$ and $B + A$ are $(m \times n)$ matrices) and that all components are equal; that is,

>**B1:** For every integer $i = 1, \ldots, m$ and for every integer $j = 1, \ldots, n$, $(A + B)_{ij} = (B + A)_{ij}$, that is, $A_{ij} + B_{ij} = B_{ij} + A_{ij}$.

Recognizing the keywords "for every" in the backward statement $B1$, you should proceed with the choose method and choose

>**A1:** Integers $i$ and $j$ with $1 \leq i \leq m$ and $1 \leq j \leq n$,

for which you must show that

>**B2:** $A_{ij} + B_{ij} = B_{ij} + A_{ij}$.

However, $B2$ is true because of the commutative property of addition of real numbers, and so the proof is complete.

**Proof.** To see that $A + B = B + A$, note that the dimensions of both $A + B$ and $B + A$ are $(m \times n)$, so let $i$ and $j$ be integers with $1 \leq i \leq m$ and $1 \leq j \leq n$. (The word "let" here indicates that the choose method is used.) But then you have that $(A + B)_{ij} = A_{ij} + B_{ij} = B_{ij} + A_{ij} =$

$(B+A)_{ij}$ by the commutative property of addition of real numbers, and so the proof is complete.  $\square$

**B.14 Analysis of Proof.** The forward-backward method is used to begin the proof because there are no keywords in the hypothesis or conclusion. A key question associated with the conclusion that $A(B+C) = AB+AC$ is, "How can I show that two matrices are equal?" By definition, one answer is to show that the two matrices have the same dimensions (which they do because both $A(B+C)$ and $AB+AC$ are $(m \times n)$ matrices) and that all components are equal; that is,

> **B1:** For every integer $i = 1, \ldots, m$ and for every integer $j = 1, \ldots, n$, $[A(B+C)]_{ij} = (AB+AC)_{ij}$.

Recognizing the keywords "for every" in the backward statement $B1$, you should proceed with the choose method and choose

> **A1:** Integers $i$ and $j$ with $1 \leq i \leq m$ and $1 \leq j \leq n$,

for which you must show that

> **B2:** $[A(B+C)]_{ij} = (AB+AC)_{ij}$.

By definition of matrix multiplication, the element in row $i$ and column $j$ of $A(B+C)$ is $A_{i*} \bullet (B+C)_{*j}$; while the element in row $i$ and column $j$ of $AB+AC$ is $(AB)_{ij} + (AC)_{ij}$. Thus, you must show that

> **B3:** $A_{i*} \bullet (B+C)_{*j} = (AB)_{ij} + (AC)_{ij}$.

However, starting with the left side of $B3$, you have

> **A2:** $A_{i*} \bullet (B+C)_{*j} = A_{i*} \bullet (B_{*j}+C_{*j}) = A_{i*} \bullet B_{*j} + A_{i*} \bullet C_{*j} = (AB)_{ij} + (AC)_{ij}$.

The proof is now complete because $A2$ is the same as $B3$.

**Proof.** To see that $A(B+C) = AB+AC$, note that both matrices have dimension $(m \times n)$, so, let $i$ and $j$ be integers with $1 \leq i \leq m$ and $1 \leq j \leq n$. (The word "let" here indicates that the choose method is used.) You then have:

$$
\begin{aligned}
[A(B+C)]_{ij} &= A_{i*} \bullet (B+C)_{*j} && \text{(def. of matrix mult.)} \\
&= A_{i*} \bullet (B_{*j}+C_{*j}) && \text{(prop. of matrix add.)} \\
&= A_{i*} \bullet B_{*j} + A_{i*} \bullet C_{*j} && \text{(prop. of dot product)} \\
&= (AB)_{ij} + (AC)_{ij} && \text{(def. of matrix mult.)} \\
&= (AB+AC)_{ij} && \text{(def. of matrix add.).}
\end{aligned}
$$

The proof is now complete.  $\square$

## Solutions to Selected Exercises in Appendix C

**C.1  Analysis of Proof.** The forward-backward method is used to begin the proof because there are no keywords in the hypothesis or conclusion. A key question associated with the conclusion $a|(-b)$ is, "How can I show that an integer (namely, $a$) divides another integer (namely, $-b$)?" Using the definition, one answer is to show that

> **B1:** There is an integer $d$ such that $-b = da$.

Recognizing the keywords "there is" in the backward statement $B1$, you should now use the construction method. Specifically, turn to the forward process to construct the desired integer $d$.

Working forward from the hypothesis that $a$ divides $b$, by definition, you know that

> **A1:** There is an integer $c$ such that $b = ca$.

Multiplying the equality in $A1$ through by $-1$ yields that

> **A2:** $-b = (-c)a$.

From $A2$ you can see that the desired value of the integer $d$ in $B2$ is $d = -c$. According to the construction method, you must still show that this value of $d$ satisfies the certain property and the something that happens in $B1$; namely, that $-b = da$, but this is clear from $A2$.

**Proof.** To show that $a|(-b)$, by definition, it is shown that there is an integer $d$ such that $-b = da$. However, from the hypothesis that $a|b$, there is an integer $c$ such that $b = ca$. Letting $d = -c$, it is easy to see that $-b = (-c)a = da$, thus completing the proof. $\square$

**C.4  Analysis of Proof.** The words "Suppose not ..." indicate that the author is using the contradiction method (see Chapter 9). Accordingly, the author assumes that the conclusion is not true; that is, that

> **A1 (NOT B):** There is an integer $a > 1$ such that $a$ cannot be written as a product of primes.

The author must now reach a contradiction (which is identified in the last sentence of the proof). To that end, from $A1$, the author constructs

> **A2:** $b =$ the first integer $> 1$ such that $b$ cannot be written as a product of primes.

The author is justified in making this statement because the author is specializing the Least Integer Principle to the following set:

$M = \{c \in Z : c > 1 \text{ and } c \text{ cannot be written as a product of primes}\}.$

To apply specialization, it is necessary to show that $M$ is a nonempty set of positive integers. However, $A1$ ensures that $M \neq \emptyset$ (because $a \in M$) and the definition of $M$ ensures that all elements are $> 0$. Hence, the author is justified in making statement $A2$. The remainder of the proof is working forward from $A2$ to show that $b$ can be written as a product of primes, which contradicts $A2$. Specifically, the author first notes from $A2$ that

> **A3:** $b$ is not prime.

Working forward from $A3$ by writing the negation of the definition of a prime, the author states that

> **A4:** There are integers $b_1$ and $b_2$ with $1 < b_1, b_2 < b$ such that $b = b_1 b_2$.

The author now uses the fact that $b$ is the smallest integer $> 1$ that cannot be written as the product of primes (see $A2$). Specifically, without telling you, the author uses a max/min technique (see Section 12.3) to realize that

> **A5:** For every integer $a$ with $1 < a < b$, $a$ can be written as a product of primes.

Recognizing the keywords "for every" in the forward statement $A5$, the author specializes $A5$ to $a = b_1$ and again to $a = b_2$ (which both satisfy $1 < b_1, b_2 < b$ according to $A4$) to obtain

> **A6:** There are primes $q_1, \ldots, q_m$ and $r_1, \ldots, r_n$ such that $b_1 = q_1 q_2 \cdots q_m$ and $b_2 = r_1 r_2 \cdots r_n$.

Multiplying corresponding sides of the equalities in $A6$ results in

> **A7:** $b = b_1 b_2 = (q_1 q_2 \cdots q_m)(r_1 r_2 \cdots r_n)$.

However, $A7$ says that $b$ can be written as a product of primes, which contradicts $A2$ and completes the proof.

**C.6 Analysis of Proof.** Recognizing the keyword "only" in the conclusion, you should use the backward uniqueness method (see Section 11.1). Accordingly, you must first construct an identity element of the group $G$. However, this is given in the hypothesis and is property (2) in the definition of a group, so you already know that

> **A1:** There is an element $e \in G$ such that, for all elements $a \in G$, $a \odot e = e \odot a = a$.

To show that there is at most one identity element, the direct uniqueness method is used here. Accordingly, you should now assume that

**A2:** There is an element $f \in G$ such that, for all elements $a \in G$, $a \odot f = f \odot a = a$.

You must show that

**B1:** $e = f$.

You can obtain $B1$ by specializing the for-all statements in $A1$ and $A2$. Specifically, specializing the for-all statement in $A1$ to $a = f$ (which is in $G$ from $A2$), it follows that

**A3:** $f \odot e = f$.

Likewise, specializing the for-all statement in $A2$ to $a = e$ (which is in $G$ from $A1$), it follows that

**A4:** $f \odot e = e$.

The desired conclusion in $B1$ now follows from $A3$ and $A4$ because both $e$ and $f$ are equal to $f \odot e$ and hence $e = f$, thus completing the proof.

**Proof.** To see that the identity element $e \in G$ is the only element with the property that, for all $a \in G$, $a \odot e = e \odot a = a$, suppose that $f \in G$ also satisfies the property that, for all $a \in G$, $a \odot f = f \odot a = a$. Then, because $e$ is an identity element of $G$, $f \odot e = f$. Likewise, because $f$ is an identity element of $G$, $f \odot e = e$. But then $f \odot e = f = e$ and so $e = f$, completing the proof. $\square$

**C.12 Analysis of Proof.** The forward-backward method is used to begin the proof because there are no keywords in either the hypothesis or the conclusion. A key question associated with the conclusion is, "How can I show that a set (namely, $H$) together with an operation (namely, $\odot$ from $G$) forms a group?" One answer is to use the definition for a group. Accordingly, you must first show that

**B1:** The operation $\odot$ satisfies the property that, for all elements $a, b \in H$, $a \odot b \in H$.

Recognizing the keywords "for all" in the backward statement $B1$, you should now use the choose method to choose

**A1:** Elements $a, b \in H$,

for which you must show that

**B2:** $a \odot b \in H$.

A key question associated with $B2$ is, "How can I show that an element (namely, $a \odot b$) belongs to a set (namely, $H$)?" Using the defining property of $H$, you must show that

**B3:** There is an integer $k$ such that $a \odot b = x^k$.

Recognizing the keywords "there is" in the backward statement $B2$, you should now use the construction method. Accordingly, the author turns to the forward process. Working forward from $A1$ using the defining property of the set $H$, it follows that

**A2:** There are integers $i, j$ such that $a = x^i$ and $b = x^j$.

The author now works forward from $A2$ using Proposition 35 to state that

**A3:** $a \odot b = x^i \odot x^j = x^{i+j}$.

You can now see from $A3$ that the desired value of the integer $k$ in $B2$ is $k = i + j$.

To complete the proof that $(H, \odot)$ is a group, it remains to show that the following three properties hold:

**B4:**
(1) For all elements $a, b, c \in H$, $(a \odot b) \odot c = a \odot (b \odot c)$.
(2) There is an element $e \in H$ such that, for all elements $a \in H$, $a \odot e = e \odot a = a$.
(3) For all elements $a \in H$, there is an element $a^{-1} \in H$ such that $a \odot a^{-1} = a^{-1} \odot a = e$.

The author now uses the following three properties of the group $(G, \odot)$ to establish the corresponding properties in $B4$ for $(H, \odot)$:

**A4:**
(1) For all elements $a, b, c \in G$, $(a \odot b) \odot c = a \odot (b \odot c)$.
(2) There is an element $e \in G$ such that, for all elements $a \in G$, $a \odot e = e \odot a = a$.
(3) For all elements $a \in G$, there is an element $a^{-1} \in G$ such that $a \odot a^{-1} = a^{-1} \odot a = e$.

Recognizing the keywords "for all" in property (1) of the backward statement $B4$, the author uses the choose method to choose

**A5:** Elements $a, b, c \in H$,

for which it must be shown that

**B5:** $(a \odot b) \odot c = a \odot (b \odot c)$.

The author states that $B2$ is true by specializing the corresponding for-all statement in property (1) of $A4$ to $a, b, c$ in $A5$. However, to specialize that statement in $A4$, it must be that $a, b, c \in G$. The author states that this is the case without justification. The justification is that, because $a, b, c \in H$, by the defining property of $H$, there are integers $i, j, k$ such that $a = x^i$, $b = x^j$, $c = x^k$. Each of these elements is in $G$ because $\odot$ combines elements of $G$ and produces an element of $G$.

Turning now to proving property (2) in $B4$, the author recognizes the keywords "there is" and uses the construction method. Specifically, the author constructs the identity element of $H$ as the identity element $e$ of $G$. According to the construction method, the author must show that the value of $e$ satisfies the certain property and the something that happens in property (2) in $B4$; that is, the author must show that

**B6:** $e \in H$ and for all elements $a \in H$, $a \odot e = e \odot a = a$.

The author notes that $e \in H$ because $e = x^0$. Recognizing the keywords "for all" in the backward statement $B6$, the author now uses the choose method to choose

**A6:** An element $a \in H$,

for which it must be shown that

**B7:** $a \odot e = e \odot a = a$.

The author states that $B7$ is true, which is obtained by specializing the for-all statement in property (2) of $A4$ to $a$, noting that $a \in H$ and so $a \in G$.

Turning to proving property (3) in $B4$, the author recognizes the keywords "for all" in the backward statement $B4$ and uses the choose method to choose

**A7:** An element $x^k \in H$,

for which it must be shown that

**B8:** There is an element $(x^k)^{-1} \in H$ such that
$x^k \odot (x^k)^{-1} = (x^k)^{-1} \odot x^k = e$.

Recognizing the keywords "there is" in the backward statement $B8$, the author uses the construction method. Specifically, the author constructs $(x^k)^{-1} = x^{-k}$. According to the construction method, the author must show that this value of $(x^k)^{-1} = x^{-k}$ satisfies the certain property and the something that happens in $B8$; namely, that

**B9:** $(x^k)^{-1} \in H$ and $x^k \odot (x^k)^{-1} = (x^k)^{-1} \odot x^k = e$.

Now $(x^k)^{-1} = x^{-k} \in H$ by the defining property of $H$. Finally, the author uses Proposition 35 to note that

$$x^k \odot (x^k)^{-1} = x^k \odot x^{-k} = x^0 = e \quad \text{and}$$
$$(x^k)^{-1} \odot x^k = x^{-k} \odot x^k = x^0 = e.$$

The proof is now complete.

## Solutions to Selected Exercises in Appendix D

**D.1** a. A common key question is, "How can I show that a sequence converges to a number?"

    b. Using the definition, one answer to the key question in part (a) applied to the specific problem results in having to show that

> **B1:** For every real number $\epsilon > 0$, there is an integer $j \in N$ such that, for all $k \in N$ with $k > j$, $|x_k y_k - xy| < \epsilon$.

    c. Recognizing "for every" as the first of the nested quantifiers in the backward statement $B1$, you should use the choose method next.

**D.2** a. Use the set $T = \{s \in R : s > 0 \text{ and } s^n > 2\}$.

    b. **Analysis of Proof.** You can show that the set $T$ in part (a) has an infimum by specializing the Infimum Property of real numbers. To do so, you must show that

> **B1:** $T \neq \emptyset$ and $T$ is bounded below.

You can see that $T \neq \emptyset$ because $2 \in T$ (recall that $n > 1$). Finally, $0$ is a lower bound for $T$. To see that this is so, according to the definition of a lower bound, you must show that

> **B2:** For every element $x \in T, x \geq 0$.

Recognizing the keywords "for every" in the backward statement $B2$, you should now use the choose method to choose

> **A1:** An element $x \in T$,

for which it must be shown that

> **B3:** $x \geq 0$.

However, $B3$ is true because $x \in T$ and by the defining property of $T$, $x > 0$. The proof is now complete.

> **Proof.** To see that $T$ has an infimum, note that $2 \in T$ and so $T \neq \emptyset$. Also, $0$ is a lower bound for $T$ because, for $x \in T$, by the defining property of $T$, $x > 0$. The fact that $T$ has an infimum now follows from the Infimum Property of real numbers. □

**D.6** **Analysis of Proof.** Recognizing the keywords "for every" as the first quantifier in the conclusion, the choose method is used to choose

> **A1:** A real number $\epsilon > 0$,

for which it must be shown that

> **B1:** There is an element $x \in T$ such that $x < t + \epsilon$.

Recognizing the keywords "there is" in the backward statement $B1$, the author uses the construction method to produce the element $x \in T$. To do so, the author works forward from the hypothesis that $t$ is the infimum of $T$ which, by definition, means $t$ is a lower bound for $T$ and

**A2:** For every lower bound $s$ for $T$, $s \le t$.

The author then states that

**A3:** $t + \epsilon$ is not a lower bound for $T$.

Now $A3$ is true because, if $t + \epsilon$ is a lower bound for $T$, then you could specialize $A2$ to $s = t + \epsilon$ and hence conclude that $t + \epsilon \le t$; that is, $\epsilon \le 0$, which contradicts $A1$. The author then works forward from $A3$ by writing the $NOT$ of the definition of a lower bound for a set to obtain

**A4:** There is an element $x \in T$ such that $x < t + \epsilon$.

The proof is now complete because $A4$ is the same as $B1$.

**D.8** The sequence $X = (x_1, x_2, \ldots)$ does not converge to the real number $x$ means that there is a real number $\epsilon > 0$ such that, for every integer $j \in N$, there is an integer $k \in N$ with $k > j$ such that $|x_k - x| \ge \epsilon$. The statement is obtained by using the rules in Chapter 8 to negate the definition of $X$ converging to $x$, as follows:

NOT(for every real number $\epsilon > 0$, there is an integer $j \in N$ such that, for all $k \in N$ with $k > j$, $|x_k - x| < \epsilon$).

There is a real number $\epsilon > 0$ such that NOT(there is an integer $j \in N$ such that, for all $k \in N$ with $k > j$, $|x_k - x| < \epsilon$).

There is a real number $\epsilon > 0$ such that, for every integer $j \in N$, NOT(for all $k \in N$ with $k > j$, $|x_k - x| < \epsilon$).

There is a real number $\epsilon > 0$ such that, for every integer $j \in N$, there is an integer $k \in N$ with $k > j$ such that NOT($|x_k - x| < \epsilon$).

There is a real number $\epsilon > 0$ such that, for every integer $j \in N$, there is an integer $k \in N$ with $k > j$ such that $|x_k - x| \ge \epsilon$.

**D.10 Analysis of Proof.** The forward-backward method is used to begin the proof because there are no keywords in the hypothesis or the conclusion. An associated key question is, "How can I show that a sequence [namely, $X = (\frac{1}{1}, \frac{1}{2}, \frac{1}{3}, \ldots)$] converges to a real number (namely, $0$)?" Using the definition of convergence, one answer is to show that

**B1:** For every real number $\epsilon > 0$, there is an integer $j \in N$ such that, for all $k \in N$ with $k > j$, $|\frac{1}{k} - 0| = \frac{1}{k} < \epsilon$.

Recognizing the keywords "for every" as the first quantifier in the backward statement $B1$, you should use the choose method to choose

**A1:** A real number $\epsilon > 0$,

for which it must be shown that

**B2:** There is an integer $j \in N$ such that, for all $k \in N$ with $k > j$, $\frac{1}{k} < \epsilon$.

Recognizing the keywords "there is" as the first quantifier in the backward statement $B2$, you should now consider using the construction method. You can turn to the forward process in an attempt to do so. However, another approach is to assume that you have *already* constructed the desired value of $j \in N$. To complete the construction method, you would then have to show that your value of $j$ satisfies the something that happens in $B2$; namely, that

**B3:** For all $k \in N$ with $k > j$, $\frac{1}{k} < \epsilon$.

The idea now is to try to prove $B3$, and in so doing to discover what value of $j \in N$ allows you to do so. Proceeding with the backward process and recognizing the keywords "for all" in $B3$, you should now use the choose method to choose

**A2:** An integer $k \in N$ with $k > j$,

for which you must show that

**B4:** $\frac{1}{k} < \epsilon$.

Now, from $A2$, you know that $k > j$, so

**A3:** $\frac{1}{k} < \frac{1}{j}$.

You can obtain $B4$ from $A3$ provided that

**B5:** $\frac{1}{j} < \epsilon$.

Indeed, this tells you that you need to construct $j$ so that $B5$ holds. Solving the inequality in $B5$ for $j$ using the fact that $\epsilon > 0$ leads to

**B6:** $j > \frac{1}{\epsilon}$.

In other words, constructing $j$ to be any integer satisfying $B6$ enables you to show that $B5$, and hence $B4$, is true, thus completing the proof.

**Proof.** To see that the sequence $X = (\frac{1}{1}, \frac{1}{2}, \frac{1}{3}, \ldots)$ converges to 0, let $\epsilon > 0$. (The word "let" here indicates that the choose method is used.) Now let $j$ be any integer with $j > \frac{1}{\epsilon}$. (The word "let" here indicates that the construction method is used.) Now let $k \in N$ with $k > j$. (The word "let" here indicates that the choose method is used.) You then have that

$$|x_k - 0| = \frac{1}{k} < \frac{1}{j} < \epsilon.$$

The proof is now complete. □

# Glossary of Math Terms and Symbols

**and**—For two statements $A$ and $B$, the statement $A$ *AND* $B$ (written $A \bigwedge B$) is true when both $A$ and $B$ are true, and false otherwise.

**axiom**—A statement whose truth is accepted without a proof.

**backward process**—The process of deriving from a statement, $B$, a new statement, $B1$, with the property that, if $B1$ is true, then so is $B$. You do this by asking and answering a key question.

**backward uniqueness method**—A technique for proving that there is one and only one object with a certain property such that something happens. To do so, first use the construction or contradiction method to show that there is at least one such object. Then use either the direct or the indirect uniqueness method to show that there is at most one such object.

**bounded above function**—A real-valued function $f$ of one real variable for which there is a real number $y$ such that, for every real number $x$, $f(x) \le y$.

**bounded set of real numbers**—A set of real numbers $S$ for which there is a real number $M > 0$ such that, for every element $x \in S$, $|x| < M$.

**choose method**—A technique for proving that, for every object with a certain property, something happens. To do so, choose a generic object with the certain property. Then show that, for this chosen object, the something happens.

**codomain of a function**—Any set that contains all possible values that can result from evaluating the function at an element in the domain of the function.

**column vector**—A matrix consisting of one column.

**complement of a set**—The set of all elements in the universal set that are not in the set.

**component of a vector**—Any of the individual values in a vector.

**conclusion**—The statement $B$ in the implication "$A$ implies $B$." When proving "$A$ implies $B$," your job is to show that the conclusion $B$ is true.

**conditional statement**—A statement whose truth depends on the value of one or more variables.

**construction method**—A technique for proving that there is an object with a certain property such that something happens. To do so, construct, guess, produce, or devise an algorithm to produce the desired object. Then show that the object you constructed has the certain property and satisfies the something that happens.

**continuous function at a point**—A function $f$ of one variable such that, at a given point $x$, for every real number $\epsilon > 0$, there is a real number $\delta > 0$ such that, for all real numbers $y$ with $|x - y| < \delta$, $|f(x) - f(y)| < \epsilon$.

**contradiction method**—A technique for proving that "$A$ implies $B$" in which you work forward from the assumption that $A$ and $NOT\ B$ are true to reach a contradiction to some statement that you know is true.

**contrapositive method**—A technique for proving that "$A$ implies $B$" in which you prove that "$NOT\ B$ implies $NOT\ A$" by working forward from $NOT\ B$ and backward from $NOT\ A$.

**contrapositive statement**—The contrapositive of the statement "$A$ implies $B$" is the statement "$NOT\ B$ implies $NOT\ A$."

**convergence of a sequence of real numbers to a given real number**—A sequence of real numbers $x_1, x_2, \ldots$ for which, at a given real number $x$, for every real number $\epsilon > 0$, there is an integer $j \geq 1$ such that, for every integer $k$ with $k > j$, $|x_k - x| < \epsilon$.

**converse statement**—The converse of the statement "$A$ implies $B$" is the statement "$B$ implies $A$."

**convex function**—A function $f$ of one variable such that, for all real numbers $x$ and $y$ and for all real numbers $t$ with $0 \leq t \leq 1$, it follows that $f(tx + (1 - t)y) \leq tf(x) + (1 - t)f(y)$.

**convex set**—A set $C$ of real numbers such that, for all elements $x, y \in C$ and for all real numbers $t$ with $0 \leq t \leq 1$, $tx + (1 - t)y \in C$.

**corollary**—A proposition whose truth follows almost immediately from a theorem.

**decreasing sequence of real numbers**—A sequence $x_1, x_2, \ldots$ of real numbers such that, for every integer $k = 1, 2, \ldots$, $x_k > x_{k+1}$.

**definition**—An agreement, by all parties concerned, as to the meaning of a particular term.

**derivative of a function at a point**—The real number $f'(\bar{x})$ is the derivative of the function $f$ at the point $\bar{x}$ if and only if, for every real number $\epsilon > 0$, there is a real number $\delta > 0$ such that, for every real number $x$ with $0 < |x - \bar{x}| < \delta$, it follows that $\left| \frac{f(x)-f(\bar{x})}{x-\bar{x}} - f'(\bar{x}) \right| < \epsilon$.

**dimension of a vector**—The number of components in a vector.

**dimension of a matrix**—The number of rows and columns in a matrix.

**direct uniqueness method**—A technique for proving that there is a unique object with a certain property such that something happens. To do so, first construct the desired object $X$. Then assume that there is also an object $Y$ with the certain property and for which the something happens. Work forward to show that $X$ and $Y$ are the same.

**divides**—An integer $a$ divides an integer $b$ (written $a|b$) if and only if there is an integer $c$ such that $b = ca$.

**domain of a function**—The set of all allowable values at which the function can be evaluated.

**dot product of two $n$-vectors**—The dot product of the two $n$-vectors $\mathbf{x} = (x_1, \ldots, x_n)$ and $\mathbf{y} = (y_1, \ldots, y_n)$ is $\mathbf{x} \bullet \mathbf{y} = x_1 y_1 + \cdots + x_n y_n$.

**either/or methods**—Techniques for proving that "$A$ implies $B$" when $A$ and/or $B$ contain the key words "either/or" (see **proof by elimination** and **proof by cases**).

**element of a matrix**—Any entry in the matrix.

**element of a set**—An item that belongs to a given set.

**empty set**—The set with no elements, written $\emptyset$.

**equal matrices**—Two $(m \times n)$ matrices $A$ and $B$ are equal (written $A = B$) if and only if, for each row $i = 1, \ldots, n$ and column $j = 1, \ldots, m$, $A_{ij} = B_{ij}$.

**equal pairs of real numbers**—Two pairs of real numbers $(x_1, y_1)$ and $(x_2, y_2)$ for which $x_1 = x_2$ and $y_1 = y_2$.

**equal sets**—Two sets $S$ and $T$ are equal (written $S = T$) if and only if $S$ is a subset of $T$ and $T$ is a subset of $S$.

**equal vectors**—Two $n$-vectors $\mathbf{x} = (x_1, \ldots, x_n)$ and $\mathbf{y} = (y_1, \ldots, y_n)$ are equal (written $\mathbf{x} = \mathbf{y}$) if and only if, for each integer $i = 1, \ldots, n$, $x_i = y_i$.

**equilateral triangle**—A triangle, all of whose sides have the same length.

**equivalent statements**—Two statements $A$ and $B$ for which "$A$ implies $B$" and "$B$ implies $A$."

**even integer**—An integer $n$ whose remainder on dividing by 2 is 0. Equivalently, an integer $n$ for which there is an integer $k$ such that $n = 2k$.

**existential quantifier**—The key words "there is" ("there are," "there exists").

**forward-backward method**—The technique for proving that "$A$ implies $B$" in which you assume that $A$ is true and try to show that $B$ is true. To do so, you apply the forward process to $A$ and the backward process to $B$.

**forward process**—The process of deriving from a statement, $A$, a new statement, $A1$, with the property that $A1$ is true because $A$ is true.

**forward uniqueness method**—The technique of working forward from a statement of the form, "there is a unique object with a certain property such that something happens." To do so, look for two such objects, $X$ and $Y$. Then conclude, as a new statement in the forward process, that $X$ and $Y$ are the same; that is, that $X = Y$.

**greater-than-or-equal-to functions**—For functions $f$ and $g$ of one variable, $g \geq f$ on the set $S$ of real numbers if and only if, for every element $x \in S$, $g(x) \geq f(x)$.

**greatest common divisor**—An integer $d$ such that, for two given integers $a$ and $b$, (1) $d$ divides $a$ and $d$ divides $b$ and (2) whenever $c$ is an integer for which $c$ divides $a$ and $c$ divides $b$, it follows that $c$ divides $d$.

**greatest lower bound for a set of real numbers**—A real number $t$ is the greatest lower bound for a set $T$ of real numbers if and only if (1) $t$ is a lower bound for $T$ and (2) for any lower bound $s$ for $T$, $s \leq t$.

**group**—A set $G$ and an operation $\odot$ on $G$ with the property that, for all $a, b \in G$, $a \odot b \in G$ that together satisfy the following three properties:

1. (*Associative Law*) For all $a, b, c \in G$, $(a \odot b) \odot c = a \odot (b \odot c)$.

2. (*Existence of an Identity Element*) There is an element $e \in G$ such that, for all $a \in G$, $a \odot e = e \odot a = a$. ($e$ is called the **identity element**.)

3. (*Existence of an Inverse Element*) For each $a \in G$, there is an element $a^{-1} \in G$ such that $a \odot a^{-1} = a^{-1} \odot a = e$. (The element $a^{-1}$ is called the **inverse** of $a$.)

**hypothesis**—The statement $A$ in the implication "$A$ implies $B$." When proving "$A$ implies $B$," you can assume that the hypothesis $A$ is true.

**identity element of a group**—An element $e$ of a group $(G, \odot)$ with the property that, for every element $a \in G$, $a \odot e = e \odot a = a$.

**identity matrix**—An $(n \times n)$ matrix $I$ all of whose values are 0 except that each diagonal element $I_{ii}$ is 1.

**implication**—A statement of the form, "If $A$ is true, then $B$ is true," where $A$ and $B$ are given statements.

**increasing function**—A real-valued function $f$ of one real variable such that, for all real numbers $x$ and $y$ with $x \leq y$, $f(x) \leq f(y)$.

**increasing sequence of real numbers**—A sequence $x_1, x_2, \ldots$ of real numbers such that, for every integer $k = 1, 2, \ldots$, $x_k < x_{k+1}$.

**indirect uniqueness method**—A technique for proving that there is a unique object with a certain property such that something happens. To do so, first construct the desired object $X$. Then assume that there is an object $Y$, different from $X$, with the certain property and for which the something happens. Work forward to reach a contradiction.

**induction**—A technique for proving that, for every integer $n \geq$ some initial integer $n_0$, some statement $P(n)$ is true. To do so, show that $P(n_0)$ is true. Then assume that $P(n)$ is true and show that $P(n+1)$ is true by relating $P(n+1)$ to $P(n)$.

**infimum of a set of real numbers**—A real number $t$ is the infimum of a set $T$ of real numbers if and only if (1) $t$ is a lower bound for $T$ and (2) for any lower bound $s$ for $T$, $s \leq t$.

**infimum property of the real numbers**—The property that any nonempty set of real numbers that has a lower bound has an infimum.

**injective function**—A real-valued function $f$ of one real variable such that, for all real numbers $x$ and $y$ with $x \neq y$, $f(x) \neq f(y)$.

**integers**—The set of numbers $\{\ldots, -2, -1, 0, 1, 2, \ldots\}$.

**intersection of two sets**—The set of all elements that are in both sets.

**inverse element of a group**—For an element $a$ of a group $(G, \odot)$, an element $a^{-1} \in G$ is the inverse element of $a$ if and only if $a \odot a^{-1} = a^{-1} \odot a = e$, where $e$ is the identity element of $G$.

**inverse statement**—The inverse of the statement "$A$ implies $B$" is the statement "$NOT\ A$ implies $NOT\ B$."

**invertible matrix**—An $(n \times n)$ matrix $A$ for which there is an $(n \times n)$ matrix $C$ such that $AC = CA = I$, the $(n \times n)$ identity matrix.

**isosceles triangle**—A triangle, two of whose sides have equal length.

**key answer**—An answer to a key question.

**key question**—The specific question obtained by asking how you can show that a given statement $B$ is true.

**least integer principle**—The property that a nonempty set of positive integers always has a smallest element.

**least upper bound for a set**—A real number $u$ such that, for a given set $S$ of real numbers, (1) $u$ is an upper bound for $S$ and (2) for every upper bound $v$ for $S$, $u \leq v$.

**lemma**—A proposition that is used in the proof of a subsequent theorem.

**linear function**—A real-valued function $f$ of one real variable for which there are real numbers $m$ and $b$ such that, for all real numbers $x$, $f(x) = mx + b$.

**linearly independent vectors**—The $n$-vectors $\mathbf{x}^1, \ldots, \mathbf{x}^k$ are linearly independent if and only if, for all real numbers $t_1, \ldots, t_k$ with the property that $t_1\mathbf{x}^1 + \cdots + t_k\mathbf{x}^k = \mathbf{0}$, it follows that $t_1 = \cdots = t_k = 0$.

**lower bound for a set of real numbers**—A real number $t$ is a lower bound for a set $T$ of real numbers if and only if, for every element $x \in T$, $t \leq x$.

**matrix**—A rectangular table of numbers arranged in $m$ rows and $n$ columns.

**matrix multiplication**—Given an $(m \times p)$ matrix $A$ and a $(p \times n)$ matrix $B$, $AB$ is the $(m \times n)$ matrix in which, for each row $i = 1, \ldots, m$ and column $j = 1, \ldots, n$, $(AB)_{ij} = A_{i*} \bullet B_{*j}$ = row $i$ of $A$ times column $j$ of $B$.

**maximizer of a function**—A real number $x^*$ such that, for every real number $x$, $f(x) \leq f(x^*)$ (where $f$ is a real-valued function of one real variable).

**max/min methods**—Methods for proving that the maximum or minimum of a given set $S$ of real numbers is $\leq$ or $\geq$ a given real number $x$ by converting the statement to an equivalent statement containing the quantifier "there is" or "for all," as follows:

- $\min\{s \in S\} \geq x$ is equivalent to for all $s \in S$, $s \geq x$.

- $\min\{s \in S\} \leq x$ is equivalent to there is an $s \in S$ such that $s \leq x$.

- $\max\{s \in S\} \geq x$ is equivalent to there is an $s \in S$ such that $s \geq x$.

- $\max\{s \in S\} \leq x$ is equivalent to for all $s \in S$, $s \leq x$.

**member of a set**—An item that belongs to a given set.

**neighborhood of radius $\epsilon$ around a real number $x$**—The set $N_\epsilon(x) = \{y \in R : |x - y| < \epsilon\}$.

**nested quantifiers**—A statement that contains more than one quantifier.

**nonsingular matrix**—An $(n \times n)$ matrix $A$ for which there is an $(n \times n)$ matrix $C$ such that $AC = CA = I$, the $(n \times n)$ identity matrix.

**not**—For a statement $A$, the statement $NOT\ A$ is true when $A$ is false and is false when $A$ is true.

**odd integer**—An integer $n$ for which there is an integer $k$ such that $n = 2k + 1$.

**one-to-one function**—A real-valued function $f$ of one real variable such that, for all real numbers $x$ and $y$ with $x \neq y$, $f(x) \neq f(y)$.

**onto function**—A real-valued function $f$ of one real variable such that, for every real number $y$, there is a real number $x$ such that $f(x) = y$.

**or**—For two statements $A$ and $B$, the statement $A\ OR\ B$ (written $A \bigvee B$) is false when $A$ is false and $B$ is false, and true otherwise.

**prime**—An integer $p > 1$ whose only positive integer divisors are 1 and $p$.

**proof**—A convincing argument expressed in the language of mathematics.

**proof by cases**—A technique for proving that "$A$ implies $B$" when $A$ has the form "either $C$ or $D$." To do so, first assume that $C$ is true and show that $B$ is true. Then assume that $D$ is true and show that $B$ is true.

**proof by elimination**—A technique for proving that "$A$ implies $B$" when $B$ has the form "either $C$ or $D$." To do so, assume that $A$ and $NOT\ C$ are true and show that $D$ is true. Alternatively, you can assume that $A$ and $NOT\ D$ are true and show that $C$ is true.

**proof technique**—Any method for proving that the statement "$A$ implies $B$" is true.

**proposition**—A true statement of interest.

**quantifier**—One of the two groups of key words "there is" ("there are," "there exists") and "for all" ("for each," "for every," "for any").

**rational number**—A real number $r$ for which there are integers $p$ and $q$ with $q \neq 0$ such that $r = p/q$.

**rationals**—The set of all rational numbers.

**real number**—Any number that can be expressed in decimal format.

**reals**—The set of all real numbers.

**row vector**—A matrix consisting of one row.

**scalar**—A real number.

**sequence**—A function $x : N \to R$, where $N = \{1, 2, \ldots\}$, and $R$ is the set of real numbers. For $k \in N$, the notation $x_k$ is used instead of $x(k)$, and the sequence is written $X = (x_1, x_2, \ldots)$ or $X = (x_k : k \in N)$ or $X = (x_k)$.

**set**—A collection of items.

**square integer**—An integer $n$ such that there is an integer $k$ with $n = k^2$.

**standard form for "for all"**—For all objects with a certain property, something happens.

**standard form for "there is"**—There is an object with a certain property such that something happens.

**statement**—A mathematical sentence that is either true or false.

**strictly increasing function**—A real-valued function $f$ of one real variable such that, for all real numbers $x$ and $y$ with $x < y$, $f(x) < f(y)$.

**strictly monotone sequence of real numbers**—A sequence of real numbers $x_1, x_2, \ldots$ that is either decreasing or increasing.

**subset**—A set $S$ is a subset of a set $T$ (written $S \subseteq T$ or $S \subset T$) if and only if, for every element $x \in S$, $x \in T$.

**surjective function**—A function $f$ of one variable such that, for every real number $y$ there is a real number $x$ such that $f(x) = y$.

**theorem**—An important proposition.

**transpose of a matrix**—The transpose of an $(m \times n)$ matrix $A$ is the $(n \times m)$ matrix $A^t$ in which, for all $i = 1, \ldots, n$ and for all $j = 1, \ldots, m$, $(A^t)_{ij} = A_{ji}$.

**truth table**—A table that lists the truth of a complex statement (such as "$A$ implies $B$") for all possible combinations of truth values of the individual statements (in this case, $A$ and $B$).

**union of two sets**—The set of elements that are in either of the two sets.

**uniqueness methods**—Techniques for working with statements of the form, "there is a unique object with a certain property such that something happens" (see the **forward** and **backward uniqueness methods** and also the **direct** and **indirect uniqueness methods**).

**universal quantifier**—The key words "for all" ("for each," "for every," "for any").

**upper bound for a set**—A real number $u$ such that, for all $x \in S$, $x \le u$ (where $S$ is a given set of real numbers).

**vector**—An ordered list of real numbers.

**zero matrix**—A matrix, all of whose elements are 0.

## Glossary of Mathematical Symbols

| Symbol | Meaning | Page |
|---|---|---|
| $\Rightarrow$ | implies | 3 |
| $\square$ | Q. E. D. (which was to be demonstrated) | 14 |
| $\wedge$ | and | 26 |
| $\vee$ | or | 26 |
| $\Leftrightarrow$ | if and only if | 27 |
| $\sim$ | not | 33 |
| $\exists$ | there is (there are, there exists) | 42 |
| $\ni$ | such that | 42 |
| $\in$ | is an element of | 51 |
| $\emptyset$ | empty set | 52 |
| $\subseteq$ $(\subset)$ | subset | 53 |
| $\forall$ | for all (for each, for any, for every) | 53 |

# References

1. Ash, Robert B. *A Primer of Abstract Mathematics*. The Mathematical Association of America, 1998.

2. Bittinger, Marvin L. *Logic, Proof, and Sets*, 2nd ed. Reading, MA: Addison-Wesley, 1970.

3. Chartrand, Gary, Albert D. Polimeni, and Ping Zhang. *A Transition to Advanced Mathematics*, 2nd ed. Reading, MA: Addison-Wesley, 2008.

4. Cupillari, Antonella. *The Nuts and Bolts of Proofs*, 3rd ed. Burlington, MA: Elsevier Academic Press, 2005.

5. Daep, P. and P. Gorkin. *Reading, Writing, and Proving: A Closer Look at Mathematics*. New York: Springer-Verlag, 2003.

6. D'Angelo, John P. and Douglas B. West. *Mathematical Thinking: Problem-Solving and Proofs*, 2nd ed. Englewood Cliffs: Prentice-Hall, 2000.

7. Exner, George R. *An Accompaniment to Higher Mathematics*. New York: Springer-Verlag, 1996.

8. Fendel, Daniel and Diane Resek. *Foundations of Higher Mathematics: Exploration and Proof*. Reading: Addison-Wesley, 1990.

9. Fletcher, Peter and C. Wayne Patty. *Foundations of Higher Mathematics*, 3rd ed. Belmont: Brooks/Cole, 1995.

10. Gerstein, Larry J. *Introduction to Mathematical Structures and Proofs.* New York: Springer-Verlag, 1996.

11. Gibilisco, Stan. *Math Proofs Demystified.* New York: McGraw-Hill, 2005.

12. Granier, Rowan and John Taylor. *100% Mathematical Proof.* New York: John Wiley, 1996.

13. Levine, Alan. *Discovering Higher Mathematics—Four Habits of Highly Effective Mathematicians.* San Diego, CA: Harcourt Academic Press, 2000.

14. Lucas, John F. *Introduction to Abstract Mathematics*, 2$^{\text{nd}}$ ed. New York: Ardsley House, 1990.

15. Morash, Ronald P. *Bridge to Abstract Mathematics: Mathematical Proof and Structures*, 2$^{\text{nd}}$ ed. New York: McGraw-Hill, 1991.

16. Polya, George. *How to Solve It.* Garden City: Doubleday, 1957.

17. Polya, George. *Mathematical Discovery* (combined ed.). New York: John Wiley, 1981.

18. Schumacher, Carol. *Chapter 0—Fundamental Notions of Abstract Mathematics*, 2$^{\text{nd}}$. Reading: Addison-Wesley, 2001.

19. Schwartz, Diane D. *Conjecture & Proof: An Introduction to Mathematical Thinking.* Fort Worth: Harcourt Brace & Company, 1997.

20. Smith, Douglas, Maurice Eggen, and Richard St. Andree. *A Transition to Advanced Mathematics*, 5$^{\text{th}}$ ed. Monterey: Brooks/Cole, 2001.

21. Solow, Daniel. *The Keys to Advanced Mathematics.* Cleveland: Books Unlimited, 1995.

22. Solow, Daniel. *The Keys to Linear Algebra.* Cleveland: Books Unlimited, 1998.

23. Sundstrom, Ted. *Mathematical Reasoning: Writing and Proof.* Englewood Cliffs: Prentice Hall, 2003.

24. Vellman, Daniel J. *How to Prove It: A Structured Approach.* Cambridge, UK: Cambridge University Press, 1994.

25. Wicklegren, Wayne A. *How to Solve Problems.* San Francisco: W. H. Freeman, 1974.

# Index

# SUMMARY OF PROOF TECHNIQUES

| Proof Technique | When to Use It | What to Assume |
|---|---|---|
| Forward Uniqueness (page 123) | When $A$ has the key word "unique" in it. | There is such an object, $X$. |
| Direct Uniqueness (page 125) | When $B$ has the key word "unique" in it. | There are two such objects, and $A$ |
| Indirect Uniqueness (page 126) | When $B$ has the key word "unique" in it. | There are two different objects, and $A$ |
| Induction (page 131) | When a statement $P(n)$ is true for each integer $n \geq n_0$. | $P(n)$ is true for $n$. |
| Proof by Cases (page 143) | When A has the form "$C$ OR $D$." | Case 1: $C$ <br> Case 2: $D$ |
| Proof by Elimination (page 145) | When $B$ has the form "$C$ OR $D$." | $A$ and NOT $C$ <br><br> or <br><br> $A$ and NOT $D$ |
| Max/Min 1 (page 153) | When $A$ or $B$ has the form "$\max S \leq z$" or "$\min S \geq z$". | |
| Max/Min 2 (page 153) | When $A$ or $B$ has the form "$\max S \geq z$" or "$\min S \leq z$". | |